T0189229

ROBUST OPTIMIZATION-DIRECTED DESIGN

Printed in the United States of America.

9 8 7 6 5 4 3 2 1   SPIN 11539148

springeronline.com

Nonconvex Optimization and Its Applications

VOLUME 81

ROBUST OPTIMIZATION-DIRECTED DESIGN

Edited by

ANDREW J. KURDILA
University of Florida, Gainesville, Florida

PANOS M. PARDALOS
University of Florida, Gainesville, Florida

MICHAEL ZABARANKIN
Stevens Institute of Technology, Hoboken, New Jersey

 Springer

Library of Congress Cataloging-in-Publication Data

Kurdila, Andrew.
    Robust optimization-directed design / edited by Andrew J. Kurdila, Panos M. Pardalos,
Michael Zabarankin.
        p. cm. — (Nonconvex optimization and its applications ; v. 81)
    Includes bibliographical references.
    ISBN-13: 978-1-4614-9834-6
    ISBN-10: 1-4614-9834-1
    ISBN-13: 978-0-387-28654-9 (e-book)
    ISBN-10: 0-387-28654-3 (e-book)
    1. System theory. 2. Mathematical optimization. 3. Programming (Mathematics) I.
Pardalos, P.M. (Panos M.), 1954– II. Zabarankin, Michael, 1974– III. Title. IV. Series.

    Q295.K875 2006
    003—dc22

                                                                        2005051643

AMS Subject Classifications: 93C20, 93C95, 93E03, 90C15, 65K10

9 8 7 6 5 4 3 2 1   SPIN 11539148

springeronline.com

Contents

Preface vii

**1 A Multigrid Approach to Optimal Control Computations for
Navier-Stokes Flows**
E. Aulisa, S. Manservisi   3

2 Control System Radii and Robustness Under Approximation
John A. Burns, Gunther H. Peichl 25

3 Equilibrium Analysis for a Network Market Model
Juan F. Escobar, Alejandro Jofré 63

**4 Distributed Solution of Optimal Control Problems Governed by
Parabolic Equations**
Matthias Heinkenschloss, Michael Herty 73

**5 Modeling and Implementation of Risk-Averse Preferences in Stochastic
Programs Using Risk Measures**
Pavlo A. Krokhmal, Robert Murphey 95

6 Shape Optimization of Electrodes for Piezoelectric Actuators
Andrew J. Kurdila, Weijian Wang, Yunfei Feng, Richard J. Prazenica 117

**7 Robust Static Super-Replication of Barrier Options in the
Black-Scholes model**
Jan H. Maruhn, Ekkehard W. Sachs 135

8 Numerical Techniques in Relaxed Optimization Problems
Tomáš Roubíček 157

**9 Combining Model and Test Data for Optimal Determination of
Percentiles and Allowables: CVaR Regression Approach, Part I**
Stan Uryasev, A. Alexandre Trindade 179

10 Combining Model and Test Data for Optimal Determination of Percentiles and Allowables: CVaR Regression Approach, Part II
Stan Uryasev, A. Alexandre Trindade 209

11 Semidefinite Programming for Sensor Network and Graph Localization
Yinyu Ye 247

Preface

There has been and continues to be a great deal of work on optimization, multi-disciplinary optimal design (MDO), but in spite of the amount of research in MDO, there is a lack of effort in sharing new ideas aimed at addressing some of the practical issues that could lead to more widespread and effective use of optimization as a practical design tool.

Robust design or managing design uncertainties (model uncertainty, parametric uncertainty, etc.) is the unpleasant issue that is crucial in much of the MDO work. There is a lot of work in stochastic optimization, which tries to address some of the issues, and may have some promise for complex, integrated design problems, especially for many practical examples springing from this field. The "Optimization-Directed" expression in the *Robust Optimization-Directed Design* title is meant to suggest that the focus is not on agonizing over whether optimization strategies identify a true global optimum, but on whether they make significant design improvements. Recently, there has been enormous practical interest in strategies for applying optimization tools to the development of robust solutions/designs in

- Aerodynamics: airframes design, modeling and control
- Integration of sensing (laser radars, vision-based systems, millimeter-wave radars) and control
- Cooperative control with poorly modeled uncertainty
- Cascading failures in military and civilian applications
- Multi-mode seekers/sensor fusion
- Data association problems and tracking systems

In April 2004, the University of Florida (UF) Graduate Engineering and Research Center (GERC) and the Center for Agile Autonomous Flight successfully hosted the first conference on Robust Optimization-Directed Design (RODD) in Shalimar, Florida. The RODD meeting brought together outstanding researchers in RODD, researchers working on uncertainty management in complex modeling and simulation problems, stochastic optimization and mathematical modeling experts, in order to understand the state of the art in RODD, and to see what tools and techniques may be available to help in some of the many complex design issues that are arising in the joint Air Force Research Laboratory/Munitions Directorate, Air Force Office of Scientific Research and UF GERC research effort currently being conducted at the new Center for Agile Autonomous Flight. About 30 researchers from government, industry and academia attended the conference and presented their views on robust design, what it means and how it is distinct or related to other fields of research. This book contains refereed papers summarizing the participants' research in uncertainty modelling, robust design, optimal control and stochastic optimization.

We would like to take the opportunity to thank the authors of the papers, the UF Center for Agile Autonomous Flight for financial support, Dr. Marc Jacobs for the help in organizing the conference, the anonymous referees and Springer Publisher for making the publication of this volume possible.

Andrew J. Kurdila, Panos M. Pardalos and Michael Zabarankin
April 2005

Everything flows, nothing stands still. No man can cross the same river twice, because neither the man nor the river are the same.

Heraclitus, 535–475 B.C.

1

A Multigrid Approach to Optimal Control Computations for Navier-Stokes Flows

E. Aulisa[1] and S. Manservisi[2]

[1] DIENCA
    University of Bologna
    Via dei colli 16
    Bologna, Italy
[2] Departments of Mathematics and Statistics
    Texas Tech University
    Lubbock Texas 79409-1042
    smanserv@math.ttu.edu

Summary. Optimal control computations with boundary controls are presented by using a new multigrid approach for reliable and robust optimization. The multigrid implementation is based on a local Vanka-type solver for the Navier-Stokes and the adjoint system. The solution is achieved by solving and relaxing element by element the optimal control problem which is formulated by using an embedded domain approach. Numerical tests for steady and unsteady solutions are presented to prove the effectiveness and robustness of the method for flow matching and tracking.

Key words: optimal control, Navier-Stokes equations, Multigrid methods, Vanka-type solvers

1.1 Introduction

Optimal boundary control problems associated with the Navier-Stokes equations have a wide and important range of applications. Aerodynamic and hydrodynamic problems such as the design of cars, airplanes, and jet engines provide a few settings. Despite the fact that this field has been extensively studied, determining the optimal control for a system governed by the Navier-Stokes equations is still a difficult and time consuming task. The optimal control of the Navier-Stokes equations shows many challenges and has been considered by numerous authors, e.g., [1], [4], [6], [7], [8], [11], [12], [13], [16], [19] and [24]. However in many of these papers the formulation is not suitable for applications since the function spaces proposed for the solution may not allow accurate finite element implementations or may not allow a suitable numerical form for the optimality system.

In this paper, we study a class of optimal flow control problems and its multigrid implementation for which the fluid motion is controlled by velocity forcing, i.e., injection or suction, along a portion of the boundary, and the cost or objective functional is a measure of the discrepancy between the flow velocity and a given target velocity. We consider the two-dimensional incompressible flow of a viscous fluid on the do-

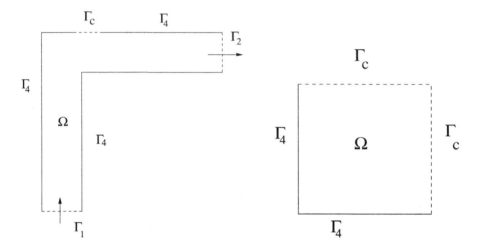

Fig. 1.1. Flow domain: Γ_1 inflow, Γ_2 outflow and Γ_c controlled boundary.

main Ω with boundary Γ. For example a typical L-shaped channel domain is shown in Figure 1.1 on the left and a square domain on the right with boundary control over Γ_c. The velocity $\mathbf{u}$ and the pressure p satisfy the Navier-Stokes system

$$\frac{\partial \mathbf{u}}{\partial t} + (\mathbf{u} \cdot \nabla)\mathbf{u} = -\nabla p + \nu \triangle \mathbf{u} + \mathbf{f} \qquad \text{in } \Omega \qquad (1.1)$$

$$\nabla \cdot \mathbf{u} = 0 \qquad \text{in } \Omega \qquad (1.2)$$

along with the Dirichlet boundary conditions

$$\mathbf{u} = \mathbf{g} = \begin{cases} \mathbf{g}_1 & \text{on } \Gamma_1 \\ \mathbf{g}_2 & \text{on } \Gamma_2 \\ \mathbf{g}_c & \text{on } \Gamma_c \\ \mathbf{0} & \text{on } \Gamma_4 \end{cases} \qquad (1.3)$$

where $\mathbf{f}$ is the given body force. In (1.3), ν denotes the inverse of the Reynolds number whenever the variables are appropriately nondimensionalized. The vectors $\mathbf{g}_1$ and $\mathbf{g}_2$ are the given velocities at the inflow Γ_1 and outflow Γ_2 of the channel, respectively. Along the boundary Γ_4 of the channel the velocity vanishes. The velocity at the outflow Γ_2 may be specified by using Neumann boundary control. The function $\mathbf{g}$ must satisfy the compatibility condition

$$\int_\Gamma \mathbf{g} \cdot \mathbf{n} \, ds = 0 \,, \qquad (1.4)$$

where $\mathbf{n}$ is the unit normal vector along the surface Γ.

There is a substantial literature discussing the set of all possible boundary controls. Clearly, the function $\mathbf{g}_c$ must belong to $H^{1/2}(\Gamma_c)$, the Sobolev space of order $1/2$. However $H^{1/2}(\Gamma_c)$ or $H^1(\Gamma_c)$ may not be sufficient to enable one to explicitly derive a first-order necessary condition. Thus in general the set of all admissible controls $\mathbf{g}$ must be restricted to more regular spaces, namely, to belong to $H^{3/2}(\Gamma_c)$. In this paper we define a set of allowable boundary controls by using distributed control on the extended domain. Let $\widehat{\Omega}$ and $\widehat{\Gamma}$ be the extended domain and boundary respectively where $\widehat{\Gamma}$ contains $\Gamma - \Gamma_c$. We define $\widehat{f}$ the extended force function such that $\widehat{f} = \mathbf{f}$ on Ω. Then, the set of admissible boundary controls can be taken as the set of all $\mathbf{g}_c \in H^{3/2}(\Gamma_c)$ such that $\mathbf{g}_c$ is the trace of the extended solution $\widehat{u}$ of (1.3) over $\widehat{\Omega}$ with distributed control $\widehat{f}$.

One could examine several practical objective functionals for determining the shape of the boundary, e.g., the reduction of the drag due to viscosity or the identification of the velocity at a fixed vertical slit downstream. To fix ideas, we focus on the minimization of the cost functional

$$\mathcal{J} = \int_{\Omega_1} |\mathbf{u} - \hat{u}|^2 \, d\mathbf{x} + \frac{\beta}{2} \int_{\widehat{\Omega}-\Omega} |\widehat{f}|^2 \, d\mathbf{x}, \qquad (1.5)$$

for stationary matching or

$$\mathcal{J} = \int_0^T \int_{\Omega_1} |\mathbf{u} - \mathbf{U}|^2 \, d\mathbf{x} + \frac{\alpha}{2} \int_\Omega |\mathbf{u}(T) - \mathbf{U}(T)|^2 \, d\mathbf{x}$$
$$+ \frac{\beta}{2} \int_0^T \int_{\widehat{\Omega}-\Omega} |\widehat{f}|^2 \, d\mathbf{x}, \qquad (1.6)$$

for time dependent control. The velocity field $\mathbf{U}$ is the desired velocity field defined on $\Omega_1 \subset \Omega$, α and β are nonnegative constants. Formally speaking, by using this formulation the boundary control problem is reduced to solving a distributed control problem over the $\widehat{\Omega} - \Omega$ domain.

In literature the standard steady optimal control problem is formulated by using the following functional (see for example [1; 13])

$$\mathcal{J} = \frac{\alpha}{2} \int_\Omega |\mathbf{u} - \mathbf{U}|^2 \, d\mathbf{x} + \frac{\beta}{2} \int_{\Gamma_c} (|\mathbf{g}|^2 + \beta_1 |\mathbf{g}_x|^2) \, d\mathbf{x}, \qquad (1.7)$$

where the minimization of the first term involving $(\mathbf{u} - \mathbf{U})$ is the real goal of the velocity matching problem and the other terms have been introduced in order to bound the control function and prove the existence of the solution of the optimal control problem and the optimality system. We may effectively limit the size of the control and prove the existence of the first order necessary condition for optimality through an appropriate choice of the positive coefficients β and β_1 but the optimal control based on this admissible set of solutions and the choice of β and β_1 is not very friendly from the numerical point of view and it turns out to be a very difficult task if injection or suction boundary velocity is required to satisfy the integral constraint (1.4).

In this paper the functional (1.5) is used and the optimal boundary control is obtained as the trace of the solution over the extended domain. Computations of the optimality system are performed by using a new multigrid approach. The implementation is based on a local Vanka-type solver for the Navier-Stokes and the adjoint system where solution is achieved by solving and relaxing element by element the optimal control problem. The multigrid smoother operator is constructed directly from the optimal control formulation and requires the iterative exact solution of the optimality system over a limited number of unknowns. Numerical tests for steady and unsteady solutions are presented. Unsteady optimal control is presented in the form of a time piecewise optimal control problem which proves to be a very effective and robust method for flow tracking. Also in this multigrid approach the solution of the unsteady optimal problem is achieved by solving iteratively several optimal problems over all the time steps but over a limited number of degrees of freedom in space. This allows us to solve the couple time-space optimality system exactly over a sufficient large number of time steps and enhances enormously the capability of solving boundary optimal control problems for complex geometries.

1.2 The Stationary Boundary Control Problem

We denote by $H^s(\mathcal{O})$, $s \in \mathbb{R}$, the standard Sobolev space of order s with respect to the set $\mathcal{O}$, which is either the flow domain Ω, or its boundary Γ, or part of its boundary. Whenever m is a nonnegative integer, the inner product over $H^m(\mathcal{O})$ is denoted by $(f, g)_m$ and (f, g) denotes the inner product over $H^0(\mathcal{O}) = L^2(\mathcal{O})$. Hence, we associate with $H^m(\mathcal{O})$ its natural norm $\|f\|_{m,\mathcal{O}} = \sqrt{(f, f)_m}$. Whenever possible, we will neglect the domain label in the norm.

For vector-valued functions and spaces, we use boldface notation. For example, $\mathbf{H}^s(\Omega) = [H^s(\Omega)]^n$ denotes the space of $\mathbb{R}^n$-valued functions such that each component belongs to $H^s(\Omega)$. Of special interest is the space

$$\mathbf{H}^1(\Omega) = \left\{ v_j \in L^2(\Omega) \ \middle| \ \frac{\partial v_j}{\partial x_k} \in L^2(\Omega) \quad \text{for } j, k = 1, 2 \right\},$$

equipped with the norm $\|\mathbf{v}\|_1 = (\sum_{k=1}^2 \|v_k\|_1^2)^{1/2}$.

For $\Gamma_s \subset \Gamma$ with nonzero measure, we also consider the subspace

$$\mathbf{H}^1_{\Gamma_s}(\Omega) = \{ \mathbf{v} \in \mathbf{H}^1(\Omega) \mid \mathbf{v} = \mathbf{0} \ \text{ on } \Gamma_s \}.$$

Also, we write $\mathbf{H}^1_0(\Omega) = \mathbf{H}^1_\Gamma(\Omega)$. For any $\mathbf{v} \in \mathbf{H}^1(\Omega)$, we write $\|\nabla\mathbf{v}\|$ for the seminorm. Let $(\mathbf{H}^1_{\Gamma_s})^*$ denote the dual space of $\mathbf{H}^1_{\Gamma_s}$. Note that $(\mathbf{H}^1_{\Gamma_s})^*$ is a subspace of $\mathbf{H}^{-1}(\Omega)$, where the latter is the dual space of $\mathbf{H}^1_0(\Omega)$. The duality pairing between $\mathbf{H}^{-1}(\Omega)$ and $\mathbf{H}^1_0(\Omega)$ is denoted by $\langle \cdot, \cdot \rangle$.

Let $\mathbf{g}$ be an element of $\mathbf{H}^{1/2}(\Gamma)$. It is well known that $\mathbf{H}^{1/2}(\Gamma)$ is a Hilbert space with norm

$$\|\mathbf{g}\|_{1/2,\Gamma} = \inf_{\mathbf{v}\in\mathbf{H}^1(\Omega);\, \gamma_\Gamma\mathbf{v}=\mathbf{g}} \|\mathbf{v}\|_1 ,$$

where γ_Γ denotes the trace mapping $\gamma_\Gamma : \mathbf{H}^1(\Omega) \to \mathbf{H}^{1/2}(\Gamma)$. We let $(\mathbf{H}^{1/2}(\Gamma))^*$ denote the dual space of $\mathbf{H}^{1/2}(\Gamma)$ and $\langle \cdot, \cdot \rangle_\Gamma$ denote the duality pairing between $(\mathbf{H}^{1/2}(\Gamma))^*$ and $\mathbf{H}^{1/2}(\Gamma)$. Let Γ_s be a smooth subset of Γ. Then, the trace mapping $\gamma_{\Gamma_s} : \mathbf{H}^1(\Omega) \to \mathbf{H}^{1/2}(\Gamma_s)$ is well defined and $\mathbf{H}^{1/2}(\Gamma_s) = \gamma_{\Gamma_s}(\mathbf{H}^1(\Omega))$.

Since the pressure is only determined up to an additive constant by the Navier-Stokes system with velocity boundary conditions, we define the space of square integrable functions having zero mean over Ω as

$$L_0^2(\Omega) = \{\, p \in L^2(\Omega) \mid \int_\Omega p \, d\mathbf{x} = 0 \,\}.$$

In order to define a weak form of the Navier-Stokes equations, we introduce the continuous bilinear forms

$$a(\mathbf{u}, \mathbf{v}) = 2\nu \sum_{i,j=1}^{2} \int_\Omega D_{ij}(\mathbf{u}) \, D_{ij}(\mathbf{v}) \, d\mathbf{x} \qquad \forall \, \mathbf{u}, \mathbf{v} \in \mathbf{H}^1(\Omega) \tag{1.8}$$

and

$$b(\mathbf{v}, q) = -\int_\Omega q \, \nabla \cdot \mathbf{v} \, d\mathbf{x} \quad \forall q \in L_0^2(\Omega), \quad \forall \mathbf{v} \in \mathbf{H}^1(\Omega) \tag{1.9}$$

and the trilinear form

$$\begin{aligned} c(\mathbf{w}; \mathbf{u}, \mathbf{v}) &= \int_\Omega \mathbf{w} \cdot \nabla \mathbf{u} \cdot \mathbf{v} \, d\mathbf{x} \\ &= \sum_{i,j=1}^{2} \int_\Omega w_j \left(\frac{\partial u_i}{\partial x_j} \right) v_i \, d\mathbf{x}, \quad \forall \mathbf{w}, \mathbf{u}, \mathbf{v} \in \mathbf{H}^1(\Omega). \end{aligned} \tag{1.10}$$

Obviously, $a(\cdot, \cdot)$ is a continuous bilinear form on $\mathbf{H}^1(\Omega) \times \mathbf{H}^1(\Omega)$, $b(\cdot, \cdot)$ is a continuous bilinear form on $\mathbf{H}^1(\Omega) \times L_0^2(\Omega)$ and $c(\cdot; \cdot, \cdot)$ is a continuous trilinear form on $\mathbf{H}^1(\Omega) \times \mathbf{H}^1(\Omega) \times \mathbf{H}^1(\Omega)$. For details concerning the function spaces we have introduced, one may consult [2; 25] and for details about the bilinear and trilinear forms and their properties, one may consult [9; 25].

We now formulate the mathematical model of the optimal boundary control problem. Let $\widehat{\Omega}$ be an extended domain and $\widehat{\Gamma}$ be the corresponding boundary. If Γ_c is the part of the boundary where we apply the control we assume that $\Gamma - \Gamma_c$ is a subset of $\widehat{\Gamma}$, namely only the controlled part of the boundary lies inside the extended domain $\widehat{\Omega}$. In the rest of the paper we denote by u the restriction to Ω of the function $\widehat{u}$ defined over the domain $\widehat{\Omega}$ and vice-versa. Some properties of an extension or a solenoidal extension of a function defined in Ω can be found in [2].

The optimal boundary control problem can then be stated by using the extended domain $\widehat{\Omega}$ and the distributed extended force $\widehat{f}$ in the following way:

find $\widehat{f} \in \mathbf{L}^2(\widehat{\Omega} - \Omega)$ such that $(\widehat{u}, \widehat{p}, \widehat{\tau})$ minimizes the functional

$$\mathcal{J} = \int_{\Omega_1} |\mathbf{u} - \mathbf{U}|^2 \, d\mathbf{x} + \frac{\beta}{2} \int_{\Omega - \widehat{\Omega}} |\widehat{f}|^2 \, d\mathbf{x}, \tag{1.11}$$

and satisfies

$$\begin{cases} a(\widehat{u},\widehat{v}) + c(\widehat{u};\widehat{u},\widehat{v}) + \langle\widehat{\tau},\widehat{v}\rangle_{\widehat{\Gamma}} + b(\widehat{v},p) \\ \qquad = \langle\widehat{f},\widehat{v}\rangle, \quad \forall\widehat{v}\in\mathbf{H}^1(\widehat{\Omega}) \\ b(\widehat{u},\widehat{q}) = 0, \quad \forall\widehat{q}\in L_0^2(\widehat{\Omega}) \\ \langle\widehat{u},\widehat{s}\rangle_{\widehat{\Gamma}} = \langle\widehat{g},\widehat{s}\rangle_{\widehat{\Gamma}}, \quad \forall\widehat{s}\in\mathbf{H}^{-1/2}(\widehat{\Gamma}) \end{cases} \tag{1.12}$$

with $\widehat{f} = \mathbf{f}$ over Ω. The domain Ω_1 is the part of the domain over which the matching is desired. The corresponding boundary control $\mathbf{g}_c \in \mathbf{H}^{3/2}(\Gamma_c)$ can be found after the solution of the above optimal control problem as the trace of the extended solution $\widehat{u}$ over Γ_c. We note that the boundary control $\mathbf{g}_c$ automatically satisfies the compatibility condition (1.4). Existence and uniqueness results for solutions of the system (1.12) are contained in [9; 25]. Note that solutions of (1.12) exists for any value of the Reynolds number. However the uniqueness can be guaranteed only for "large enough" values of ν or for "small enough" data $\mathbf{f}$ and $\mathbf{g}$. The admissible set of states and controls is given by

$$\begin{aligned} \mathcal{A}_{ad} = \{(\mathbf{u},p,\widehat{f},\mathbf{g}_c)&\in\mathbf{H}^1(\Omega)\times L_0^2(\Omega)\times\mathbf{L}_0^2(\widehat{\Omega})\times\mathbf{H}^{3/2}(\Gamma_c) \\ &\text{with } \mathbf{g}_c = \gamma_{\Gamma_c}\widehat{u} \text{ and } \widehat{f} = \mathbf{f} \text{ over } \Omega \\ &\text{such that } \mathcal{J}(\mathbf{u},\widehat{f}) < \infty \text{ and } (\widehat{u},\widehat{p}) \text{ satisfies (1.12)}\}. \end{aligned}$$

The existence of optimal solutions in this admissible set can be studied by using standard techniques (see for example [1; 8; 11; 12; 13]). Following this approach it is possible to show that optimal control solutions must satisfy a first-order necessary condition. They must satisfy the following Navier-Stokes system

$$\begin{cases} \nu a(\widehat{u},\widehat{v}) + c(\widehat{u};\widehat{u},\widehat{v}) + b(\widehat{v},\widehat{p}) = \langle\widehat{f},\widehat{v}\rangle, \quad \forall\widehat{v}\in\mathbf{H}^1_{\widehat{\Gamma}-\Gamma_2}(\widehat{\Omega}) \\ b(\widehat{u},\widehat{q}) = 0, \quad \forall\widehat{q}\in L_0^2(\widehat{\Omega}) \\ \langle\widehat{u},\widehat{s}\rangle_{\widehat{\Gamma}-\Gamma_2} = \langle\widehat{g},\widehat{s}\rangle_{\widehat{\Gamma}-\Gamma_2}, \quad \forall\widehat{s}\in\mathbf{H}^{-1/2}(\widehat{\Gamma}) \end{cases} \tag{1.13}$$

and the adjoint system

$$\begin{cases} \nu a(\widehat{w},\widehat{v}) + c(\widehat{w};\widehat{u},\widehat{v}) + c(\widehat{u};\widehat{w},\widehat{v}) + b(\widehat{v},\widehat{\sigma}) \\ \qquad = \int_{\Omega_1}(\mathbf{u}-\mathbf{U})\widehat{v}\,d\mathbf{x}, \quad \forall\widehat{v}\in\mathbf{H}^1_{\widehat{\Gamma}-\Gamma_2}(\widehat{\Omega}) \\ b(\widehat{w},\widehat{q}) = 0, \quad \forall\widehat{q}\in L_0^2(\widehat{\Omega}) \\ \langle\widehat{w},\widehat{s}\rangle_{\widehat{\Gamma}-\Gamma_2} = 0, \quad \forall\widehat{s}\in\mathbf{H}^{-1/2}(\widehat{\Gamma}) \end{cases} \tag{1.14}$$

with

$$\mathbf{g}_c = \gamma_{\Gamma_c}\widehat{u}, \tag{1.15}$$

and $\widehat{f} = \mathbf{f}$ over Ω and $\widehat{f} = \widehat{w}/\beta$ over $\widehat{\Omega} - \Omega$. Γ_2 is the part of the boundary where homogeneous Neumann boundary conditions (outflow) are imposed. The optimality system for the boundary control is reduced to a distributed optimal control problem which requires much less computational resources than other boundary control formulations. In this case there are no regularization parameters involved with the exception of β and the compatibility condition is automatically satisfied. The tangential control can be numerically achieved by using non-embedded techniques but in all these formulations the compatibility constraint is a limit to the feasibility of the normal boundary control. The normal boundary control must obey to this integral constraint reducing enormously the possibility to achieve accurate and fast numerical solutions of the necessary optimal control system with non-embedded techniques.

1.2.1 Numerical Computation of the Boundary Control Problem

The optimal boundary control problem can be solved by using a multigrid approach and the multigrid smoothing operator for each grid level can be derived directly from the optimal control problem. There is a vast class of smoothing operators for multigrid methods but we are interested in the class of Vanka-type solvers. In this class of solvers, which are well known for solving Navier-Stokes equations, the iterative solution is achieved by solving several exact systems involving blocks of variables. In particular we use the close relationship between this class of solvers and the class of solvers arising from saddle point or minimization problems

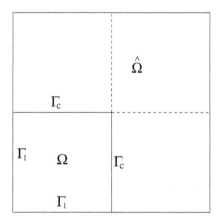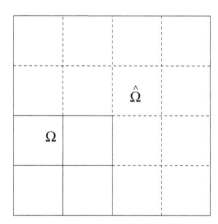

Fig. 1.2. Domain Ω and extended domain $\widehat{\Omega}$ at grid level l_0 (on the left) and at the grid level l_1 (on the right).

which allows us to use conforming isoparametric finite elements.

Let $\widehat{\Omega}_h$ be the square geometry described in Figure 1.2 on the left. Now, by starting at the multigrid coarse level l_0 we subdivide $\widehat{\Omega}_h$ into triangles or rectangles by unstructured families of meshes T_h^{i,l_0}. Based on the simple element midpoint refinement different multigrid levels can be built to reach a complete unstructured mesh $T_h^{i,l}$ for finite element over the entire domain Ω_h at the top finest multigrid level l_{n_t}. For example in Figure 1.2 on the right the mesh obtained by simple midpoint refinement is shown at the grid level l_1 (on the right).

We introduce the approximation spaces $\mathbf{X}_{h_l} \subset \mathbf{H}^1(\widehat{\Omega})$ and $S_{h_l} \subset L^2(\widehat{\Omega})$ for the extended velocity and pressure respectively at the multigrid level l. The approximate function obeys to the standard approximation properties including the LBB-condition. Let $P_{h_l} = X_{h_l}|_{\partial\widehat{\Omega}}$, i.e., P_{h_l} consists of all the restrictions, to the boundary $\partial\widehat{\Omega}$, of functions belonging to X_{h_l}. For all choices of conforming finite element space X_h we then have that $P_{h_l} \subset H^{-\frac{1}{2}}(\partial\widehat{\Omega})$. See [12] for details concerning these approximation spaces. The extended velocity and pressure fields $(\widehat{u}_{h_l}, \widehat{p}_{h_l}) \in \mathbf{X}_{h_l}(\widehat{\Omega}_h) \times S_{h_l}(\widehat{\Omega}_h)$ at the level l satisfy the Navier-Stokes equations

$$
\begin{cases}
a(\widehat{u}_{h_l}, \widehat{v}_{h_l}) + c(\widehat{u}_{h_l}; \widehat{u}_{h_l}, \widehat{v}_{h_l}) + b(\widehat{v}_{h_l}, \widehat{p}_{h_l}) = \langle \widehat{f}_{h_l}, \widehat{v}_{h_l} \rangle \\
\qquad \forall \widehat{v}_{h_l} \in \mathbf{X}_{h_l}(\widehat{\Omega}_h) \cap \mathbf{H}^1_{\widehat{\Gamma}_h - \Gamma_{2h}}(\widehat{\Omega}_h) \\
b(\widehat{u}_{h_l}, \widehat{r}_{h_l}) = 0, \quad \forall \widehat{r}_{h_l} \in S_{h_l}(\widehat{\Omega}_h) \\
\langle \widehat{u}_{h_l}, \widehat{s}_{h_l} \rangle_{\widehat{\Gamma}_h - \Gamma_{2h}} = \langle \mathbf{g}, \widehat{s}_{h_l} \rangle_{\widehat{\Gamma}_h - \Gamma_{2h}}, \quad \forall \widehat{s}_{h_l} \in \mathbf{P}_{h_l}(\widehat{\Gamma}_h)
\end{cases}
\tag{1.16}
$$

and the adjoint

$$
\begin{cases}
a(\widehat{w}_{h_l}, \widehat{v}_{h_l}) + c(\widehat{w}_{h_l}; \widehat{u}_{h_l}, \widehat{v}_{h_l}) + c(\widehat{u}_{h_l}; \widehat{w}_{h_l}, \widehat{v}_{h_l}) + b(\widehat{v}_{h_l}, \widehat{\sigma}_{h_l}) \\
\quad = \int_{\Omega_1} (\mathbf{u}_{h_l} - \mathbf{U}) \cdot \widehat{v}_{h_l}\, dx, \quad \forall \widehat{v}_{h_l} \in \mathbf{X}_{h_l}(\widehat{\Omega}_h) \cap \mathbf{H}^1_{\widehat{\Gamma}_h - \Gamma_{2h}}(\widehat{\Omega}_h) \\
b(\widehat{w}_{h_l}, \widehat{q}_{h_l}) = 0, \quad \forall \widehat{q}_{h_l} \in S_{h_l}(\widehat{\Omega}_h) \\
\langle \widehat{w}_{h_l}, \widehat{s}_{h_l} \rangle_{\widehat{\Gamma}_h - \Gamma_{2h}} = 0, \quad \forall \widehat{s}_{h_l} \in \mathbf{P}_{h_l}(\widehat{\Gamma}_h)
\end{cases}
\tag{1.17}
$$

with

$$
\mathbf{g}_{ch_l} = \gamma_{\Gamma_c} \widehat{u}_{h_l}
\tag{1.18}
$$

and $\widehat{f}_{h_l} = \mathbf{f}_h$ over Ω_{h_l} and $\widehat{f}_{h_l} = \widehat{w}_{h_l}/\beta$ over $\widehat{\Omega}_{h_l} - \Omega_{h_l}$. Existence and uniqueness results for finite element solutions of (1.16) are well known; see, e.g., [9; 11].

The unique representations of $\widehat{u}_{h_l}, \widehat{w}_{h_l}$ and $\widehat{p}_{h_l}, \widehat{\sigma}_{h_l}$ as a function of the nodal point values $\widehat{u}_l(k_1), \widehat{w}_l(k_1)$ and $\widehat{p}_l(k_2), \widehat{\sigma}_l(k_2)$ ($k_1 = 1, 2, ...nvt$ with $nvt =$ number of vertex velocity points and $k_2 = 1, 2, ...npt$ with $npt =$ number of vertex pressure points) define the finite element isomorphisms $\Phi_l : U_l \to X_{hl}$, $\Phi_l^+ : W_l \to X_{hl}$, $\Psi_l : \Pi_l \to S_{hl}$ $\Psi_l^+ : \Sigma_l \to S_{hl}$ between the vector spaces $U_l, W_l, \Pi_l, \Sigma_l$ of nvt-dimension and npt-dimension vectors and the finite element spaces X_{h_l}, S_{h_l}.

At the level l we introduce the corresponding finite element matrices A_l, B_l and $C_l(\widehat{u}_{h_l})$ for the discrete Navier-Stokes operators a, b, c defined by (1.8)–(1.10) respectively. Their corresponding finite element matrices for the adjoint operators are denoted by A_l^+, B_l^+ and $C_l^+(\widehat{u}_{h_l})$. The Navier-Stokes/adjoint coupled terms are denoted by H_l and G_l. Now the problem (1.16) is equivalent to

$$\begin{pmatrix} A_l + C_l \; B_l{}^T & H_l & 0 \\ B_l & 0 & 0 & 0 \\ G_l & 0 & A_l^+ + C_l^+ \; (B_l^+)^T \\ 0 & 0 & B_l^+ & 0 \end{pmatrix} \begin{pmatrix} \widehat{u}_{h_l,n} \\ \widehat{p}_{h_l,n} \\ \widehat{w}_{h_l,n} \\ \widehat{\sigma}_{h_l,n} \end{pmatrix} = \begin{pmatrix} \widehat{F}_{h_l} \\ 0 \\ \widehat{R}_{h_l} \\ 0 \end{pmatrix} \qquad (1.19)$$

at the multigrid level l. In the vector spaces U_l, W_l, Π_l and Σ_l we use the usual Euclidean norms which can be proved equivalent to the norms introduced to the corresponding finite element approximation spaces (see [5; 17] for details).

Essential elements of a multigrid algorithm are the velocity and pressure prolongation maps

$$P_{l,l-1}(u) : U_{l-1} \to U_l,$$
$$P_{l,l-1}(p) : \Pi_{l-1} \to \Pi_l,$$

and the velocity and restriction operators

$$R_{l-1,l}(u) = P^*_{l,l-1}(u) : U_l \to U_{l-1},$$
$$R_{l-1,l}(u) = P^*_{l,l-1}(u) : \Pi_l \to \Pi_{l-1}.$$

Since we would like to use conforming Taylor-Hood finite element approximation spaces we have the nested finite element hierarchies $X_{h_0} \subset X_{h_1} \subset \dots \subset X_{h_l}$ and $S_{h_0} \subset S_{h_1} \subset \dots \subset S_{h_l}$ and the canonical prolongation maps $P_{l,l-1}(u)$, $P_{l,l-1}(p)$ can be obtained simply by

$$P_{l,l-1}(u) = \Phi_{l-1}(\Phi_l^{-1}(u)),$$
$$P_{l,l-1}(p) = \Psi_{l-1}(\Psi_l^{-1}(p)).$$

For details and properties one can consult [17; 22] and citations therein.

We solve the coupled system (1.19) by using an iterative method. Multigrid solvers for coupled velocity/pressure systems compute simultaneously the solution for both pressure and velocity and they are known to be ones of the best class of solvers for laminar Navier-Stokes equations (see [18; 26]). An iterative coupled solution of the linearized and discretized incompressible Navier-Stokes equations requires the approximate solution of sparse saddle point problems. In this multigrid approach the most suitable class of solvers is the Vanka-type smoothers. They can be considered as block Gauss-Seidel methods where one block consists of a small number of degrees of freedom (for details see [26; 17; 18]). The characteristic feature of this type of smoother is that in each smoothing step a large number of small linear systems of equations has to be solved. In the Vanca-type smoother, a block consists of all degrees of freedom which are connected to few neighboring elements. As shown in Figure 1.3 for conforming finite elements the block could consist of all the elements containing a pressure vertex or four pressure nodes , namely 21 velocity nodes (circles and squares) with one pressure node (square) or 16 velocity nodes (circles and squares) with four pressure nodes (squares) respectively. Thus, in the first case a relaxation step with this Vanca-type smoother consists of the iterative solution of the corresponding block of equations over all the pressure nodes. In the second case a relaxation step consists of the solution of the block of equations over all the elements where the velocity

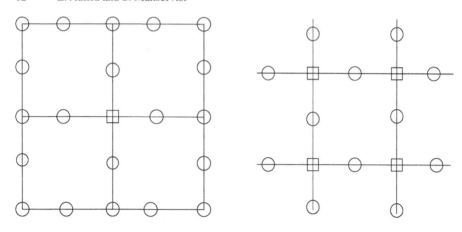

Fig. 1.3. Blocks of unknowns: $21V + 1P$ (on the left) and $16V + 4P$ (on the right).

and pressure variables are updated iteratively. Different blocks of unknowns can be solved including local constraints as they arise from the optimal control problem. For convergence and properties of this class of smoothers one can consult [26; 17; 18] and citations therein.

1.2.2 Boundary Control Test 1

We consider a unit square domain $\Omega = [0, 0.5] \times [0, 0.5]$ with boundary Γ as shown in Figure 1.2 on the left. Let $\widehat{\Omega}$ be $[0, 1] \times [0, 1]$. The boundary Γ consists of Γ_1, where homogeneous Dirichlet boundary conditions are applied, and Γ_c where the boundary control is applied. There are no Neumann boundary conditions and therefore Γ_2 is empty. The steady target velocity **U** for this test is given by

$$u(x, y) = 2(1 - x)^2(1 - \cos(4\pi x))((1 - y)(\cos(4\pi y) - 1)$$
$$+ 2\pi(1 - y)^2 \sin(4\pi y)),$$
$$v(x, y) = 2(1 - y)^2(\cos(4\pi y) - 1)((1 - x)(\cos(4\pi x) - 1)$$
$$+ 2\pi(1 - x)^2 \sin(4\pi x)).$$

In Figure 1.4 we show the results for different values of the penalty parameter β for Reynolds numbers equal to 100 and rectangular isoparametric Taylor-Hood finite elements. On the top there is the extended solution $\widehat{u}$ over the extended domain $\widehat{\Omega}$ for $\beta = 10^{-3}$. The desired solution and the controlled solution (for asymptotic β) are shown in Figure 1.4 in the middle from the left to the right respectively. We note that the boundary control can achieve some matching of the desired flow if the normal and the tangential control are combined. This embedded method can handle the normal control in a relative straightforward manner, satisfies the compatibility constraint and improves the effectiveness of the control. In these computations $\Omega_1 = \Omega$ but a better matching can be reached if the controlled area Ω_1 is tuned. On the bottom of Figure 1.4

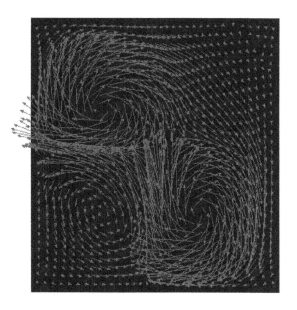

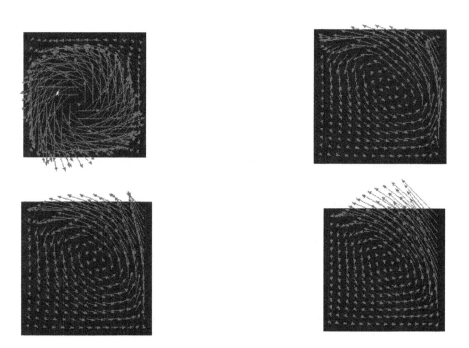

Fig. 1.4. Extended controlled flow (top) over $\widehat{\Omega} = [0,1] \times [0,1]$. Desired flow (central left) and controlled flow for $\beta = 1 \times 10^{-3}$ (central right), 1×10^{-2} (bottom left) and 1×10^{-1} (bottom right) over $\Omega = \Omega_1 = [0,0.5] \times [0,0.5]$.

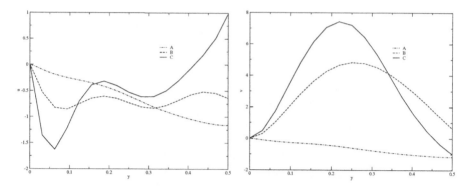

Fig. 1.5. Boundary control (u-component on the left and v-component on the right) on the vertical part of Γ_c for $\beta = 1 \times 10^{-1}$ (A), 1×10^{-2} (B) and 1×10^{-3} (C).

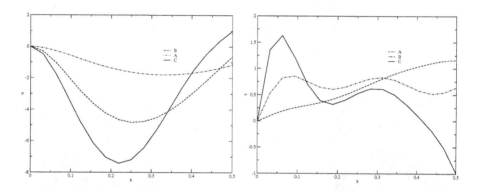

Fig. 1.6. Boundary control (u-component on the left and v-component on the right) on the horizontal part of Γ_c for $\beta = 1 \times 10^{-1}$ (A), 1×10^{-2} (B) and 1×10^{-3} (C).

we have the controlled solutions for different values of $\beta = 0.01$ (on the left) and $\beta = 0.1$ (on the right). The controlled boundary Γ_{ch} consists of a vertical and an horizontal part. Figure 1.5 shows the boundary control on the vertical part of Γ_c for $\beta = 1 \times 10^{-1}$ (A), 1×10^{-2} (B) and 1×10^{-3} (C). The u-component is shown on the left and the v-component on the right. In a similar way Figure 1.6 shows the u-component (on the left) and the v-component (on the right) of the horizontal part of the controlled boundary Γ_c for $\beta = 1 \times 10^{-1}$ (A), 1×10^{-2} (B) and 1×10^{-3} (C). We note that the controlled normal component of the boundary control may be positive and negative, namely there is injection and suction along the same portion of the boundary. If a standard non-embedded method is used the normal component of the control must satisfy the integral equation (1.4) and this may be numerically very challenging. Also this technique solves the corner point in a natural and straightforward manner while in the standard boundary control this point must be fixed by an artificial boundary condition which may limit the strength of the control.

In the second numerical experiment we would like to illustrate an example in which the boundary control can be efficiently applied to real situations. Suppose we have a velocity regulator as shown in Figure 1.1 on the left. The inflow over Γ_1 is assigned and we would like to control the fluid motion near to the output. By injection or suction along a portion of the boundary (for example Γ_c) it is possible to control accurately the velocity field. In order to model the problem we introduce, as shown in Figure 1.7 on the left, a L-shape domain with eight small cavities. The cavities are present in the real design and represent the area in which the fluid may be controlled. If a control is active in that area then we model such control as a boundary control, remove the cavity from the domain Ω and use it as a part of the extended domain. A very accurate study of this regulator can be done by taking into account all the seven cavities and the corresponding boundary controls but in this paper we investigate a simulation in which only the three parts of the boundary Γ_{1c}, Γ_{2c} and Γ_{3c} are controlled as shown in Figure 1.7 on the right. The desired velocity is a constant velocity field in the controlled area. The initial flow is a flow with parabolic velocity in Γ_1 with maximal velocity of $2.5m/s$. The Reynolds number of this initial velocity is 150 Reynolds with laminar motion everywhere. We compute the solution in two cases: constant horizontal target velocity of $0.5m/s$ (target case A) and $3.5m/s$ (target case B) in the controlled area. The controlled area, shown in Figure 1.8 on the left, is bounded by the line a and c. The vertical centerline of the controlled area is label by b. The stationary computations are performed with the penalty parameter β equal to 1000. In Figure 1.8 the controlled and desired u-component of the velocity are shown at $x = a$ (left), $x = b$ (center) and $x = c$ (right) for desired target case A and B. We note that the boundary suction and injection can control efficiently the average velocity to the target case A and B. In Figure 1.9-1.12 we see the controlled velocity for case A. In particular in Figure 1.9 the velocity field is shown in part of the domain Ω which is bounded on the right by the line c of the controlled area and from the left to the right of Figure 1.12 we have the boundary velocity over Γ_1, Γ_{2c} and Γ_{3c} respectively. In case A there is a strong suction in both boundary controls Γ_{1c} and Γ_{2c} in order to reduce the velocity in the controlled area and the boundary control in Γ_{3c} is relative small. In Figure 1.13 we see the velocity field for case B over boundary Γ_{1c} on the left, boundary Γ_{2c} in the center and boundary Γ_{3c} on the right.

In Figure 1.13 the u-component and the v-component of the velocity field are plotted as a function of the edge coordinate of the cavity. In this case there is suction in Γ_{1c} and injection in Γ_{2c} and Γ_{3c}. In case B the control in Γ_{3c} leads to a better matching in the desired velocity profile along c in the desired area close the upper boundary.

1.2.3 Boundary Control Test 2

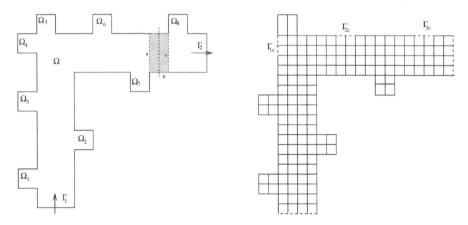

Fig. 1.7. Extended domain $\widehat{\Omega}$ (on the left) and domain Ω with boundary control over Γ_{1c}, Γ_{2c} and Γ_{3c} (on the right).

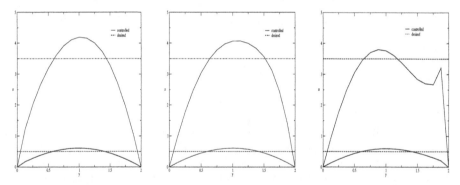

Fig. 1.8. Controlled and desired U-component of the velocity at $x = a$ (left), $x = b$ (center) and $x = c$ (right) in the controlled area for the desired target A and B.

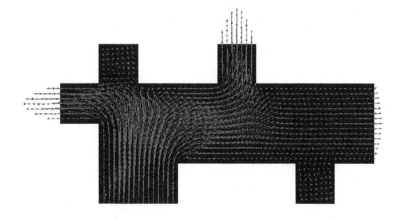

Fig. 1.9. Part of the velocity field for case A.

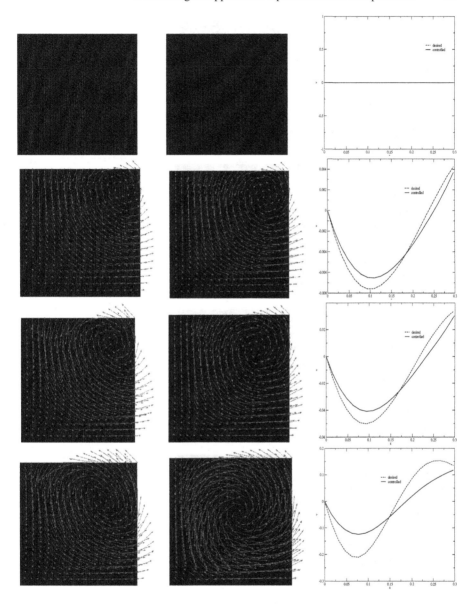

Fig. 1.10. Desired (left) and controlled (center) flow at different time $t = 0, 0.0125, 0.25$ and 0.75 from the top to the bottom. On the right the controlled and desired v-component along the x-axis at $y = 0.25$.

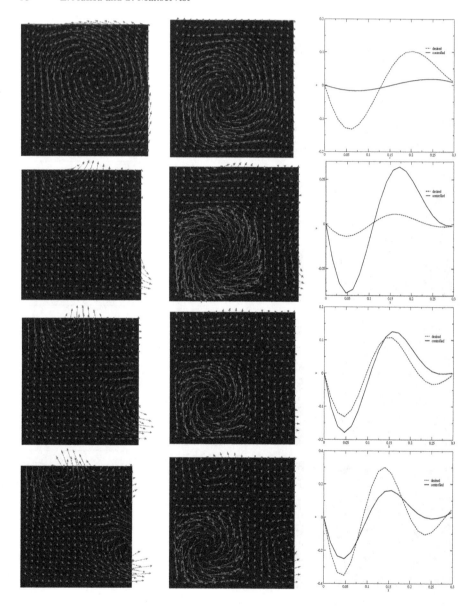

Fig. 1.11. Desired (left) and controlled (center) flow for different time $t = 1.5$, 1.25, 1.325 and 1.5 from the top to the bottom. On the right the controlled and desired v-component along the x-axis at $y = 0.25$.

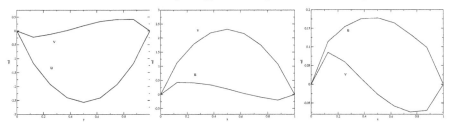

Fig. 1.12. Boundary control for case A. Boundary Γ_{1c} on the left, boundary Γ_{2c} in the middle and boundary Γ_{3c} on the right.

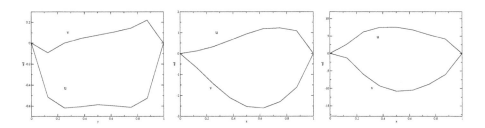

Fig. 1.13. Boundary control for case B. Boundary Γ_{1c} on the left, boundary Γ_{2c} on the center and boundary Γ_{3c} on the right.

1.3 Time Dependent Boundary Control Test

Now we test the proposed multigrid method with a time dependent boundary control problem. This problem reflects the desire to steer a candidate velocity field **u** to a given steady target velocity field **U** by appropriately controlling the velocity along a portion of the boundary of the flow domain. We consider a two-dimensional flow over a square domain $\Omega = [0, .35] \times [0, .35]$ with boundary Γ, control in $\Gamma_c \subset \Gamma$, homogeneous Dirichlet boundary condition in $\Gamma - \Gamma_c$ and extended domain $\widehat{\Omega} = [0, 1] \times [0, 1]$. The desired velocity $\mathbf{U} = (U, V)$ is given by

$$\phi(k, t, z) = (1 - \cos(2k\pi t z)) \times (1 - z)^2,$$

$$U(t, x, y) = \frac{d}{dy}\left(\phi(k, t, x)\,\phi(k, t, y)\right),$$

$$V(t, x, y) = -\frac{d}{dx}\left(\phi(k, t, x)\,\phi(k, t, y)\right)$$

with $k = 2$. Since we can use the multigrid approach to solve the time dependent optimal control problem over a block of unknowns we consider piecewise continuous boundary control in time. We assume that the time interval $[0, T]$ is divided in m equal intervals of time $\Delta t = T/m$ and over each interval we seek a continuous boundary control. This allows to discretize the optimal control system over every interval Δt in a limited number of subinterval δt and solve exactly the system over all time-space domain by the multigrid technique proposed in the previous section. We remark that if ΔT is discretized by a single time step ($\delta t = \Delta t$) we recover the linear feedback control presented in [14]. The optimal boundary problem, by using the notation of the previous section, can be stated in the following way

find $\widehat{f} \in \mathbf{L}^2([0,T], \mathbf{L}^2(\widehat{\Omega} - \Omega))$ such that $(\widehat{u}, \widehat{p}, \widehat{\tau})$ minimizes the functional

$$\mathcal{J} = \frac{\alpha_0}{2} \int_0^T \int_{\Omega_1} |\mathbf{u} - \mathbf{U}|^2 \, d\mathbf{x} \, dt + \sum_{j=1}^m \frac{\alpha_j}{2} \int_{\Omega_2} |\mathbf{u}(T_j) - \mathbf{U}(T_j)|^2 \, d\mathbf{x}$$

$$+ \frac{\beta}{2} \int_0^T \int_{\widehat{\Omega}-\Omega} |\widehat{f}|^2 \, d\mathbf{x} \, dt$$

and satisfies

$$\begin{cases} \langle \widehat{u}_t, \widehat{v} \rangle + a(\widehat{u}, \widehat{v}) + c(\widehat{u}; \widehat{u}, \widehat{v}) + \langle \widehat{\tau}, \widehat{v} \rangle_{\widehat{\Gamma}} + b(\widehat{v}, p) = \langle \widehat{f}, \widehat{v} \rangle \\ \qquad\qquad\qquad\qquad\qquad \forall \widehat{v} \in \mathbf{H}_0^1(\widehat{\Omega}) \\ b(\widehat{u}, \widehat{q}) = 0, \quad \forall \widehat{q} \in L_0^2(\widehat{\Omega}) \\ \langle \widehat{u}, \widehat{s} \rangle_{\widehat{\Gamma}} = \langle \widehat{g}, \widehat{s} \rangle_{\widehat{\Gamma}}, \quad \forall \widehat{s} \in \mathbf{H}^{-1/2}(\widehat{\Gamma}) \end{cases} \qquad (1.20)$$

with $\widehat{f} = \mathbf{f}$ over $[0,T] \times \Omega$. The term with the constant α_j is required if the adjoint function and therefore the control must not vanish at the end of the interval at the time T_j.

By using standard techniques (see for example [1; 8; 11; 12; 13]) we have that the optimal control solution must satisfy the following Navier-Stokes system over $[T_{j-1}, T_j]$

$$\begin{cases} \langle \widehat{u}_t, \widehat{v} \rangle + \nu a(\widehat{u}, \widehat{v}) + c(\widehat{u}; \widehat{u}, \widehat{v}) + b(\widehat{v}, \widehat{p}) = \langle \widehat{f}, \widehat{v} \rangle, \quad \forall \widehat{v} \in H_0^1(\widehat{\Omega}) \\ b(\widehat{u}, \widehat{q}) = 0, \quad \forall \widehat{q} \in L_0^2(\widehat{\Omega}) \\ (\widehat{u}, \widehat{s})_{\partial \widehat{\Omega}} = 0, \quad \forall \mathbf{s} \in H^{-1/2}(\partial \widehat{\Omega}) \\ \widehat{u}(T_{j-1}, \mathbf{x}) = \widehat{u}(T_{j-2}, \mathbf{x}), \quad \forall \mathbf{x} \in \widehat{\Omega} \end{cases}$$

the adjoint system over $[T_{j-1}, T_j]$

$$\begin{cases} -\langle \widehat{w}_t, \widehat{v} \rangle + \nu a(\widehat{w}, \widehat{v}) + c(\widehat{w}; \widehat{u}, \widehat{v}) + c(\widehat{u}; \widehat{w}, \widehat{v}) \\ \qquad + b(\widehat{v}, \widehat{\sigma}) = \alpha_0 (\mathbf{u} - \mathbf{U}, \widehat{v})_{\Omega_1}, \quad \forall \widehat{v} \in H_0^1(\widehat{\Omega}) \\ b(\widehat{w}, \widehat{q}) = 0, \quad \forall \widehat{q} \in L_0^2(\widehat{\Omega}) \\ \mathbf{w} = 0, \quad \forall \mathbf{x} \in \partial \widehat{\Omega} \\ \widehat{w}(T_j, \mathbf{x}) = \alpha_j (\widehat{u}(T_j) - \mathbf{U}(T_j)), \quad \forall \mathbf{x} \in \Omega_2 \end{cases}$$

for $j = 0, 1, \ldots, m$ and $\widehat{f} = \mathbf{f}$ over Ω and $\widehat{f} = \widehat{w}/\beta$ over $\widehat{\Omega} - \Omega$. The boundary control can be computed as $\mathbf{g} = \gamma_\Gamma \mathbf{u}$ with

$$\mathbf{g}_c = \gamma_{\Gamma_c} \widehat{u} \tag{1.21}$$

$\int_\Gamma \mathbf{g} \cdot \mathbf{n}\, d\Gamma = 0$. The optimality system for the boundary control can be computed in a very straightforward way if compared to other boundary control formulations.

In Figures 1.10-1.11 we show the desired (left) and the controlled (center) vector field for different time. All the figures are normalized with respect to the maximal velocity. From the left top to the bottom right of Figures 1.10-1.11 we have the velocity field at time $t = 0, 0.0125, 0.25, 0.75, 1.0, 1.25, 1.325$ and 1.5 respectively. The interval Δt is 0.025 with $\beta = 1000$ and $\alpha_j = 100$ with $j = 0, 1, \ldots$. Each interval has been divided in four time steps ($\delta = \Delta t/4$) and the complete optimality system solved with an exact method over all the time-space domain over the block of unknowns in Figure 1.3 on the right with a V-multigrid cycle. Since $\widehat{w}$ should not vanish at any time we set $\Omega_2 = [0, 0.5] \times [0, 0.5]$, namely $\Omega = \Omega_1 \subset \Omega_2 \subset \widehat{\Omega}$. On the right of Figures 1.10-1.11 the controlled and desired v-component along the x-axis at $y = 0.25$. It is clear that the matching is achieved essentially on the boundary and the boundary velocity cannot control the interior of the domain if the desired flow moves rapidly. However, this represents the optimum that can be achieved with the energy available.

1.4 Conclusions

We introduced an embedded method for boundary control which allows tracking and matching velocity field very efficiently. It is accurate and avoids the cumbersome regularizations of the standard boundary control. Also this methods allows to solve the problem for normal boundary control which must obey to the compatibility condition. A particular class of multigrid solvers is used in this paper to solve exactly the optimal control problem at the element level producing accurate and robust solutions. All this leads to improved computability and reliability for the numerical solution of steady and time dependent boundary control.

References

[1] F. ABERGEL AND R. TEMAM, *On some control problems in fluid mechanics*, Theoretical and Computational Fluid Dynamics **1**, 1990, pp. 303-326
[2] R. ADAMS, *Sobolev Spaces*, Academic Press, New York, 1975.
[3] G. ARMUGAN AND O. PIRONNEAU, On the problem of riblets as a drag reduction device, *Optim. Control Appl. Meth.* **10** 1989, pp. 93–112.
[4] M. BERGGREN, *Numerical solution of a flow control problem: Vorticity reduction by dynamic boundary action* , Siam J. Sci. Comput., **19** (3), 1998

[5] Z. CHEN, *On the convergence of non-nested multigrid methods with nested spaces on coarse grids* , Numer. Methods in Partial Differential Equations, pp. 265-284, **16**, 2000

[6] T. BEWLEY, R. TEMAM, AND M. ZIANE, *A robust framework for robust control in fluid mechanics*, Report, Center for Turbulence Research, Stanford University, Stanford, 1998.

[7] M. DESAI AND K. ITO, *Optimal control of Navier-Stokes equations*, SIAM J. Cont. Optim. **32**, 1994, pp. 1428–1446.

[8] A. FURSIKOV, M. GUNZBURGER AND L. HOU, *Boundary value problems and optimal boundary control for the Navier-Stokes system: the two-dimensional case* , SIAM J. Control Optim. **3** 36, 1998.

[9] V. GIRAULT AND P. RAVIART, *The Finite Element Method for Navier-Stokes Equations: Theory and Algorithms*, Springer-Verlag, New York, 1986.

[10] M. GUNZBURGER, L. HOU AND T. SVOBODNY, *Analysis and finite element approximation of optimal control problems for the stationary Navier-Stokes equations with Dirichlet controls*, Math. Mod. Num. Anal. **25**, 1991, pp. 771-748.

[11] M. GUNZBURGER AND S. MANSERVISI, *Analysis and approximation of the velocity tracking problem for Navier-Stokes flows with distributed control,* to appear in SIAM J. Numer. Anal.

[12] M. GUNZBURGER AND S. MANSERVISI, *The velocity tracking problem for Navier-Stokes flows with boundary control,* in SIAM J. Contr. Optim., Vol. 39 Number 2 pp. 594-634 (2000)

[13] M. GUNZBURGER AND S. MANSERVISI, *On a shape control problem for the stationary Navier-Stokes equations,* Mathematical Modelling and Numerical Analysis, Vol. 34, N 6, pp. 1233-1258 (2000)

[14] M. GUNZBURGER AND S. MANSERVISI, *The velocity tracking problem for Navier-Stokes flow with linear feedback control,* Journal of Comput. Meth. Appl. Mech. Eng., 189 (3) pp. 803-823 (2000)

[15] J. HASLINGER, K.H. HOFFMANN, AND M. KOCVARA, Control fictitious domain method for solving optimal shape design problems, *Math. Model. Num. Analy.* **27** 1993, pp. 157–182.

[16] M. HINZE AND K. KUNISH, *Control strategies for fluid flows - optimal versus suboptimal control*, Preprint n.573/1998, Technische Universitat, Germany 1998

[17] V. JOHN, P. KNOBLOCH, G. MATTHIES, AND L. TOBISKA, *Non-nested Multilevel Solvers for Finite Element Discretizations of Mixed Problems* , MGnet, preprint (2000)

[18] V. JOHN AND L. TOBISKA, *Numerical performance of smoother in coupled multigrid methods for parallel solution of the incompressible Navier-Stokes equations*, Int. J. Numer. Meth. Fluids, **33**, pp.453-473 (2000) MGnet, preprint (2000)

[19] P. KOUMOUTSAKOS, *Active control of vortex-wall interactions*, Phys. Fluids, **9**, 1997, pp. 3808–3816.

[20] J.L. LIONS AND E. MAGENES, *Problemes aux Limites Non Homogenes et Applications*, Vol.1 et 2, Dunod, Paris, 1968.

[21] O. PIRONNEAU, *Optimal Shape Design for Elliptic Systems,* Springer, Berlin, 1984.

[22] V. SHAIDUROV, *Multigrid Methods for finite elements,* Kluwer Academic Publisher, 1989.

[23] T. SLAWIG, *Domain Optimization for the Stationary Stokes and Navier-Stokes Equations by Embedding Domain Technique,* Thesis, TU Berlin, Berlin, 1998.

[24] S. SRITHARAN, *Dynamic programming of Navier-Stokes equations*, System and Control Letters, **16**, 1991, pp. 229-307.

[25] R. TEMAM, *Navier-Stokes Equations*, North-Holland, Amsterdam, 1979.

[26] S. TUREK, *Efficient solvers for incompressible flow problems*, Lectures notes in Comput. Sci. Engrg. **6**, Springer-Verlag, New York 1999.

Control System Radii and Robustness Under Approximation

John A. Burns [*][1] and Gunther H. Peichl [**][2]

[1] Center for Optimal Design and Control
   Interdisciplinary Center for Applied Mathematics
   Virginia Tech
   Blacksburg, VA 24061-0531
   `burns@icam.vt.edu`
[2] Institut für Mathematik und Wissenschaftliches Rechnen
   Karl-Franzens-Universität Graz
   A-8010 Graz, Austria
   `gunther.peichl@uni-graz.at`

Summary. The purpose of this paper is twofold. First, we provide a short review and summarize results on the robustness of controllability and stabilizability for finite dimensional control problems. We discuss the computation of system radii which provide a measure of robustness. Second, we consider systems which arise as finite difference and finite element approximations to control systems defined by partial differential equations. In particular, we derive controllability criteria for approximations of the controlled heat equation which are easy to check numerically. For a particular example we establish tight theoretical upper and lower bounds on the controllability radii for the finite difference and finite element models and compare these bounds with numerical results. Finally, we present numerical results on stabilizability radii which suggests that conditioning of the LQR control problem may be measured by this radii.

Key words: control system radii, numerical methods, robustness

Introduction

The analysis of mathematical models used in control design and optimization often requires several stages of approximation. Also, in the area of distributed parameter

[*] This research was supported in part by the Air Force Office of Scientific Research under grant F49620-02-C-0048 and the Special Projects Office at DARPA. Parts of the research were carried out while the author was a visitor at the Institut für Mathematik und Wissenschaftliches Rechnen at Karl-Franzens-Universität Graz, Austria, supported by the Fonds zur Förderung der wissenschaftlichen Forschung, Austria, under project SFB03-18 "Optimierung und Kontrolle".
[**] This research was supported in part by the Fonds zur Förderung der wissenschaftlichen Forschung under P7522-PHY, P8146-PHY, and SFB03-18 "Optimierung und Kontrolle".

(DP) control, numerical approximation must be introduced at some point in the modelling process. Finite element, Galerkin and finite difference schemes are typically used to "discretize" continuum models, while in the frequency domain one might construct rational approximations of non-rational transfer functions. For computing purposes, state space models offer certain advantages in that there are numerous computational algorithms well suited for the matrix-linear algebra problems that occur in control design. Direct discretization of continuum models usually produce state space models as do frequency domain approximations (followed by realization schemes) and model reduction methods such as proper orthogonal decomposition (POD). It is fair to say that all approaches have advantages and disadvantages and each approach leads to its own characteristic set of problems. However, to achieve robustness in a design based on approximate models one needs to take into account the robustness of the approximate model with respect to system properties.

Given that there are several approaches to constructing finite dimensional state space models, it is reasonable to ask if there is some "measure" that can be used to select the "best" approach for a given system with a specific control design objective. In order to study this problem, it is clear that one must first decide what criteria will be used to determine which state space model is "best" for the particular problem at hand. It is very important to remember that such criteria may change it the system changes, if the control design objective changes or if the numerical method used to solve the corresponding control problem changes. Since the finite dimensional approximate/reduced order model will be used to design and optimize the infinite dimensional system, it is important that the finite dimensional model inherits the essential control system properties and that the finite dimensional control problem is numerically well-conditioned. For example, it has been observed [6; 7] that numerical conditioning of the LQR problem can be negatively influenced when non-uniform meshes are used to approximate a DP system governed by a partial differential equation.

In this paper we investigate these ideas for distributed parameter systems. We shall focus on a specific subset of these problems. Our goal is to illustrate how one can use system measures to aid in the selection of model reduction and discretization algorithms. In particular, we shall use the concept of "system radii" to measure the "quality" of finite dimensional state space models constructed by direct discretization of continuum models. The motivation for this choice lies in the observation that numerical algorithms for control design can be (numerically) unstable if applied to systems that are not controllable (observable, stabilizable, etc.). Moreover, numerical ill-conditioning can result even if the system is controllable and observable but "near" an uncontrollable or unobservable system. This idea is certainly not new and there exist many examples of this type. Demmel [13] has developed a rather nice theory of ill-conditioning and established that numerous problems in numerical linear algebra (matrix inversion, eigenvalue calculations) and control design (pole-placement, robust control) all become ill-conditioned if the state space models used in the calculations are close to an ill-posed problem. Laub and his co-workers have established similar results for the LQR problem [18; 29; 30]. Since one of the often noted "advantages" of state space models is their usefulness for computational purposes, it is reasonable to use the condition number of the problem as one measure to help select a discretization scheme. One can find a nice presentation of these ideas in the recent book [12] by Datta.

Although there are several issues that need to be addressed in the overall approximation process, we shall limit most of our discussion to the study of system radii for systems that typically occur when partial differential equations are discretized by finite element and finite difference schemes. These finite dimensional systems often have nice symmetry properties that can be exploited in the computation of the system radii. The basic problem of preserving system properties under approximation has been addressed by other authors [4; 17; 32] and is crucial to any method. However, we concentrate on the problem of selecting a "good" approximation from the class of all schemes that preserve the appropriate system properties.

The paper is organized as follows. In Section 2.1 we review the basic definitions of system (radii) measures for finite dimensional systems and present examples to illustrate some relationships between these measures and typical control problems. We also summarize a few known results concerning these measures and give some new results on computing these measures. In Section 2.2 we discuss the problem of approximating infinite dimensional systems and use finite element and finite difference approximations of the heat equation to illustrate the ideas. This simple example is rich enough to provide some indication of the difficulties one encounters in developing theoretical and computational results for such problems. In Section 2.3 we provide a case study and compare theoretical bounds to computed values. Finally, we close with a short summary and a simple numerical example to illustrate the potential use of system radii to estimate the numerical condition number of an LQR problem.

2.1 Measures of Robustness

As noted in [1] numerical algorithms which assume a specific system property such as controllability or stabilizability can be expected to be numerically ill-conditioned if the system model is nearly uncontrollable (or nearly unstabilizable). The following simple example illustrates the type of difficulties that one can encounter.

Motivating Example. Consider the control system governed by the second order system

$$\frac{d}{dt}\begin{bmatrix} x_1(t) \\ x_2(t) \end{bmatrix} = \begin{bmatrix} 1 & 0 \\ 0 & -1 \end{bmatrix} \begin{bmatrix} x_1(t) \\ x_2(t) \end{bmatrix} + \begin{bmatrix} \delta \\ \epsilon \end{bmatrix} u(t).$$

Observe that this system is controllable (and stabilizable) if and only if $\delta \neq 0$ and $\epsilon \neq 0$. It becomes uncontrollable if $\epsilon = 0$ and unstabilizable if and only if $\delta = 0$. Moreover, the system becomes "nearly uncontrollable" as $\epsilon \to 0$ (and "nearly unstabilizable" as $\delta \to 0$) in the sense that a perturbation of order ϵ (δ) in the input matrix $\mathrm{col}(\delta, \epsilon)$ may result in an uncontrollable (unstabilizable) system.

In order to demonstrate the effect of near unstabilizability on control design, we consider the problem of minimizing the quadratic functional

$$J(u) = \int_0^{+\infty} [(x_1(t))^2 + (x_2(t))^2 + u^2(t)]dt.$$

The optimal feedback gain is given by

$$k^*(\delta, \epsilon) = [k_1^*, k_2^*] = -[\delta, \epsilon]\Pi(\delta, \epsilon)$$

where $\Pi(\delta, \epsilon)$ is the solution to the Riccati equation

$$A^*\Pi + \Pi A - \Pi B r^{-1} B^* \Pi + Q = 0,$$

and $Q = I_2$ and $r = 1$.

It is straightforward to show that

$$\Pi(\delta, \epsilon) = \begin{bmatrix} \frac{2+\epsilon^2+2\sqrt{1+\epsilon^2+\delta^2}}{2\delta^2} & \frac{-\epsilon}{2\delta} \\ \frac{-\epsilon}{2\delta} & \frac{1}{2} \end{bmatrix}$$

and hence

$$k^*(\delta, \epsilon) = -\Big[\frac{1+\sqrt{1+\epsilon^2+\delta^2}}{\delta}, 0\Big].$$

Note that as $\delta \to 0$, $\|\Pi(\delta, \epsilon)\| \to +\infty$ and $\|k^*\| \to +\infty$. Here, $\|\cdot\|$ denotes any suitable matrix or vector norm. Thus, as the system approaches an unstabilizable system, the Riccati equation becomes ill-conditioned. As expected, the conditioning of the Riccati equation is not affected by the loss of controllability. However, consider the problem of finding a feedback operator $k_p(\delta, \epsilon) = [k_p^1, k_p^2]$ that places the closed-loop poles at -2 and -4. In particular, if $\delta\epsilon \neq 0$ then the unique solution to this problem exists and is given by

$$k_p = \begin{bmatrix} \frac{-15}{2\delta}, \frac{3}{2\epsilon} \end{bmatrix}.$$

Observe that as δ or ϵ approach 0, the system becomes nearly uncontrollable and $\|k_p\| \to +\infty$.

The previous example illustrates the need for a device to measure nearness of a system to uncontrollability, respectively unstabilizability. In order to make these ideas precise we introduce the following notation. We identify the control system Σ

$$\dot{x} = Ax + Bu, \tag{Σ}$$

where $A \in \mathbb{R}^{n\times n}$, $B \in \mathbb{R}^{n\times m}$, $m \leq n$, with the matrix $[A, B] \in \mathbb{R}^{n\times(n+m)}$. For any $\lambda \in \mathbb{C}$ we introduce

$$H(\lambda) = [A - \lambda I, B] \in \mathbb{C}^{n\times(n+m)}.$$

The Hautus - test (see [28]) for controllability is based on embedding $[A, B]$ in the set of complex systems

$$\Gamma = \{[A, B] : A \in \mathbb{C}^{n\times n}, B \in \mathbb{C}^{n\times m}\}.$$

The distance between two systems Σ_1, Σ_2 is defined by

$$\delta(\Sigma_1, \Sigma_2) = \|[A_1 - A_2, B_1 - B_2]\|,$$

where $\| \cdot \|$ denotes any suitable matrix norm on $\mathbb{C}^{n \times (n+m)}$. Given $\Sigma \in \Gamma$ and a subset $S \subset \Gamma$ the distance between Σ and S is defined by

$$d(\Sigma, S) = \inf\{\delta(\Sigma, \Sigma_\alpha) : \Sigma_\alpha \in S\}.$$

Let $N_c \subset \Gamma$ be the set of all complex systems that are not controllable and $N_s \subset \Gamma$ be the set of all complex systems that are not stabilizable, i.e.

$$N_c = \{[A, B] \in \Gamma : [A, B] \text{ is not controllable}\},$$
$$N_s = \{[A, B] \in \Gamma : [A, B] \text{ is not stabilizable}\}.$$

Given $\Sigma \in \Gamma$ one defines the measure of controllability by

$$\gamma_c = d(\Sigma, N_c) \tag{2.1}$$

and the measure of stabilizability by

$$\gamma_c = d(\Sigma, N_s). \tag{2.2}$$

These definitions may be found in [14; 29; 34]. There are several reasons that these measures are useful. First, they can provide an estimate of the condition number for control design algorithms (see [13] and the example on robust pole placement therein). Moreover, they provide numerical bounds on the errors that can be tolerated in the data defining the system matrices A and B.

To obtain more explicit formulae for γ_c and γ_s the norm in $\mathbb{C}^{n \times (n+m)}$ has to be related to the norms in $\mathbb{C}^n$ and $\mathbb{C}^{n+m}$. We shall modify a concept which was used in [21] to calculate the stability radius of a matrix. If we choose the Euclidean norm in $\mathbb{C}^n$ and in $\mathbb{C}^{n+m}$, and the spectral norm in $\mathbb{C}^{n \times (n+m)}$, then γ_c and γ_s are determined by the singular values of $H(\lambda)$. The singular values $\sigma_1, \ldots, \sigma_p$, $p = \min(r, s)$ of a matrix $H \in \mathbb{C}^{r \times s}$ (see [19] for basic definitions) will be ordered in the standard fashion $\sigma_1 \geq \cdots \geq \sigma_p$. We shall also use $\sigma_{min}(H)$ to denote the smallest singular value σ_p of a matrix H.

Definition 1. *Let $\| \cdot \|_n$, $\| \cdot \|_m$ be norms on $\mathbb{C}^n$ and $\mathbb{C}^m$ respectively and let $\| \cdot \|_n^*$ ($\| \cdot \|_m^*$) denote the norm dual to $\| \cdot \|_n$ ($\| \cdot \|_m$). A matrix norm $\| \cdot \|_{n,m}$ on $\mathbb{C}^{n \times m}$ is said to be strongly compatible with $\| \cdot \|_n^*$ and $\| \cdot \|_m^*$ if the following two conditions hold*

(C1) $\|x^* A\|_m^* \leq \|A\|_{n,m} \|x^*\|_n^*$ *for all $A \in \mathbb{C}^{n \times m}$, $x^* \in (\mathbb{C}^n)^*$.*
(C2)For any pair of vectors $x^* \in (\mathbb{C}^n)^*$, $x^* \neq 0$, $y^* \in (\mathbb{C}^m)^*$ there exists $H \in \mathbb{C}^{n \times m}$ satisfying

$$y^* = x^* H \quad and \quad \|H\|_{n,m} \|x^*\|_n^* = \|y^*\|_m^*.$$

In the above definition $(\mathbb{C}^n)^*$ denotes the dual space of $\mathbb{C}^n$ which is (algebraically) identified with $\mathbb{C}^{1 \times n}$. Easy modifications of the proofs in [21] establish that the operator norms on $\mathbb{C}^{n \times m}$ as well as the Hölder norms are strongly compatible with $\| \cdot \|_n^*$ and $\| \cdot \|_m^*$. In particular, this implies that the spectral norm and the Frobenius norm (see [19]) are strongly compatible with the Euclidean norms in $\mathbb{C}^n$ and $\mathbb{C}^m$.

Theorem 1. *Choose any vector norms* $\| \cdot \|_n$, $\| \cdot \|_{n+m}$ *in* $\mathbb{C}^n$ *and* $\mathbb{C}^{n+m}$ *and any matrix norm in* $\mathbb{C}^{n \times (n+m)}$ *strongly compatible with* $\| \cdot \|_n^*$ *and* $\| \cdot \|_{n+m}^*$. *If the system* $[A, B] \in \Gamma$ *is controllable, then the measure of controllability is given by*

$$\gamma_s = \min_{\lambda \in \mathbb{C}} \min_{\substack{\|x\|_n^* = 1 \\ x^* \in (\mathbb{C}^n)^*}} \|x^*[A - \lambda I, B]\|_{n+m}^* . \tag{2.3}$$

Proof. Let α denote the number on the righthand side of (2.3) and let $[\delta A_0, \delta B_0] \in \Gamma$ satisfy $\|[\delta A_0, \delta B_0]\|_{n,n+m} = \gamma_c$ and $[A + \delta A_0, B + \delta B_0] \in N_c$. Hence, there exists $\lambda \in \mathbb{C}$, $x^* \in (\mathbb{C}^n)^*$, $\|x^*\|_n^* = 1$ with

$$x^*[A + \delta A_0 - \lambda I, B + \delta B_0] = 0.$$

Using (C1) this implies

$$\|x^*[A - \lambda I, B]\|_{n+m}^* \leq \|x^*\|_n^* \|[\delta A_0, \delta B_0]\|_{n,n+m}$$

and a fortiori

$$\alpha \leq \gamma_c.$$

On the other hand one can argue the existence of $\lambda_0 \in \mathbb{C}$, $x_0^* \in (\mathbb{C}^n)^*$, $\|x_0^*\|_n^* = 1$ with

$$\alpha = \|x_0^*[A - \lambda_0 I, B]\|_{n+m}^*.$$

Condition (C2) applied to $x^* = x_0^*$, $y^* = x_0^*[A - \lambda_0 I, B]$ ensures the existence of $[\delta \hat{A}, \delta \hat{B}] \in \Gamma$ with

$$x_0^*[A - \lambda_0 I, B] = x_0^*[\delta \hat{A}, \delta \hat{B}] \tag{2.4}$$

$$\|[\delta \hat{A}, \delta \hat{B}]\|_{n+m} \cdot \|x_0^*\|_n^* = \|x_0^*[A - \lambda_0 I, B]\|_{n+m}^*. \tag{2.5}$$

The identity (2.4) shows that $[A - \delta \hat{A}, B - \delta \hat{B}] \in N_c$, and (2.5) implies that

$$\gamma_c \leq \alpha.$$

$\square$

Corollary 1. *Let the norms be chosen as in Theorem 1. If the system* $[A, B] \in \Gamma$ *is stabilizable, then the measure of stabilizability is given by*

$$\gamma_s = \min_{\substack{\lambda \in \mathbb{C} \\ 0 \leq \mathrm{Re}\,\lambda}} \min_{\substack{\|x\|_n^* = 1 \\ x^* \in (\mathbb{C}^n)^*}} \|x^*[A - \lambda I, B]\|_{n+m}^* . \tag{2.6}$$

2 Control System Radii and Robustness Under Approximation

Corollary 2. *Choose the Euclidean norms in $\mathbb{C}^n$ and $\mathbb{C}^{n+m}$ and the spectral norm (or Frobenius norm) in $\mathbb{C}^{n \times (n+m)}$.*

i) *If $[A, B] \in \Gamma$ is controllable, then*

$$\gamma_c = \min_{\lambda \in \mathbb{C}} \sigma_{\min}(H(\lambda)). \tag{2.7}$$

ii) *If $[A, B] \in \Gamma$ is stabilizable, then*

$$\gamma_s = \min_{Re\lambda \geq 0} \sigma_{\min}(H(\lambda)). \tag{2.8}$$

The characterizations (2.7) and (2.8) were first derived in [14] using a different argument.

Throughout the remaining part of this chapter we shall use the Euclidean norms in $\mathbb{C}^n$ and $\mathbb{C}^{n+m}$ and the Frobenius norm $\| \cdot \|_F$ in $\mathbb{C}^{n \times (n+m)}$. In [15; 44; 45] an alternative solution to (2.1) was given.

Theorem 2. *If $[A, B] \in \Gamma$ is controllable, then*

$$\gamma_c = \min_{\substack{\|q\|=1 \\ q \in \mathbb{C}^n}} \|q^* [A(I - qq^*), B]\|. \tag{2.9}$$

Although (2.9) is equivalent to an optimization problem given in [15], the approach to establish (2.9) given in [44; 45] is entirely different. The minimal perturbation $[\delta A_0, \delta B_0]$ is determined by using a state transformation that produces a certain canonical form. In [44; 45] the identity (2.9) is established by showing its equivalence with (2.7). For completeness we want to give a direct proof based on the above motivation.

Proof of Theorem 2. Let $[\delta A, \delta B] \in \Gamma$ satisfy $[A + \delta A, B + \delta B] \in N_c$. Then there exists $Q = [Q_1, q] \in \mathbb{C}^n$, $Q_1 \in \mathbb{C}^{n,n-1}$, $q \in \mathbb{C}^n$, $Q^*Q = I$ such that in the transformed system

$$Q^* (A + \delta A) Q = \begin{pmatrix} Q_1^* (A + \delta A) Q_1 & Q_1^* (A + \delta A) q \\ q^* (A + \delta A) Q_1 & q^* (A + \delta A) q \end{pmatrix},$$

$$Q^* (B + \delta B) = \begin{pmatrix} Q_1^* (B + \delta B) \\ q^* (B + \delta B) \end{pmatrix},$$

$$q^*(A + \delta A)Q_1 = 0 \quad \text{and} \quad q^*(B + \delta B) = 0 \tag{2.10}$$

hold. Fix $q \in \mathbb{C}^n$. The minimal norm perturbation δB, in (2.10) is given by

$$\delta B_0 = -qq^* B.$$

Since the columns of Q_1 are orthogonal, the first equation in (2.10) implies the existence of $\lambda \in \mathbb{C}$ with

$$q^*(A + \delta A) = \lambda q^*$$

which gives the minimal norm perturbation δA_0

$$\delta A_0 = qq^*\lambda - qq^*A.$$

Also $\|\delta A_0\|_F = \|q\|_2 \|q^*\lambda - q^*A\|_2$ minimizing $\|q^*\lambda - q^*A\|_2^2$ with respect to λ so that the minimum is attained at $\lambda_0 = q^*Aq$. Hence, the minimum norm perturbation δA_0 is given by

$$\delta A_0 = qq^*(q^*Aq) - qq^*A = q(q^*Aq)q^* - qq^*A = -qq^*A(I - qq^*).$$

Thus we obtain

$$[\delta A_0, \delta B_0] = -qq^*[A(I - qq^*), B].$$

and therefore

$$\|[\delta A_0, \delta B_0]\|_F = \|q^*[A(I - qq^*), B]\|_2.$$

Minimizing with respect to q, yields the desired result.   □

Remark. As a consequence of the above proof we note that if $q_0 \in \mathbb{C}^n$ is optimal in (2.9), then

$$[\delta A_0, \delta B_0] = -q_0 q_0^*[A(I - q_0 q_0^*), B]$$

yields a minimal norm perturbation of $[A, B]$ destroying controllability. It is shown in [44, Theorem 4.3] that if $\lambda_0 \in \mathbb{C}$ is optimal in (2.7) and $U = [u_1, \ldots, u_n] \in \mathbb{C}^{n \times n}$, $V = [v_1, \ldots, v_{n+m}] \in \mathbb{C}^{(n+m) \times (n+m)}$ determine a singular value decomposition of $H(\lambda_0)$, then $q_0 = u_n$ minimizes (2.9) and

$$[\delta A_0, \delta B_0] = \gamma_c u_n v_n^*, \quad \lambda_0 = u_n^* A u_n. \tag{2.11}$$

holds. Conversely if q_0 is a minimizer for (2.9) then $\lambda_0 = q_0^* A q_0$ minimizes (2.7) having q_0 as a left singular vector. In particular (2.11) reveals that for real systems $[A, B] \in \mathbb{R}^{n \times (n+m)}$ the closest uncontrollable (unstabilizable) system will in general be complex.

In order to study the effect of real perturbations, real measures have been introduced (see [1; 15; 45; 18]). In particular, let

$$\Omega = \{\Sigma = [A, B] : A \in \mathbb{R}^{n \times n}, B \in \mathbb{R}^{n \times m}\}$$

denote the set of real systems. Then one defines the real measure of controllability by

$$\omega_c = d(\Sigma, N_c \cap \Omega)$$

and the real measure of stabilizability by

$$\omega_s = d(\Sigma, N_s \cap \Omega).$$

In general it is clear that

$$\gamma_c \leq \omega_c \quad \text{and} \quad \gamma_s \leq \omega_s$$
$$\gamma_c \leq \gamma_s \quad \text{and} \quad \omega_c \leq \omega_s$$

hold and it can happen that there is a significant difference between the various measures. It is tempting to assume that ω_c could be found by computing the quantity

$$\omega_{c,1} = \min_{\lambda \in \mathbb{R}} \sigma_{\min}(H(\lambda)). \tag{2.12}$$

In general, however, $\omega_{c,1}$ yields just an upper bound for ω_c. This is a consequence of the following theorem which was established in [45]:

Theorem 3. *If $[A, B] \in \Omega$ is controllable, then*

$$\omega_c = \min(\omega_{c,1}, \omega_{c,2}),$$

where $\omega_{c,1}$ is given by (2.12) and $\omega_{c,2}$ is defined by

$$\omega_{c,2} = \min_{\substack{Q \in \mathbb{R}^{n \times 2} \\ Q^T Q = I_n}} \left\| Q^T \left[A - QQ^T, Q^T B \right] \right\|_F. \tag{2.13}$$

Corollary 3. *If $[A, B] \in \Omega$ is controllable and $n = 2$, then*

$$\omega_c = \min(\omega_{c,1}, \|B\|_F).$$

The proof of this result comes from the observation that $QQ^T = I_2$ holds for $n = 2$. We illustrate the above discussion by means of the following example.

Example 1. Let $\varepsilon < 2$ and define

$$A = \begin{pmatrix} 0 & -1 \\ 1 & 0 \end{pmatrix} \quad \text{and} \quad B = \begin{pmatrix} \varepsilon \\ 0 \end{pmatrix}.$$

It follows that $\omega_{c,1} = 1$, hence $\omega_c = \varepsilon$ and $[\delta A, \delta B] = [0, -B]$ is a real minimal norm perturbation destroying controllability. A short calculation shows that $\gamma_c = \varepsilon\sqrt{\frac{1}{2} - \frac{\varepsilon^2}{16}}$, the minimum in (2.7) being attained at $\lambda_0 = \pm i\sqrt{1 - \frac{\varepsilon^4}{16}}$. Observe that although $\omega_c = \omega_{c,2}$, it does not necessarily follow that the corresponding minimal norm perturbation destroying controllability $[\delta A, \delta B] = [QQ^T A(I - QQ^T), QQ^T B]$ has rank 2. This should be kept in mind in interpreting the corresponding results in [18]. For a more detailed discussion of ω_c we refer to [18; 45]. We complement these results by a characterization of the equality $\omega_c = \gamma_c$ which is adapted from an analogous result concerning the calculation of stability radii in [33]. (In the next two results we use the spectral norm in $\mathbb{C}^{n \times (n+m)}$.)

Theorem 4. *Let* $[A, B] \in \Omega$ *be controllable and define*

$$\Lambda_0 = \{\lambda^* \in \mathbb{C} : \sigma_{\min}(H(\lambda^*)) = \gamma_c\}.$$

Then $\gamma_c = \omega_c$ *holds if and only if there exist* $\lambda_0 \in \Lambda_0$ *and a singular value decomposition of* $H(\lambda_0)$,

$$H(\lambda_0) = \sum_{i=1}^{n} \sigma_i u_i v_i^*, \ u_i \in \mathbb{C}^n, \ v_i \in \mathbb{C}^{n+m}, \ i = 1, \dots, n \qquad (2.14)$$

satisfying

$$u_n^T u_n = v_n^T v_n. \qquad (2.15)$$

Proof. First we show the necessity of (2.15). It suffices to discuss $\Lambda_0 \cap \mathbb{R} = \varnothing$. Assume that $\gamma_c = \omega_c$ holds and that the minimum in (2.7) is attained for some $\lambda_0 \in \mathbb{C} \setminus \mathbb{R}$. Hence, there is a minimal norm (with respect to the spectral norm) real perturbation $E \in \mathbb{R}^{n \times (n+m)}$ of $[A, B]$ satisfying

$$\|E\|_2 = \gamma_c = \sigma_n = \sigma_{\min} H(\lambda_0).$$

Thus, using (2.11) E admits the representation

$$E = -\sigma_n u_n v_n^*. \qquad (2.16)$$

It follows that

$$u_n^*([A, B] + E) = u_n^* H(\lambda_0) + u_n^* E + \lambda_0 u_n^* [I, 0] = \lambda_0 u_n^* [I, 0].$$

Multiplying on the right by $\bar{v}_n$ and taking complex conjugates (note that $E \in \mathbb{R}^{n \times (n+m)}$) we arrive at

$$u_n^T([A, B] + E)v_n = \bar{\lambda}_0 \, u_n^T [I, 0] v_n. \qquad (2.17)$$

Similarly, starting with $([A, B] + E)v_n$ it follows that

$$u_n^T([A, B] + E)v_n = \lambda_0 \, u_n^T [I, 0] v_n. \qquad (2.18)$$

Comparing (2.17) and (2.18) we obtain

$$u_n^T [I, 0] v_n = 0. \qquad (2.19)$$

Although the remaining part of the proof is identical to the one in [33] we present it here for the sake of completeness. The decomposition (2.14) together with (2.19) imply

$$u_n^T [A, B] v_n = \sigma_n u_n^T u_n,$$

while

$$[A, B] - \bar{\lambda}_0 [I, 0] = \sum_{i=1}^{n} \sigma_i \bar{u}_i v_i^T,$$

the complex conjugate of (2.14), combined with (2.19) yields

$$u_n^T[A, B]v_n = \sigma_n v_n^T v_n.$$

This completes the necessity part of the proof. Sufficiency of (2.15) may be shown in exactly the same way as in [33].   □

A consequence of Theorem 4 is the following easily checked sufficient condition for $\gamma_c = \omega_c$ to hold. We present it for the sake of completeness and refer to [33] for the proof.

Proposition 1. *Let* $[A, B] \in \Omega$ *be controllable and* Λ_0 *be as defined in Theorem 4. If for some* $\lambda_0 \in \Lambda_0$ $\sigma_n = \sigma_{n-1}$ *holds in the SVD* (2.14) *of* $H(\lambda_0)$, *then the real and complex controllability measures coincide, i.e.* $\omega_c = \gamma_c$.

It is apparent from Corollary 2, Theorem 2 and Theorem 3 that computing these measures is a difficult numerical problem and various algorithms have been developed for this purpose. Most of them rely on Corollary 2 (see [4; 44; 18]). For an alternative approach based on Theorems 2 and 3 see [45]. In order to reduce the required computational effort in minimizing (2.7) it is certainly advantageous to have a priori information on the location of the minimizing frequency λ^*.

Theorem 5. *Assume that* $[A, B] \in \Gamma$ *is controllable.*

(i) *If* $A = A^*$, *then*
$$\gamma_c = \min_{\lambda \in \mathbb{R}} \sigma_{\min}(H(\lambda)).$$

(ii) *If* $A = -A^*$, *then*
$$\gamma_c = \min_{\lambda \in \mathbb{R}} \sigma_{\min}(H(i\lambda)).$$

(iii) *If* $[A, B] \in \Omega$ *and* $A = A^T$, *then*
$$\gamma_c = \omega_c = \min_{\lambda \in \mathbb{R}} \sigma_{\min}(H(\lambda)).$$

Proof.   (i) The square of $\sigma_{\min}[H(\lambda)]$ is equal to the minimum eigenvalue of $H(\lambda)H(\lambda)^*$. If $\lambda = \alpha + i\beta$, then

$$H(\lambda)H(\lambda)^* = [A - \lambda I, B]\begin{bmatrix} A^* - \bar{\lambda}I \\ B^* \end{bmatrix}$$
$$= I|\lambda|^2 - 2\mathrm{Re}(\lambda)A + A^2 + BB^*$$
$$= \{\alpha^2 I - 2\alpha A + A^2 + BB^*\} + \beta^2 I$$
$$= \{(\alpha I - A)^2 + BB^*\} + \beta^2 I.$$

The Hermitian matrix $G(\alpha) = \{(\alpha I - A)^2 + BB^*\}$ has real eigenvalues $\lambda_1(\alpha)$, $\lambda_2(\alpha), \ldots, \lambda_n(\alpha)$ and the spectral theorem [31, p. 312] implies that the eigenvalues of

$$H(\lambda)H^*(\lambda) = G(\alpha) + \beta^2 I$$

are given by $\lambda_i(\alpha) + \beta^2, i = 1, 2, \ldots, n$. Therefore, for each $\lambda = \alpha + i\beta$, the minimum eigenvalue of $H(\lambda)H^*(\lambda)$ occurs at $\beta = 0$ and hence

$$\gamma_c = \min_{\lambda \in \mathbb{C}} \sigma_{\min}[H(\lambda)] = \min_{\lambda \in \mathbb{R}} \sigma_{\min}[H(\lambda)]$$

which completes the proof of (i).

The proof of (ii) follows from (i) by observing that $A = -A^*$ if and only if $[iA] = [iA]^*$.

If $[A, B] \in \Omega$ and $A = A^T$, then part (i) implies that

$$\gamma_c = \min_{\lambda \in \mathbb{R}} \sigma_{\min}[H(\lambda)] = \sigma_{\min}[H(\hat{\lambda})]$$

for some real $\hat{\lambda}$. Since $H(\hat{\lambda})$ is real, the singular vectors of $H(\lambda)$ are real and hence the minimum norm rank reducing perturbation is real (see [19, p. 19]). Therefore, $\gamma_c = \omega_c$, part (iii) is established and this completes the proof. $\square$

We note that Theorem 5 could be deduced exploiting (2.11). However, the proof presented above also establishes the following result.

Theorem 6. *Assume that $[A, B] \in \Gamma$ is stabilizable.*

(i) If $A = A^*$, then

$$\gamma_s = \min_{\lambda \geq 0} \sigma_{\min}(H(\lambda)).$$

(ii) If $A = -A^*$, then

$$\gamma_s = \min_{\lambda \in \mathbb{R}} \sigma_{\min}(H(i\lambda)).$$

(iii)If $[A, B] \in \Omega$ and $A = A^T$, then

$$\gamma_s = \omega_s = \min_{\lambda \geq 0} \sigma_{\min}(H(\lambda)).$$

Corollary 4. *If $[A, B] \in \Omega$ is stabilizable, $A = A^T$ and $x^T A x \leq 0$ for all $x \in \mathbb{R}^n$, then*

$$\omega_s = \gamma_s = \sigma_{\min}(H(0)) = \sigma_{\min}([A, B]).$$

Proof. It follows from Theorem 6 that

$$\omega_s = \gamma_s = \min_{\lambda \in \mathbf{R}^+} \sigma_{\min}[H(\lambda)]$$

so we need only show that $\lambda^* = 0$ provides such a minimum. Since $(\lambda I - A)^2 + BB^T$ is positive semi-definite and $\sigma_{\min}^2[H(\lambda)]$ is the minimum eigenvalue $\hat{\sigma}(\lambda)$ of

$$H(\lambda)H^T(\lambda) = (A - \lambda I)^2 + BB^T,$$

it follows that

$$\hat{\sigma}(\lambda) = \min_{\|x\|=1} [x^T(A - \lambda I)^2 x + x^T BB^T x]$$

$$\geq \min_{\|x\|=1} [x^T A^2 x + x^T BB^T x] + \min_{\|x\|=1} [\lambda^2 \|x\|^2 - 2\lambda x^T Ax].$$

The last term in this inequality is non-negative so that

$$\hat{\sigma}(\lambda) \geq \min_{\|x\|=1} [x^T(A^2 + BB^T)x] = \hat{\sigma}(0)$$

and this completes the proof. □

We conclude this section with an example demonstrating that symmetry of A is sufficient for $\gamma_c = \min_{\lambda \in \mathbb{R}} \sigma_{\min}(H(\lambda))$ to hold, but the symmetry of A is not necessary.

Example 2. Let $\varepsilon \leq \frac{1}{2}$ and

$$A = \begin{pmatrix} 1 & 0 \\ \varepsilon & 1 \end{pmatrix} \quad \text{and} \quad B = \begin{pmatrix} 1 \\ 0 \end{pmatrix}.$$

It follows that

$$\sigma_{\min}^2(H(\lambda)) = |1 - \lambda|^2 + \frac{1}{2}(1 + \varepsilon^2) - [\varepsilon^2 |1 - \lambda|^2 + \frac{1}{4}(1 - \varepsilon^2)^2]^{\frac{1}{2}}.$$

Therefore, $\sigma_{\min}(H(\lambda))$ attains its minimum at $\lambda = 1$ which implies $\gamma_c = w_c = \varepsilon$. Note that in this case we also have $\gamma_c = \gamma_s$.

2.2 Infinite Dimensional Systems and Approximations

In this section we consider the control system

$$\dot{z}(t) = \mathcal{A}z(t) + \mathcal{B}u(t), \quad z(0) = z_0 \tag{2.20}$$

with output

$$y(t) = \mathcal{C}z(t). \tag{2.21}$$

We assume $\mathcal{A}$ generates a C_0-semigroup $\mathcal{S}(t)$ on the Hilbert space Z, $\mathcal{B} : U \to Z$, $\mathcal{C} : Z \to Y$ are bounded linear operators and U, Y are Hilbert spaces. Solutions of (2.20) will be mild solutions defined by

$$z(t) = \mathcal{S}(t)z_0 + \int_0^t \mathcal{S}(t - s)\mathcal{B}u(s)ds. \tag{2.22}$$

For $t > 0$ the reachable set at time t is given by

$$\mathcal{R}(t) = \left\{ \int_0^t \mathcal{S}(t-s)\mathcal{B}u(s)ds \middle| u \in L_2(0,t;U) \right\}.$$

System (2.20) is said to be exactly controllable in time t, if $\mathcal{R}(t) = Z$ and exactly controllable if $\cup_{t>0}\mathcal{R}(t) = Z$. Also (2.20) is called approximately controllable in time t, if $\overline{\mathcal{R}(t)} = Z$ and approximately controllable if $\overline{\cup_{t>0}\mathcal{R}(t)} = Z$. For analytic semigroups it is known (see [16]) that $\overline{\cup_{t>0}\mathcal{R}(t)} = Z$ if and only if there is a finite time $\hat{t}$ such that $\overline{\mathcal{R}(\hat{t})} = Z$. This is also true for semigroups generated by finite delay differential equations (see [2; 41]).

System (2.20) is said to be (exponentially) stabilizable if there is a bounded linear operator $\mathcal{F} : Z \to U$ such that the closed-loop operator $\mathcal{A}_c = \mathcal{A} + \mathcal{B}\mathcal{F}$ generates a C_0-semigroup $S(t)$ satisfying

$$\|\mathcal{S}(t)\| \le Me^{-\beta t}$$

for some $M > 0$ and $\beta > 0$. There are analogous definitions of observability and detectability and various other types of controllability (i.e. null controllability). A good summary of these definitions and topics may be found in [11]. However, we shall concentrate primarily on controllability questions and make some comments about stabilizability. It will be clear that dual results will exist for observability and detectability.

The first problem one faces when trying to define system measures for infinite dimensional systems is that in general most of the system properties are not generic. Consider the following simple example.

Example 3. Let $U = Z = \ell_2$ and define $\mathcal{A} = I$ and $\mathcal{B} : \ell_2 \to \ell_2$ by $\mathcal{B}(u_1, u_2, u_3, \dots)$ $= (u_1, u_2/2, u_3/3, \dots, u_i/i, \dots)$. The operators $\mathcal{A}$ and $\mathcal{B}$ are bounded and the system $(\mathcal{A}, \mathcal{B})$ is approximately controllable. Define the perturbed systems $\mathcal{A}^N = \mathcal{A}$ and $\mathcal{B}^N(u_1, u_2, \dots) = (u_1, u_2/2, \dots, u_{N-1}/(N-1), 0, u_{N+1}/(N+1), \dots)$. Observe that $\|\mathcal{A}^N - \mathcal{A}\| = 0$ and $\|\mathcal{B}^N - \mathcal{B}\| \le 1/N$ and yet the system $(\mathcal{A}^N, \mathcal{B}^N)$ is not controllable for all $N \ge 1$. If we choose for U the Banach space $\ell_2^{-1} = \{(\xi_i) | \sum_{i=1}^\infty i^{-2}|\xi_i|^2 < \infty\}$, then $(\mathcal{A}, \mathcal{B})$ as defined above is exactly controllable and stabilizable and yet $(\mathcal{A}^N, \mathcal{B}^N)$ is neither controllable nor stabilizable for all $N \ge 1$.

Example 4. Let $Z = \ell_2$ and define $\mathcal{A}$ and $\mathcal{A}^N$ by $\mathcal{A} = -I$ and $\mathcal{A}^N x = \mathcal{A}^N(x_1, x_2, \dots) = (-x_1, -x_2, \dots, -x_{N-1}, 2x_N, -x_{N+1}, \dots)$. Observe that $\|e^{\mathcal{A}t}\| = e^{-t}$ and $\|\mathcal{A}^N x - \mathcal{A}x\| = 3|x_N| \to 0$. Moreover, $\|\mathcal{A}^N\| = 2$ so that $\mathcal{A}^N$ is a numerically stable and consistent approximation of $\mathcal{A}$. If e^N denotes the unit vector $e^N = (0, 0, \dots, 1, 0, 0, \dots) \in \ell_2$, then $\mathcal{A}^N e^N = 2e^N$. $\mathcal{A}^N$ has an unstable eigenvalue $\lambda = 2$ for all $N \ge 1$.

These examples indicate that there are no such things as stability, controllability or stabilizability measures for general infinite dimensional systems. For a more detailed discussion see also [37]. In order to define a reasonable measure it is essential to limit the set of allowable perturbations to a specific (structured) set. Initial results on structured perturbations that preserve stability have been established in [40]. In [8] it was shown that, for certain delay systems, approximate controllability is preserved under small perturbations of the system coefficients (including the delay).

From a design point of view it is worthwhile to think of finite dimensional approximating systems as "structured" perturbations of infinite dimensional systems. In particular, one can use finite element and finite difference schemes to construct very special (perturbed) approximating control systems for distributed parameter models governed by partial and functional differential equations. Therefore, one question of interest is that of determining those numerical schemes that preserve the various system properties and then finding among such schemes the ones that maximize the measures of the finite dimensional models.

This problem was considered for approximations of control systems with delays in [8]. For single-input systems it was possible to give sufficient conditions for an approximation scheme to preserve controllability. Moreover, it was shown in [8] that several of the standard numerical schemes for delay equations satisfy these conditions and hence preserve controllability under approximation. The situation becomes much more complex when the system is governed by parabolic and hyperbolic partial differential equations.

Two specific numerical schemes for approximating differential operators are the finite difference and finite element methods. Both approaches have certain advantages and limitations. In many cases (not always) these schemes lead to system matrices with a very special structure. For example, it is typical that such schemes produce matrices that are symmetric and sparse (banded, block tridiagonal, etc.). When such methods are used to approximate control systems governed by partial differential equations it is possible to use this structure in the design and analysis of the control problem. Moreover, since there are often several methods for approximating a particular control problem it is important to identify those schemes that produce models that are robust. More precisely, we are interested in determining the schemes that have good system measures γ_c, γ_s, etc. We shall consider this problem for a one dimensional heat equation. We focus on the standard finite (central) difference and (piecewise linear) finite element schemes.

Consider the system governed by the heat equation

$$y_t(t, x) = y_{xx}(t, x) + b(x)u(t), \quad 0 \le x \le 1, \quad t > 0, \tag{2.23}$$

with Dirichlet boundary conditions $y(t, 0) = y(t, 1) = 0$. Here we assume that $b(\cdot) \in L_2(0, 1)$. Let $\mathcal{A}$ be the operator defined on $L_2(0, 1)$ by

$$\mathcal{D}(\mathcal{A}) = \{\phi \in L_2(0, 1) | \phi \in H^2(0, 1), \ \phi(0) = \phi(1) = 0\}, \tag{2.24}$$

and for $\phi \in \mathcal{D}(\mathcal{A})$

$$\mathcal{A}\phi = \frac{d^2}{dx^2}\phi. \tag{2.25}$$

The operator $\mathcal{A}$ generates a C_0-semigroup $\mathcal{S}(t)$ on $L_2(0, 1)$ given by

$$[\mathcal{S}(t)\phi](x) = \sum_{k=1}^{\infty} e^{-\lambda_k t} \langle \phi, \phi_k \rangle \phi_k(x), \tag{2.26}$$

where $\lambda_k = k^2\pi^2$, $\phi_k(x) = \sqrt{2}\sin(k\pi x)$ and $\langle\cdot,\cdot\rangle$ denotes the standard inner product on $L_2(0,1)$. We define $\mathcal{B} : \mathbb{R} \to L_2(0,1)$ by

$$[\mathcal{B}u](x) = b(x)u, \tag{2.27}$$

and consider the equation (2.23) as a control problem in $L_2(0,1)$ governed by

$$\dot{z}(t) = \mathcal{A}z(t) + \mathcal{B}u(t). \tag{2.28}$$

We denote by $\Sigma^H = (\mathcal{A},\mathcal{B})$ the system operators defined by (2.24) - (2.25) and (2.27). Recall that (2.28) is approximately controllable in $L_2(0,1)$ if and only if

$$\langle\phi_k, b\rangle \neq 0 \quad \text{for all} \quad k = 1, 2, \ldots. \tag{2.29}$$

We note that many of the results below can be extended to problems in more than one space variable and to some general parabolic systems. However, this introduces so many technical details that many of the basic ideas get lost. Also, it will become clear that even this "simple" one dimensional heat equation leads to difficult problems. We refer the reader to [25] for a discussion of controllability for more general problems.

The system (2.28) will be approximated by using finite difference and finite element schemes for (2.23). Divide the interval $(0,1)$ into $N+1$ equal subintervals $[x_i, x_{i+1}]$ where $x_i = \frac{i}{N+1}$, $i = 0, 1, \ldots, N+1$. Assuming that $b(\cdot) \in H^1(0,1)$, then applying the central difference approximation of $\frac{d^2}{dx^2}$ leads to the N dimensional system

$$\dot{z}^N(t) = A_D^N z^N(t) + B_D^N u(t), \tag{2.30}$$

where

$$A_D^N = (N+1)^2 \begin{bmatrix} -2 & 1 & 0 & & \\ 1 & -2 & 1 & & 0 \\ & \ddots & \ddots & \ddots & \\ 0 & & 1 & -2 & 1 \\ & & & 1 & -2 \end{bmatrix} = (N+1)^2 \tilde{A}_D^N, \tag{2.31}$$

$$B_D^N = \operatorname{col}\big(b(x_1), b(x_2), \ldots, b(x_N)\big), \tag{2.32}$$

and $z^N(t)$ is identified with the vector

$$z^N(t) = \operatorname{col}\big(y(t, x_1), y(t, x_2), \ldots, y(t, x_N)\big). \tag{2.33}$$

The system $\Sigma_D^N = (A_D^N, B_D^N)$ is called the finite difference model.

We turn now to the finite element scheme. For each $i = 1, 2, \ldots, N$ let $h_i^N(x)$ denote the hat function

$$h_i^N(x) = \begin{cases} (N+1)(x - x_{i-1}) & x_{i-1} \leq x \leq x_i, \\ -(N+1)(x - x_{i+1}) & x_i \leq x \leq x_{i+1}, \\ 0, & \text{elsewhere.} \end{cases} \tag{2.34}$$

If $y(t, x)$ is approximated by

$$y^N(t, x) = \sum_{i=1}^{N} z_i^N(t) h_i^N(x),$$ (2.35)

then a standard Galerkin procedure leads to the finite element approximation

$$E_E^N \dot{z}^N(t) = F_E^N z^N(t) + G_E^N u(t).$$ (2.36)

Moreover

$$E_E^N = [\langle h_i^N, h_j^N \rangle] = \frac{1}{6(N+1)} \begin{bmatrix} 4 & 1 & 0 & & & \\ 1 & 4 & 1 & & & 0 \\ & \ddots & \ddots & \ddots & & \\ & & \ddots & \ddots & \ddots & \\ 0 & & & 1 & 4 & 1 \\ & & & & 1 & 4 \end{bmatrix},$$ (2.37)

$$F_E^N = -[\langle \dot{h}_i^N, h_j^N \rangle] = (N+1)\tilde{A}_D^N,$$ (2.38)

and

$$G_E^N = \mathrm{col}\left(\langle b, h_1^N \rangle, \langle b, h_2^N \rangle, \ldots, \langle b, h_N^N \rangle \right).$$ (2.39)

Let

$$A_E^N = [E_E^N]^{-1} F_E^N, \quad B_E^N = [E_E^N]^{-1} G_E^N,$$

and define the finite element model $\Sigma_E^N = (A_E^N, B_E^N)$ by

$$\dot{z}^N(t) = A_E^N z^N(t) + B_E^N u(t).$$ (2.40)

It is obvious for this simple case that both schemes preserve stabilizability under approximation uniformly in N (i.e. have property (POES) as defined in [4]). In fact, it is shown in [3] that the same is true for finite element schemes applied to more general parabolic problems in several space dimensions. Their approach can also be extended to certain (but not all) finite difference schemes for such systems. It is *not* obvious that these schemes preserve controllability (even for the particular model considered here) and in fact this problem is not yet resolved. Therefore, it is worthwhile to have some conditions on the system that can be used to determine the controllability properties of the models Σ_D^N and Σ_E^N.

First consider the finite difference model Σ_D^N. The tridiagonal matrix A_D^N has eigenvalues (see [42])

$$\lambda_{D,k}^N = -4(N+1)^2 \sin^2 \frac{k\pi}{2(N+1)}, \quad k = 1, 2, \ldots, N,$$ (2.41)

and associated eigenvectors

$$z_{D,k}^N = \mathrm{col}\left(\sin(k\alpha_N), \sin(2k\alpha_N), \ldots, \sin(Nk\alpha_N) \right),$$ (2.42)

where $\alpha_N = \pi/(N+1)$. Consequently, it follows (see the identity 1.351 in [20]) that

$$\|z_{D,k}^N\|^2 = \sum_{j=1}^{N} \sin^2(jk\alpha_N) = \frac{N}{2} - (-1)^k \frac{\sin(Nk\pi/(N+1))}{2\sin(k\pi/(N+1))} = \frac{1}{2}(N+1).$$
(2.43)

In view of (2.41) - (2.43), it is clear that

$$\Phi^N = \sqrt{\frac{2}{(N+1)}} \begin{bmatrix} \sin\alpha_N & \sin 2\alpha_N & \cdots & \sin N\alpha_N \\ \sin 2\alpha_N & \sin 4\alpha_N & \cdots & \sin 2N\alpha_N \\ \vdots & & & \\ \sin N\alpha_N & \sin 2N\alpha_N & \cdots & \sin N^2\alpha_N \end{bmatrix}$$
(2.44)

is the orthogonal transformation that diagonalizes A_D^N, i.e. $[\Phi^N]^T = [\Phi^N]^{-1}$ and

$$[\Phi^N]^T A_D^N \Phi^N = \Lambda_D^N = \text{diag}(\lambda_{D,1}^N, \lambda_{D,2}^N, \ldots, \lambda_{D,N}^N).$$
(2.45)

Observe that Φ^N is also symmetric so that $[\Phi^N]^T = \Phi^N = [\Phi^N]^{-1}$. Let ϕ_k^N denote the k-th column of Φ^N,

$$\phi_k^N = \sqrt{\frac{2}{(N+1)}} \, \text{col}\,(\sin(k\alpha_N), \sin(2k\alpha_N), \ldots, \sin(Nk\alpha_N)).$$
(2.46)

Lemma 1. *The finite difference model $\Sigma_D^N = (A_D^N, B_D^N)$ is controllable if and only if*

$$\langle \phi_k^N, B_D^N \rangle \neq 0 \quad \text{for all} \quad k = 1, 2, \ldots, N.$$
(2.47)

Proof. Let $\xi_k = \langle \phi_k^N, B_D^N \rangle$, $k = 1, 2, \ldots, N$ and note that

$$\Phi^N B_D^N = \text{col}(\xi_1^N, \xi_2^N, \ldots, \xi_N^N).$$

The system Σ_D^N is controllable if and only if the controllability matrix

$$\mathcal{K}_D^N = [B_D^N, A_D^N B_D^N, \ldots, [A_D^N]^{N-1} B_D^N]$$

has maximal rank N. However,

$$\text{rank}\,\mathcal{K}_D^N = \text{rank} \begin{bmatrix} \xi_1^N & & & \\ & \xi_2^N & & 0 \\ & & \ddots & \\ 0 & & & \ddots \\ & & & \xi_N^N \end{bmatrix} \begin{bmatrix} 1 & \lambda_{D,1}^N & \cdots & (\lambda_{D,1}^N)^{N-1} \\ 1 & \lambda_{D,2}^N & \cdots & (\lambda_{D,2}^N)^{N-1} \\ \vdots & & & \vdots \\ \vdots & & & \vdots \\ 1 & \lambda_{D,N}^N & \cdots & (\lambda_{D,N}^N)^{N-1} \end{bmatrix}$$

Since the eigenvalues of A_N^D are all distinct, the Vandermonde matrix is non-singular. Hence, $\mathcal{K}_D^N$ has rank N if and only if (2.47) holds. □

If $b(\cdot) \in H^1(0,1)$, let $g(x) = b(x + \frac{1}{2})$ for $-\frac{1}{2} \le x \le \frac{1}{2}$. In the cases where g is odd or even, (2.29) does not hold. If $g(x) = g(-x)$, then

$$B_D^N = \text{col}\left(b(x_1), \ldots, b(x_\ell), b(x_\ell), \ldots, b(x_1)\right), \quad \text{if } N = 2\ell$$

and

$$B_D^N = \text{col}\left(b(x_1), \ldots, b(x_\ell), b(x_{\ell+1}), b(x_\ell), \ldots, b(x_1)\right), \quad \text{if } N = 2\ell + 1.$$

If $g(x) = -g(-x)$, then

$$B_D^N = \text{col}\left(b(x_1), \ldots, b(x_\ell), -b(x_\ell), \ldots, -b(x_1)\right), \quad \text{if } N = 2\ell$$

and

$$B_D^N = \text{col}\left(b(x_1), \ldots, b(x_\ell), 0, -b(x_\ell), \ldots, -b(x_1)\right), \quad \text{if } N = 2\ell + 1.$$

Proposition 2. *If $g(x)$ is odd or even, then Σ^H and Σ_D^N are not controllable for all $N \ge 1$.*

Proof. As noted above, Σ^H is not controllable since (2.29) fails. Let $\xi_k^N = \langle \phi_k^N, B_D^N \rangle$. If $N = 2\ell$, then a direct calculation yields

$$\xi_k^N = \sqrt{\frac{2}{N+1}} \sum_{i=1}^\ell b(x_i)(1 \mp (-1)^k) \sin \frac{ik\pi}{2\ell + 1}.$$

If $N = 2\ell + 1$, then it follows just as above that

$$\xi_k^N = \sqrt{\frac{2}{N+1}} \sum_{i=1}^\ell b(x_i)(1 \mp (-1)^k) \sin \frac{ik\pi}{2(\ell+1)} + b(x_{\ell+1}) \sin \frac{k\pi}{2}.$$

Above, the minus sign is valid if g is even, the plus sign if g is odd. Hence we conclude that the even (odd) numbered coordinates of ξ^N vanish if g is symmetric (skew symmetric). □

A close look at the above proof yields a clear relationship between (2.29) and (2.47) in the special cases above where (2.29) fails because of the special form of $g(\cdot)$. This form is also present in B_D^N and it is precisely this form that causes (2.47) to fail. As we shall see below, the same structure is preserved by the finite element scheme. In particular, let

$$\tilde{E}_E^N = 6(N+1)E_E^N, \qquad \tilde{F}_E^N = (N+1)^{-1}F_E^N,$$
$$\tilde{A}_E^N = [\tilde{E}_E^N]^{-1}\tilde{F}_E^N = \frac{1}{6(N+1)^2}A_E^N,$$
$$\tilde{B}_E^N = [\tilde{E}_E^N]^{-1}G_E^N = \frac{1}{6(N+1)}B_E^N \quad \text{and}$$
$$\tilde{A}_D^N = (N+1)^{-2}A_D^N.$$

Clearly, $\tilde{\Sigma}_E^N = (\tilde{A}_E^N, \tilde{B}_E^N)$ is controllable if and only if $\Sigma_E^N = (A_E^N, B_E^N)$ is controllable.

Observe that the "stiffness" matrix $(N+1)^2 \tilde{F}_E^N$ is the system matrix for the finite difference equation, i.e.

$$\tilde{F}_E^N = \tilde{A}_D^N,$$

and that the "mass" matrix $\tilde{E}_E^N$ can be written as

$$\tilde{E}_E^N = \tilde{F}_E^N + 6I^N = \tilde{A}_D^N + 6I^N, \tag{2.48}$$

where I^N is the $N \times N$ identity matrix.

Lemma 2. *Let $\tilde{A}_E^N$ and $\tilde{A}_D^N$ be as given above. Then $\lambda \in \mathbb{C}$ is an eigenvalue of $\tilde{A}_E^N$ with eigenvector z_λ^N if and only if $6\lambda/(1-\lambda)$ is an eigenvalue of $\tilde{A}_D^N$ with eigenvector z_λ^N.*

The proof of Lemma 2 follows immediately from (2.48). Moreover, as a consequence of Lemma 2 and (2.48) it follows that Φ^N defined by (2.44) diagonalizes $\tilde{A}_D^N$ and $\tilde{A}_E^N$. We use these observations to establish the following result.

Proposition 3. *The finite element model $\Sigma_E^N = (A_E^N, B_E^N)$ is controllable if and only if*

$$\langle \phi_k^N, G_E^N \rangle \neq 0 \quad for\ all \quad k = 1, 2, \dots, N. \tag{2.49}$$

Proof. The system Σ_E^N is controllable if and only if $\tilde{\Sigma}_E^N$ is controllable, i.e. if and only if

$$\tilde{\mathcal{K}}_E^N = \left[\tilde{B}_E^N, \tilde{A}_E^N \tilde{B}_E^N, \dots, [\tilde{A}_E^N]^{N-1} \tilde{B}_E^N \right]$$

has rank N. However, $\tilde{E}_E^N \tilde{F}_E^N = \tilde{F}_E^N \tilde{E}_E^N$ so that

$$\begin{aligned}
\operatorname{rank} \tilde{\mathcal{K}}_E^N &= \operatorname{rank}\big[[\tilde{E}_E^N]^{-1} G_E^N, [\tilde{E}_E^N]^{-1}([\tilde{E}_E^N]^{-1}\tilde{F}_E^N) G_E^N, \\
&\qquad \dots, [\tilde{E}_E^N]^{-1}([\tilde{E}_E^N]^{-1}\tilde{F}_E^N)^{N-1} G_E^N \big] \\
&= \operatorname{rank}[\Phi^N G_E^N, \tilde{\Lambda}^N \Phi^N G_E^N, \dots, (\tilde{\Lambda}^N)^{N-1} \Phi^N G_E^N],
\end{aligned}$$

where $\tilde{\Lambda}^N$ is a non-singular diagonal matrix. Hence, $N = \operatorname{rank} \tilde{\mathcal{K}}_E^N$ if and only if (2.49) holds and this completes the proof. ∎

Proposition 4. *If $g(x)$ is odd or even, then Σ^H and Σ_E^N are not controllable for all $N \geq 1$.*

As seen above there is a nice relationship between the system matrices for Σ_D^N and Σ_E^N. Moreover, one has the following

Proposition 5. *Assume that there exists a non-singular transformation T^N such that*

(1) $\tilde{A}_D^N = T^N \tilde{A}_D^N (T^N)^{-1}$
(2) $G_E^N = T^N B_D^N.$

Then, Σ_D^N is controllable if and only if Σ_E^N is controllable.

Proof. Since scalar factors do not affect the controllability, it suffices to consider the

$$\text{rank}\left[[\tilde{E}_E^N]^{-1}\tilde{F}_E^N - \lambda I^N, [\tilde{E}_E^N]^{-1}G_E^N\right] = \text{rank}[\tilde{F}_E^N - \lambda \tilde{E}_E^N, G_E^N]$$

$$= \text{rank}\left[\tilde{A}_D^N - \frac{6\lambda}{1-\lambda}I^N, G_E^N\right] = \text{rank}\left[T^N\tilde{A}_D^N(T^N)^{-1} - \frac{6\lambda}{1-\lambda}I^N, T^N B_D^N\right].$$

The conclusion follows from the Hautus test and Lemma 2. $\square$

Example 5. Let $b(x) = x, 0 \leq x \leq 1$. It follows from (2.29) that Σ^H is controllable and $\langle \phi_k^N, B_D^N \rangle$ can be calculated. In particular, $B_D^N = \frac{1}{N+1}\text{col}(1, 2, \ldots, N)$ and

$$\xi_k^N = \langle \phi_k^N, B_D^N \rangle = \sqrt{\frac{2}{N+1}} \sum_{i=1}^{N} \frac{i}{N+1} \sin ki\alpha_N.$$

Applying the identity 1.352 (i) in [20] to the expression

$$\sqrt{\frac{2}{N+1}}\frac{1}{N+1}\left[\frac{\sin[(N+1)k\alpha_N]}{4\sin^2\frac{k\alpha_N}{2}} - \frac{N+1}{2}\frac{\cos[\frac{2N+1}{2}k\alpha_N]}{\sin\frac{k\alpha_N}{2}}\right],$$

and performing a few elementary manipulations it follows that

$$\xi_k^N = (-1)^{k+1}\sqrt{\frac{1}{2(N+1)}}\cot\frac{k\pi}{2(N+1)}, \quad k = 1, 2, \ldots, N. \tag{2.50}$$

Since

$$0 < |\xi_N^N| < |\xi_{N-1}^N| < \cdots < |\xi_1^N|, \tag{2.51}$$

it follows that the finite difference model Σ_D^N is controllable for all $N \geq 1$. If the finite element scheme is applied, then

$$G_E^N = \frac{1}{(N+1)^2}\text{col}(1, 2, \ldots, N), \tag{2.52}$$

and $G_E^N = T^N B_D^N$ where $T^N = (N+1)I^N$. Therefore, it follows from Proposition 2.8 that the finite element scheme Σ_E^N is also controllable for all $N \geq 1$.

Remark. It is important to note that checking for controllability of finite dimensional systems is in general a subtle numerical task [36; 30]. In contrast, verification of (2.49) is numerically accurate and straightforward. In the next section we shall concentrate on the robustness of the system measures for this example.

2.3 System Measures and Case Studies

In this section we consider the system measures γ_c and γ_s for each of the two schemes applied to Example 5. Also, we use the simple model problem defined in Example 5 to present a comparison of the condition number for the Riccati equation that comes from an LQR problem and the stabilizability radii.

Approximations for the Heat equation

In particular, we consider the problem governed by

$$y_t(t, x) = y_{xx}(t, x) + xu(t), \quad 0 \le x \le 1, \quad t > 0 \tag{2.53}$$

with Dirichlet boundary conditions $y(t, 0) = y(t, 1) = 0$. The finite difference system $\Sigma_D^N = (A_D^N, B_D^N)$ and the finite element system $\Sigma_E^N = (A_E^N, B_E^N)$ are defined as above, where for this problem

$$B_D^N = \frac{1}{N+1} \, \text{col} \, (1, 2, \ldots, N) \tag{2.54}$$

and

$$B_E^N = [E_E^N]^{-1} G_E^N = [E_E^N]^{-1} \frac{1}{(N+1)^2} \, \text{col} \, (1, 2, \ldots, N). \tag{2.55}$$

Let $\gamma_{DC}^N (\gamma_{DS}^N)$ and $\gamma_{EC}^N (\gamma_{ES}^N)$ denote the controllability (stabilizability) measures of Σ_D^N and Σ_E^N, respectively. From the previous section we know that Σ_D^N and Σ_E^N are controllable. Moreover, since $A_D^N = [A_D^N]^T$ and $A_E^N = [A_E^N]^T$ we have from Theorem 5 that

$$\gamma_{DC}^N = \omega_{DC}^N = \min_{\lambda \in \mathbf{R}} \sigma_{\min}[\lambda I^N - A_D^N, B_D^N], \tag{2.56}$$

and

$$\gamma_{EC}^N = \omega_{EC}^N = \min_{\lambda \in \mathbf{R}} \sigma_{\min}[\lambda I^N - A_E^N, B_E^N]. \tag{2.57}$$

Corollary 4 implies that

$$\gamma_{DS}^N = \omega_{DS}^N = \sigma_{\min}[A_D^N, B_D^N], \tag{2.58}$$

and

$$\gamma_{ES}^N = \omega_{ES}^N = \sigma_{\min}[A_E^N, B_E^N]. \tag{2.59}$$

These formulas can be used to compute the various measures. Moreover, it is possible to obtain explicit upper and lower bounds on the controllability measures. The following theorems provide such bounds. The proofs are given in the appendix.

Theorem 7. *Let* $\gamma_{DC}^N = \omega_{DC}^N$ *denote the controllability measure of the finite difference approximation of* (2.53). *If* $N \ge 8$, *then*

$$\delta_D^N - \epsilon_D^N < (\gamma_{DC}^N)^2 \le \delta_D^N, \tag{2.60}$$

where

$$\delta_D^N = |\xi_N^N|^2 = \frac{1}{2(N+1)}\cot^2\frac{N\pi}{2(N+1)}, \tag{2.61}$$

and

$$\epsilon_D^N = \frac{1}{\beta_D} \cdot \frac{\pi^2\left(3\ln\frac{2(N+1)}{3} + \pi\right)}{24N^2(N^2-1)}, \tag{2.62}$$

and

$$\beta_D = 726.$$

Theorem 8. *Let $\Sigma_E^N = (A_E^N, B_E^N)$ denote the finite element approximation of (2.53). If $\gamma_{EC}^N = \omega_{EC}^N$ is the controllability measure for Σ_E^N, then for $N \geq 8$*

$$\delta_E^N - \epsilon_E^N < (\gamma_{EC}^N)^2 \leq \delta_E^N, \tag{2.63}$$

where

$$\delta_E^N = \frac{9}{[3 - 2\sin^2\frac{N\pi}{2(N+1)}]^2} \cdot \delta_D^N, \tag{2.64}$$

and

$$\epsilon_E^N = \frac{1}{\beta_E} \cdot \frac{\pi^2(3\ln\frac{2(N+1)}{3} + \pi)}{24N^2(N^2-1)}, \tag{2.65}$$

and

$$\beta_E = 550.$$

Corollary 5. *If γ_{DC}^N and γ_{EC}^N are as above, then*

$$\lim_{N \to +\infty} \gamma_{DC}^N = \lim_{N \to \infty} \gamma_{EC}^N = 0.$$

The proof of Corollary 5 is an easy consequence of Theorems 7 and 8. Note that the asymptotic rates of δ_D^N and δ_E^N are the same. However, for "small" values of N, there are considerable differences. Moreover, the rate at which δ_D^N (and δ_E^N) converges to zero decreases significantly for $N \geq 10$. If one had no theoretical estimates (such as (2.61) and (2.64)) then numerically calculated values for $N \leq 10$ might lead one to believe that the decay of controllability occurs so rapidly that the high order models would be numerically uncontrollable on most digital computers.

In Tables 2.1, 2.2 we compare computed values of $(\gamma_{DC}^N)^2$ and $(\gamma_{EC}^N)^2$ with the theoretical upper and lower bounds given in Theorem 7 and Theorem 8. In our computations we used Algorithm II in [44]. The eigenvalues of A_D^N (A_E^N) served as initial guesses for the location of the minimum in (2.7). The proofs of Theorems 7 and 8 show that in the particular example one can choose as initial guess the eigenvalue with largest modulus which significantly reduces the computational effort. We also report the observation supported by numerous examples [35] that the minimum in (2.7) is usually attained close to an eigenvalue of A_D^N (A_E^N).

The most important point to make is that the finite element model is roughly one order of magnitude more robust than the corresponding finite difference model. Both systems provide order $\mathcal{O}(h^2)$, $h = 1/(N+1)$ approximation of the generator $\mathcal{A}$ and for fixed N, Σ_D^N and Σ_E^N are of the same dimension.

Condition Number and Stabilizability Radius

Here we discuss the relationship between the condition number for the algebraic Riccati equation and the stabilizability radius γ_s. We compute the actual condition number K_{RIC} as defined on page 533 (Theorem 13.4.2) in Datta's book [12] and compare this to the inverse of γ_s. To illustrate the point we return to the first example given as the motivating problem. In particular, we consider the

$$\frac{d}{dt}\begin{bmatrix} x_1(t) \\ x_2(t) \end{bmatrix} = \begin{bmatrix} 1 & 0 \\ 0 & -1 \end{bmatrix}\begin{bmatrix} x_1(t) \\ x_2(t) \end{bmatrix} + \begin{bmatrix} \delta \\ \epsilon \end{bmatrix} u(t),$$

with $\epsilon = 1$. In this case the system becomes unstabilizable as δ approaches zero. In Table 2.5 below we see that the Riccati equation condition number K_{RIC} increases as δ approaches zero. In addition, ratio $K_{RIC}*\gamma_s$ approaches a constant value 1.0797 which shows that the inverse of the stabilizability radius γ_s provides a good estimate of the conditioning number K_{RIC}. Therefore, it might also be possible to use system radii to help construct "well-conditioned" approximating control systems. This topic needs further study.

2.4 Concluding Remarks

In this paper we discussed a measure of controllability (stabilizability) quantifying the distance of a controllable (stabilizable) system to the set of uncontrollable (unstabilizable) systems. This measure was applied to finite dimensional systems which arise by approximating an infinite dimensional system. This process leads to several important questions that are often not fully addressed. In particular one should consider the following issues:

1. If the original distributed parameter system is controllable (stabilizable) are the finite dimensional approximation also controllable (stabilizable)? For the heat equation this leads to a very specific problem:
   (P1) Find sufficient conditions for $b \in H^1(0,1)$ which ensure that $\langle \varphi_k, b \rangle_{L^2} \neq 0$, $k = 1, 2, \ldots$, implies that $\langle \phi_k^N, B_D^N \rangle \neq 0$ for all $k = 1, \ldots, N$ (where ϕ_k^N, B_D^N are defined in (2.46) and (2.32)).
2. If the finite dimensional approximate model is controllable (stabilizable) how robust is the model with respect to the measure γ_c (γ_s)?
3. How does γ_c (γ_s) vary as the approximation scheme is refined?

   It is not unexpected that the controllability margin in the example discussed in this paper deteriorates as the approximation is refined. On the one hand the finite dimensional systems converge to the distributed parameter system for which (approximate) controllability is not robust with respect to bounded perturbations [37; 10]. On the other hand, the dimension of the finite dimensional approximating systems grows as the discretization is refined making it more likely that small perturbations will destroy controllability [1]. In addition, the measures considered here do not take into account the typical structure of the finite dimensional approximating systems. In general therefore, it will not be possible to interpret the closest uncontrollable system as an approximation of a perturbed distributed parameter system.

The problem of preserving stabilizability under approximation has been considered by serval authors [8; 24; 27]. This problem can be rather complex or quite simple depending on the particular distributed parameter system and the choice of approximating scheme. "Standard finite element" schemes applied to hyperbolic systems often preserve stabilizability while spline based schemes for delay equations are much more difficult to analyze [24], [27]. The problem of preserving controllability is much more complex and only a few results exist for this problem (see [8], [26]).

In addition to the system dynamics (Σ) we consider an output equation

$$y(t) = Cx(t)$$

and discuss the observability of the pair $[A, C]$. By duality one can define observability measures γ_o, ω_o analogous to γ_c, ω_c. It is known that none of these measures is invariant under an arbitrary change of basis. The effect of similarity transformations in the state space has been discussed in [38] and [39]. In particular, it was shown that for a generic controllable system, γ_c (γ_s) may attain any value in $(0, \infty)$ by choosing an appropriate basis. Furthermore, in [18] the authors suggest a sequence of similarity transformations based on the successive solution of a Riccati equation which tends to increase significantly the margin of stabilizability (controllability).

Let the output operator for the heat equation be $\mathcal{C} = \mathcal{I}$, the identity operator on $L_2(0, 1)$. If one uses the finite element scheme then the approximate output matrix is given by the mass matrix

$$C_E^N = E_E^N.$$

It is straightforward to use the ideas above to calculate detectability and observability measures for this problem. Also, if one balances the system (see [35]) one has another realization of the form

$$\tilde{A}_E^N = TA_E^N T^{-1}, \quad \tilde{B}_E^N = TB_E^N, \quad \tilde{C}_E^N = TC_E^N.$$

Observe that this system is a single input / multiple output system. In Table 2.3 we see that the stabilizability and detectability measures for the finite element model and the balanced model are bounded away from zero. Also, for the balanced system $\tilde{\gamma}_{Es}^N = \tilde{\gamma}_{Eo}^N$. However, as illustrated by Table 2.4, balancing increases the controllability measure and decreases the observability measure. This was observed in [35] for finite element approximations of various parabolic, hyperbolic and mixed distributed parameter systems. In closing we list some open problems that if solved in a satisfactory way could be useful in constructing approximate models for control design and in estimating condition numbers of numerical algorithms used in control. In particular, since in general $\gamma_o \neq \gamma_c$ and (even balancing) state transformations that increase γ_c can decrease γ_o, we pose the following problems:

(P2)   Given that $[A, B]$ is controllable and $[A, C]$ is observable, find a non-singular $T^* \in \mathbf{R}^{n \times n}$ that maximizes

$$J_{co}(T) = \min\{\tilde{\gamma}_c(T), \tilde{\gamma}_o(T)\}$$

over the set of non-singular matrices $T \in \mathbb{R}^{n \times n}$, where

$$\tilde{\gamma}_c(T) = \tilde{\gamma}_c(TAT^{-1}, TB) \quad \text{and} \quad \tilde{\gamma}_o(T) = \gamma_o(TAT^{-1}, CT^{-1}).$$

(P3) Find T^* maximizing $J_{co}(T)$ subject to the additional constraint that T is "well conditioned", i.e. that $J_{co}(T)$ is maximized over a set of the form

$$K(\delta) = \{T \in \mathbb{R}^{n \times m} : \sigma_{\min}(T) \geq \delta\} \quad \text{for some } \delta > 0.$$

These problems could also be stated for the stabilizability and detectability measures.

Appendix

In this section we present the proofs for Theorems 7 and 8. It is easy to see that the minimization problem defined by (2.9) is equivalent to the problem

$$\gamma_c^2 = \min_{\substack{\|z\|=1 \\ z \in C^n}} (z^* AA^T z - |z^* Az|^2 + \|z^* B\|^2).$$

Since

$$z^* AA^T z - |z^* Az|^2 \geq 0 \quad \text{for} \quad \|z\| = 1,$$

with equality holding if and only if z^* is a left eigenvector of A, we deduce the following upper bound for γ_c^2

$$\gamma_c^2 \leq \|z^* B\|_2^2, \tag{2.66}$$

where z^* is any normalized left eigenvector of A. (2.66) together with (2.50), (2.51) gives the upper bound for $[\gamma_{DC}^N]^2$ as presented in (2.61).

In order to get *lower* bounds for $[\gamma_{DC}^N]^2$ we intensively exploit the rich structure of Example 5. Since the measure of controllability is invariant under unitary transformations we may equivalently discuss the dependence of the eigenvalues of

$$(\lambda I^N - \Lambda_D^N)^2 + \Phi^N B_D^N (\Phi^N B_D^N)^T \tag{2.67}$$

on the real parameter λ, where Λ_D^N and Φ^N have been defined in (2.44) and (2.45). However, any eigenvalue of (2.67) is contained in one of the Gershgorin discs (see [43])

$$\mathcal{D}_i^N = \{\mu \in | |\mu - ((\lambda - \lambda_{D,i}^N)^2 + [\xi_i^N]^2)| \leq r_i^N\}$$
$$r_i^N = \sum_{\substack{j=1 \\ j \neq i}}^{N} |\xi_i^N \xi_j^N|, \quad i = 1, \dots, N,$$
$$\Phi^N B_D^N = \text{col}\,(\xi_1^N, \dots, \xi_N^N).$$

Note that λ affects only the centers of the Gershgorin discs but not their radii. Therefore, one of the Gershgorin discs is shifted as close as possible to the origin if we choose $\lambda = \lambda_{D,i_0}^N$, where i_0 is the index satisfying

$$[\xi_{i_0}^n]^2 = \min_{i=1,\dots,N} [\xi_i^N]^2.$$

In view of (2.51) $i_0 = N$ for Example 2.28.

In the following we first establish tight bounds on r_k^N and $[\xi_N^N]^2$ and show in a second step that $\mathcal{D}_N^N \cap \mathcal{D}_k^N = \varnothing$, $k = N - 1, \dots, 1$. As a consequence we infer that $\mathcal{D}_N^N$ contains exactly one eigenvalue of (2.67) [43, p. 71]. In order to enhance the accuracy of the estimates we scale the system by multiplying the last column of (2.67) by $\alpha_N > 0$ and accordingly the last row by $\frac{1}{\alpha_N}$. α_N will be appropriately chosen in the course of the proof. Consequently, the radii of the Gershgorin discs are given by

$$r_N^N = \frac{1}{\alpha_N} \frac{1}{2(N+1)} \cot \frac{N\pi}{2(N+1)} \sum_{i=1}^{N-1} \cot \frac{i\pi}{2(N+1)},$$

$$r_k^N = \frac{1}{2(N+1)} \cot \frac{k\pi}{2(N+1)} \sum_{i=1}^{N-1} \cot \frac{i\pi}{2(N+1)}$$

$$- \frac{1}{2(N+1)} \cot^2 \frac{k\pi}{2(N+1)} + \frac{\alpha_N}{2(N+1)} \cot \frac{k\pi}{2(N+1)} \cot \frac{N\pi}{2(N+1)},$$

$$k = 1, \dots, N - 1.$$

$$(2.68)$$

We will also make frequent use of the estimates

$$\cot \frac{\pi}{2(N+1)} \leq \frac{2}{3}(N+1), \qquad N \geq 2, \qquad (2.69)$$

$$\cot \frac{k\pi}{2(N+1)} \leq \frac{\pi}{2} \frac{N-k+1}{k}, \qquad k = 1, \dots, N, \qquad (2.70)$$

$$\cot \frac{k\pi}{2(N+1)} \geq \frac{2}{\pi} \frac{N+1-k}{k}, \qquad k = 1, \dots, N. \qquad (2.71)$$

With regard to r_N^N we have the estimate

$$\sum_{i=1}^{N-1} \cot \frac{i\pi}{2(N+1)} \leq \int_1^{N-1} \cot \frac{\pi x}{2(N+1)} dx + \cot \frac{\pi}{2(N+1)}$$

$$= \frac{2}{\pi}(N+1) \left(\ln \sin \frac{N-1}{N+1} \frac{\pi}{2} - \ln \sin \frac{\pi}{2(N+1)} \right) + \cot \frac{\pi}{2(N+1)}.$$

Since $\sin x \geq \frac{3}{\pi} x$ for $x \in [0, \frac{\pi}{6}]$, it follows from (2.69) that

$$\sum_{i=1}^{N-1} \cot \frac{i\pi}{2(N+1)} \le \frac{2}{\pi}(N+1)\ln\frac{2(N+1)}{3} + \frac{2}{3}(N+1).$$

This inequality combined with (2.70) gives

$$r_N^N \le \frac{1}{\alpha_N}\left[\frac{1}{2N}\ln\frac{2(N+1)}{3} + \frac{\pi}{6N}\right], \quad N = 2, 3, \ldots \quad . \quad (2.72)$$

A similar argument shows that

$$r_{N-1}^N \le \frac{1}{N-1}\ln\frac{2(N+1)}{3} + \frac{\pi}{3(N-1)} + \frac{\pi^2}{4N(N^2-1)}\alpha_N, \quad (2.73)$$

$N = 2, 3, \ldots$, and using (2.69) - (2.71) again yields

$$r_k^N \le \frac{N-k+1}{k}\left[\frac{1}{2}\ln\frac{2(N+1)}{3} + \frac{\pi}{6}\right] - \frac{1}{N+1}\frac{2}{\pi^2}\left(\frac{N+1-k}{k}\right)^2$$
$$+ \alpha_N\frac{\pi^2}{8}\frac{(N-k+1)}{N(N+1)k} \quad k = 1, \ldots, N-2. \quad (2.74)$$

Next we estimate the gap between $\lambda_{D,i}^N$ and $\lambda_{D,N}^N$, $i = 1, \ldots, N-1$. This is based on the identity

$$|\lambda_{D,N}^N - \lambda_{D,i}^N| = 4(N+1)^2 \cdot \sin\frac{(N+i)\pi}{2(N+1)}\sin\frac{(N-i)\pi}{2(N+1)}. \quad (2.75)$$

In particular, we have

$$|\lambda_{D,N}^N - \lambda_{D,N-1}^N| = 4(N+1)^2 \cdot \sin\frac{3\pi}{2(N+1)}\sin\frac{\pi}{2(N+1)},$$

and estimating the sine as above we find that

$$|\lambda_{D,N}^N - \lambda_{D,N-1}^N| \ge 27, \quad N \ge 8. \quad (2.76)$$

The estimate (2.76) is rather tight. This is clear when it is compared to the exact limit

$$\lim_{N\to\infty}|\lambda_{D,N}^N - \lambda_{D,N-1}^N| = 3\pi^2.$$

Employing the estimates

$$\sin x \ge \frac{2\sqrt{2}}{\pi}x, \qquad x \in \left[0, \frac{\pi}{4}\right] \qquad \text{and}$$
$$\sin x \ge 2\sqrt{2}\left(1 - \frac{x}{\pi}\right), \quad x \in \left[\frac{3\pi}{4}, \pi\right],$$

we obtain the bound

$$|\lambda_{D,N}^N - \lambda_{D,N-2}^N| \geq 64, \quad N \geq 8, \tag{2.77}$$

and making a simple linear estimate of $\sin x$ we finally have

$$|\lambda_{D,N}^N - \lambda_{D,k}^N| \geq 4(N-k)(N-k+2), \quad k = 1, \ldots, N-2. \tag{2.78}$$

It is easily checked that

$$r_N^N + [\xi_N^N]^2 < 1, \quad N \geq 8,$$

which is an upper bound for $\mathcal{D}_N^N$. In the following step we choose α_N as large as possible so that $\mathcal{D}_N^N \cap \mathcal{D}_{N-1}^N = \varnothing$. This is possible since

$$[\xi_{N-1}^N]^2 + |\lambda_{D,N}^N - \lambda_{D,N-1}^N|^2 - r_{N-1}^N > |\lambda_{D,N}^N - \lambda_{D,N-1}^N|^2 - r_{N-1}^N \geq 1.$$

If we insert (2.73) and (2.76) into the above inequality, it is clear that we may choose α_N so that

$$\alpha_N \leq \frac{4}{\pi^2} N(N^2 - 1) \cdot 726. \tag{2.79}$$

A similar argument using (2.74) and (2.77) shows that $\mathcal{D}_N^N \cap \mathcal{D}_{N-2}^N = \varnothing$. Hence it remains to show that $\mathcal{D}_N^N \cap \mathcal{D}_k^N = \varnothing$, $k = 1, \ldots, N-3$, $N \geq 8$. Observe that in view of (2.78), (2.74) and (2.79)

$$|\lambda_{D,N}^N - \lambda_{D,k}^N|^2 + [\xi_k^N]^2 - r_k^N \geq 1$$

is a consequence of the inequality

$$16(N-k)^2(N-k+2)^2 - \frac{N-k+1}{k} \left[\frac{1}{3} e^{-1}(N+1) + \frac{\pi}{6} \right]$$

$$- \frac{1}{2} \cdot 726(N-1) \frac{N-k+1}{k} \geq 1.$$

Therefore it suffices to establish the estimate

$$16(N-k)^2(N-k+2)k - \frac{1}{3} e^{-1}(N+1) - \frac{\pi}{6}$$
$$> \frac{k}{N-k+1} + 363(N-1), \quad k = 1, \ldots, N-3, \quad N \geq 8. \tag{2.80}$$

The estimate

$$\frac{k}{N-k+1} \leq \frac{N-3}{4}, \qquad k = 1, \ldots, N-3$$

provides an upper bound for the right hand side of (2.80). Combined with $\frac{4}{3} e^{-1} < .5$ and $\frac{2\pi}{3} < 2.5$ one can show that (2.80) follows from the stronger inequality

$$64(N-k)^2(N-k+2)k - \left(\frac{3}{2} + 726 \cdot 2 \right) N + 726 \cdot 2 > 0, \tag{2.81}$$
$$k = 1, \ldots, N-3, \quad N \geq 8.$$

We now establish (2.81). Define for fixed $N \geq 8$

$$G(N, x) = (N - x)^2(N - x + 2)x, \quad x \in [0, N].$$

Since $G(N, x)$ is nonnegative and possesses a unique maximum on $[0, N]$ it is obvious that

$$G(N, k) \geq \min\big(G(N, 1), G(N, N - 3)\big).$$

Since $G(N, 1) \geq G(N, N - 3)$, $N \geq 8$ it suffices to verify (2.81) for $k = N - 3$. This completes the proof of Theorem 7.

The proof of Theorem 8 is very similar. Therefore we restrict ourselves to a brief sketch and leave the details to the reader.

Let $\tilde{\lambda}_{D,k}^N$, $k = 1, \ldots, N$ denote the eigenvalues of $\tilde{A}_D^N$ and

$$\tilde{\Lambda}_D^N = \text{diag}\,(\tilde{\lambda}_{D,1}^N, \ldots, \tilde{\lambda}_{D,N}^N). \tag{2.82}$$

Defining

$$\tilde{\lambda}_{E,k}^N = \frac{\tilde{\lambda}_{D,k}^N}{6 + \tilde{\lambda}_{D,k}^N}, \quad k = 1, \ldots, N,$$

Lemma 2.5 implies that $\tilde{\lambda}_{E,k}^N$ is an eigenvalue of $\tilde{A}_E^N$. We collect them in

$$\begin{aligned}
\tilde{\Lambda}_E^N &= \text{diag}(\tilde{\lambda}_{E,1}^N, \ldots, \tilde{\lambda}_{E,N}^N), \\
\Lambda_E^N &= 6(N + 1)^2 \cdot \tilde{\Lambda}_E^N = \text{diag}(\lambda_{E,1}^N, \ldots, \lambda_{E,N}^N).
\end{aligned} \tag{2.83}$$

As a consequence of (2.48) we note

$$\Phi^N \tilde{E}_E^N \Phi^N = \tilde{\Lambda}_D^N + 6I^N. \tag{2.84}$$

Applying the orthogonal transformation Φ^n to Σ_E^N, it follows that

$$[w_{EC}^N] = \min_{\lambda \in \mathbf{R}} \sigma_{\min}\big([\lambda I - \Lambda_D^N, \phi^N [E_E^N]^{-1} G_E^N]\big).$$

Taking into account (2.84) and (2.52) we conclude

$$\hat{B}^N := \Phi^N [E_E^N]^{-1} G_E^N = 6(\tilde{\Lambda}_E^N + 6I^N)^{-1} \Phi^N B_D^N. \tag{2.85}$$

Consequently, analogous to (2.67) we consider

$$(\lambda I - \Lambda_E^N)^2 + \hat{B}^N [\hat{B}^N]^T.$$

Since the k-th coordinate of $\hat{B}^N$ satisfies

$$|(\hat{B}^N)_k| = \frac{6|\xi_k|}{6 - 4\sin^2 \frac{k\pi}{2(N+1)}} = \frac{6}{6 - 4\sin^2 \frac{k\pi}{2(N+1)}} \sqrt{\frac{1}{2(N+1)}} \cot \frac{k\pi}{2(N+1)},$$

and since $f(x) = (6 - 4\sin^2 x)^{-1}\cot x$ is strictly decreasing on $(0, \frac{\pi}{2})$ we obtain the estimate

$$|(\hat{B}^N)_N| < |(\hat{B}^N)_k|, \quad k = 1, \dots, N - 1.$$

This establishes the upper bound for $[w_{EC}^N]^2$ given in Theorem 8 and also shows that the choice $\lambda = \lambda_{E,N}^N$ shifts the configuration of Gershgorin discs as far as possible to the left. Analogous to the finite difference approximation the centers of the Gershgorin discs $\mathcal{E}_k^N$ for the finite element scheme are given by

$$(\lambda_{E,N}^N - \lambda_{E,k}^N)^2 + (\hat{B}^N)_k^2,$$

and their radii by

$$R_N^N = \frac{1}{\tilde{\alpha}_N} \sum_{i=1}^{N-1} |(\hat{B}^N)_N||(\hat{B}^N)_k|$$

and

$$R_k^N = |(\hat{B}^N)_k| \sum_{i=1}^{N-1} |(\hat{B}^N)_i| - |(\hat{B}^N)_k|^2 + \tilde{\alpha}_N |(\hat{B}^N)_k(\hat{B}^N)_N|.$$

In view of the result for the finite difference approximation we set

$$\tilde{\alpha}_N = \frac{4}{\pi^2} N(N^2 - 1)\beta. \tag{2.86}$$

Therefore we get by (2.68) and (2.72)

$$R_N^N \leq \frac{9}{\tilde{\alpha}_N} \left[\frac{1}{2N} \ln \frac{2(N+1)}{3} + \frac{\pi}{6N} \right] \tag{2.87}$$

where we also used the simple estimate

$$1 \leq \frac{6}{6 + \tilde{\lambda}_{D,k}^N} \leq 3. \tag{2.88}$$

Analogously, one may derive the bounds

$$R_{N-1}^N \leq \frac{18}{(6 + \tilde{\lambda}_{D,N-1}^N)} \left[\frac{1}{N-1} \ln \frac{2(N+1)}{3} + \frac{\pi}{3(N-1)} \right]$$
$$+ \frac{36\beta}{(6 + \tilde{\lambda}_{D,N}^N)(6 + \tilde{\lambda}_{D,N-1}^N)}$$

and

$$R_k^N \leq 9\frac{N-k+1}{k} \left[\frac{1}{2} \ln \frac{2(N+1)}{3} + \frac{\pi}{6} \right] - \frac{1}{N+1}\frac{2}{\pi^2} \left(\frac{N+1-k}{k} \right)^2$$
$$+ \frac{N-k+1}{k}(N-1) \cdot \frac{18\beta}{(6 + \tilde{\lambda}_{D,N}^N)(6 + \tilde{\lambda}_{D,k}^N)}.$$

We also need estimates of $|\lambda_{E,N}^N - \lambda_{E,k}^N|$. These will follow from

$$|\lambda_{E,N}^N - \lambda_{E,k}^N| = 36(N+1)^2 \frac{|\tilde{\lambda}_{D,N}^N - \tilde{\lambda}_{D,k}^N|}{(6 + \tilde{\lambda}_{D,N}^N)(6 + \tilde{\lambda}_{D,k}^N)}. \tag{2.89}$$

Therefore, (2.78) implies

$$|\lambda_{E,N}^N - \lambda_{E,k}^N| \geq 36 \cdot 4 \cdot \frac{(N-k)(N-k+2)}{(6 + \tilde{\lambda}_{D,N}^N)(6 + \tilde{\lambda}_{D,k}^N)}, \tag{2.90}$$

for $k = 1, \ldots, N-2$. Next we establish

$$|(\hat{B}^N)_N|^2 + R_N^N < 2, \quad N \geq 8.$$

In view of (2.87), (2.88) and the definition of $\hat{B}^N$ one obtains

$$|(\hat{B}^N)_N|^2 + R_N^N < 9 \left(\frac{\pi^2}{4\beta N(N^2 - 1)} \left(\frac{1}{2N} \ln \frac{2(N+1)}{3} + \frac{\pi}{6N} \right) \right. $$
$$\left. + \frac{1}{2(N+1)} \frac{\pi^2}{4} \frac{1}{N^2} \right),$$

which using $\frac{1}{x} \ln x \leq e^{-1}$ for $x \geq 1$, $N \geq 8$ and $\beta > 1$ yields the desired bound. At this point we want to choose β so that

$$|\lambda_{E,N}^N - \lambda_{E,k}^N|^2 + (\hat{B}^N)_k^2 - R_k^N > 2,$$

for $k = 1, \ldots, N-4$. An argument similar to the one employed for the finite difference approximation shows that this can be done if one can choose β satisfying

$$36^2 \cdot 16 \frac{(N-k)^2(N-k+2)k}{(6 + \tilde{\lambda}_{D,N}^N)(6 + \tilde{\lambda}_{D,k}^N)} - \left(3(N+1)e^{-1} + \frac{2\pi}{3} \right) (6 + \tilde{\lambda}_{D,N}^N)(6 + \tilde{\lambda}_{D,k}^N)$$
$$-18\beta(N-1) > \frac{2}{5}(N-4)(6 + \tilde{\lambda}_{D,N}^N)(6 + \tilde{\lambda}_{D,k}^N),$$

for $k = 1, \ldots, N-4$. It is easy to see that the right hand side is bounded above by

$$6(N-4) \quad \text{for} \quad N \geq 8,$$

and the left hand side is bounded below by

$$6 \cdot 36 \cdot 8 \left(1 + \frac{\pi^2}{2 \cdot 81} \right)^{-1} \cdot (N-k)^2(N-k+2)k$$
$$-9 \left(\frac{e^{-1}}{3}(N+1) + \frac{\pi}{6} \right) \cdot 12 \cdot (1 + \frac{\pi^2}{2 \cdot 81}) - 18\beta(N-1).$$

Combining these two estimates and simplifying we find that it is sufficient to show that

$$1628(N-k)^2(N-k+2)k - 21N - 51 - 18\beta N$$
$$+ 18\beta > 0, \quad \text{for} \quad k = 1, \ldots, N-4, \quad N \geq 8. \tag{2.91}$$

An argument similar to the one used for the finite difference approximation yields that the left hand side of (2.91) achieves its minimum for $k = N - 4$. Therefore, (2.91) is satisfied if we choose β such that

$$1628 \cdot 96 \cdot (N-4) - 21N - 51 - 18\beta N - 18\beta > 0, \quad N \geq 8.$$

Since the left hand side is increasing in N (if β is not too big) we infer that an appropriate choice of β is given by

$$\beta = 4950.$$

Up to now we have shown that $\mathcal{E}_N^N \cap \mathcal{E}_k^N = \varnothing, k = 1, \ldots, N-4$. It is somewhat tedious but straightforward (using the estimates derived so far) to show that $\mathcal{E}_N^N \cap \mathcal{E}_k^N = \varnothing$ for $k = N - 3, N - 2$ and $N - 1$. This completes the proof of Theorem 8.

Table 2.1. Robustness measure of controllability for $y_t = y_{xx} + x \cdot u(t)$. Finite difference approximation.

N	lower bound	upper bound	$\left(\gamma_{Dc}^N\right)^2$ calculated
2	0.55470×10^{-1}	0.55556×10^{-1}	0.55470×10^{-1}
3	0.21428×10^{-1}	0.21447×10^{-1}	0.21434×10^{-1}
4	0.10551×10^{-1}	0.10557×10^{-1}	0.10554×10^{-1}
5	0.59801×10^{-2}	0.59831×10^{-2}	0.59821×10^{-2}
6	0.37195×10^{-2}	0.37211×10^{-2}	0.37207×10^{-2}
7	0.24720×10^{-2}	0.24729×10^{-2}	0.24727×10^{-2}
8	0.17261×10^{-2}	0.17273×10^{-2}	0.17272×10^{-2}
9	0.12535×10^{-2}	0.12543×10^{-2}	0.12542×10^{-2}
10	0.93912×10^{-3}	0.93965×10^{-3}	0.93962×10^{-3}
20	0.13367×10^{-3}	0.13371×10^{-3}	0.13371×10^{-3}
30	0.41474×10^{-4}	0.41483×10^{-4}	0.41483×10^{-4}
40	0.17915×10^{-4}	0.17918×10^{-4}	0.17918×10^{-4}
50	0.93050×10^{-5}	0.93062×10^{-5}	0.93062×10^{-5}
60	0.54370×10^{-5}	0.54377×10^{-5}	0.54377×10^{-5}
70	0.34477×10^{-5}	0.34481×10^{-5}	0.34481×10^{-5}
80	0.23218×10^{-5}	0.23220×10^{-5}	0.23220×10^{-5}
90	0.16373×10^{-5}	0.16375×10^{-5}	0.16375×10^{-5}
100	0.11975×10^{-5}	0.11976×10^{-5}	0.11976×10^{-5}

Table 2.2. Robustness measure of controllability for $y_t = y_{xx} + x \cdot u(t)$. Finite difference approximation.

N	lower bound	upper bound	$\left(\gamma_{Dc}^N\right)^2$ calculated
2	0.222136×10^0	0.222222×10^0	0.222137×10^0
3	0.115457×10^0	0.115472×10^0	0.115459×10^0
4	0.066981×10^0	0.066986×10^0	0.066983×10^0
5	0.418736×10^{-1}	0.418755×10^{-1}	0.418745×10^{-1}
6	0.277254×10^{-1}	0.277263×10^{-1}	0.277259×10^{-1}
7	0.192182×10^{-1}	0.192187×10^{-1}	0.192186×10^{-1}
8	0.138272×10^{-1}	0.138275×10^{-1}	0.138274×10^{-1}
9	0.102595×10^{-1}	0.102597×10^{-1}	0.102596×10^{-1}
10	0.078111×10^{-1}	0.078112×10^{-1}	0.078112×10^{-1}
20	0.117693×10^{-2}	0.117698×10^{-2}	0.117698×10^{-2}
30	0.369533×10^{-3}	0.369544×10^{-3}	0.369544×10^{-3}
40	0.160314×10^{-3}	0.160318×10^{-3}	0.160317×10^{-3}
50	0.834376×10^{-4}	0.834393×10^{-4}	0.834393×10^{-4}
60	0.488086×10^{-4}	0.488094×10^{-4}	0.488094×10^{-4}
70	0.309715×10^{-4}	0.309720×10^{-4}	0.309720×10^{-4}
80	0.208664×10^{-4}	0.208667×10^{-4}	0.208667×10^{-4}
90	0.147195×10^{-4}	0.147196×10^{-4}	0.147196×10^{-4}
100	0.107680×10^{-4}	0.107681×10^{-4}	0.107681×10^{-4}

Table 2.3. Stabilizability and detectability radii.

N	γ_{Es}^N	$\tilde{\gamma}_{Es}^N$	γ_{Ed}^N	$\tilde{\gamma}_{Ed}^N$
2	10.8333	10.6325	10.8036	10.6325
4	10.2512	9.9300	10.2001	9.9300
8	10.0631	9.6515	9.9708	9.6515
10	10.0499	9.6098	9.9373	9.6098
12	10.0509	9.5857	9.9180	9.5857
14	10.0590	9.5705	9.9060	9.5705
16	10.0711	–	9.8979	–

Table 2.4. Controllability and observability radii.

N	γ_{Ec}^N	$\tilde{\gamma}_{Ec}^N$	γ_{Eo}^N	$\tilde{\gamma}_{Eo}^N$
2	0.47131	0.33469	0.16667	0.22899
4	0.25881	0.22885	0.07940	0.06555
8	0.11759	0.15311	0.03927	0.00663
10	0.08838	0.13509	0.03153	0.00196
12	0.06942	0.12229	0.02639	0.00060
14	0.05634	0.11264	0.02271	0.00020
16	0.04688	–	0.01994	–

Table 2.5. Stabilizability radii and Riccati condition number. Model Problem.

δ	γ_s	$1/\gamma_s$	K_{RIC}	$K_{RIC} * \gamma_s$
$1e^0$	$8.6603e^{-1}$	$1.1547e^0$	$2.2788e^0$	1.9735
$1e^{-1}$	$8.9425e^{-2}$	$1.1183e^{+1}$	$1.2716e^{+1}$	1.1371
$1e^{-2}$	$8.9443e^{-3}$	$1.1180e^{+2}$	$1.2130e^{+2}$	1.0850
$1e^{-3}$	$8.9443e^{-4}$	$1.1180e^{+3}$	$1.2072e^{+3}$	1.0802
$1e^{-4}$	$8.9443e^{-5}$	$1.1180e^{+4}$	$1.2071e^{+4}$	1.0797
$1e^{-5}$	$8.9433e^{-6}$	$1.1180e^{+5}$	$1.2071e^{+5}$	1.0797
$1e^{-6}$	$8.9433e^{-7}$	$1.1180e^{+6}$	$1.2071e^{+6}$	1.0797

References

[1] Arnold, W. F. and Laub, A. J., *Generalized algorithms and software for algebraic Riccati equations*, Proc. IEEE **12** (1984), 1746–1754.

[2] Banks, H. T., Jacobs, M. Q. and Langenhop, C. E., Characterization of the controlled states in $W_2^{(1)}$ of linear hereditary systems, SIAM J. Control Opt. **13** (1975), 611–649.

[3] Banks, H. T. and Kunisch, K., *The linear regulator problem for parabolic systems*, SIAM J. Control Opt. **22** (1984), 684–697.

[4] Boley, D. L., *Computing rank-deficiency of rectangular matrix pencils*, Systems & Control Letters **9** (1987), 207–214.

[5] Boley, D. L. and Wu-Sheng Lu, *Measuring how far a controllable system is from an uncontrollable one*, IEEE Trans. Automat. Control **AC-31** (1986), 249–251.

[6] Borggaard, J., Burns, J. A. and Zietsman, L., *Computational challenges in control of partial differential equations*, Proc. High Performance Computing (2004), 51–56.

[7] Burns, J. A., King, B. B., Krueger, D. and Zietsman, L., *Computation of feedback operators for distributed parameter systems with non-normal linearizations*, 2003 American Control Conference (2003), Paper TA04-2.

[8] Burns, J. A. and Peichl, G. H., *Preservation of controllability under approximation*, Differential and Integral equations **2** (1989), 439–452.

[9] Burns, J. A. and Peichl, G. H., *A note on the asymptotic behavior of controllability radii for a scalar hereditary system*, Proceedings of the 4th International Conference on Control of Distributed Parameter Systems, (Kappel F., Kunisch K., Schappacher W., eds.), Birkhäuser Verlag, 1989, 27–39.

[10] Burns, J. A. and Peichl, G. H., *On robustness of controllability for finite dimensional approximations of distributed parameter systems*, Proceedings of the IMACS/IFAC International Symposion on modeling and simulation of distributed parameter systems, (Sunahara, T., Tzafestas, S.G. and Futagami T., eds.), Hiroshima Institute of Technology, 1987, 491–496.

[11] Curtain, R. C. and Pritchard, A. J., *Infinite dimensional linear systems theory*, Lecture Notes in Control and Information Sciences 7, Springer, Berlin, 1978.

[12] Datta, B., *Numerical Methods for Linear Control Systems Design and Analysis*, Elsevier Academic Press, 2003.

[13] Demmel, J. W., *On condition numbers and the distance to the nearest ill posed problem*, Numer. Math. **51** (1987), 251–289.

[14] Eising, R., *Between controllable and uncontrollable*, Systems & Control Lett. **4** (1984), 263–264.

[15] Eising, R., *The distance between a system and the set of uncontrollable systems*, Lecture Notes in Control and Information Sciences, Proc. MTNS-83, Beer Sheva, vol. 58, Springer, New York, 1984, 303–314.

[16] Fattorini, H. O., *On complete controllability of linear systems*, JDE **3** (1967), 391–462.

[17] Gibson, J. S., *Linear quadratic optimal control of hereditary differential systems: Infinite dimensional Riccati equations and numerical approximations*, SIAM J. Control Opt. **21** (1983), 95–139.

[18] Gahinet, P. and Laub A. J., *Algebraic Riccati equations and the distance to the nearest uncontrollable pair*, SIAM J. Control Opt. **30** (1992), 765–786.

[19] Golub, G. H. and van Loan, C. F., *Matrix Computations*, The John Hopkins University Press, Baltimore, 1984.

[20] Gradstein, I. S. and Ryshik, I. R., *Tables of Series, Products and Integrals*, Vol I Verlag Harri Deutsch, Frankfurt, 1981.

[21] Hinrichsen, D. and Motscha, M., *Optimization problems in the robustness analysis of linear state space systems*, Approximation and Optimization, (Gomez, A., Guerra, F., Jimenez, M. A. and Lopez, G., eds.), Lecture Notes in Mathematics, 1354 Springer, 1988, 54–78.

[22] Hinrichsen, D. and Pritchard, A. J., *Stability radii for linear systems*, Systems & Control Letters **7** (1986), 1–10.

[23] Hinrichsen, D. and Pritchard, A. J., *Stability radiius for structured perturbations and the algebraic Riccati equation*, Systems & Control Letters **8** (1986), 105–113.

[24] Ito, K. and Kappel, F., *A uniformly differentiable approximation scheme for delay systems using splines*, J. Appl. Math. Opt. **23** (1991) 217–262.

[25] El Jai, A. and Pritchard, A. J., *Sensors and Controls in the Analysis of Distributed Systems*, J. Wiley, 1988.

[26] Kappel, F. and Peichl, G. H., *Preservation of controllability under approximation for delay systems*, Matematica Aplicada e Computacional **8** (1989), 23–47.

[27] Kappel, F. and Salamon, D., *On the stability properties of spline approximations for retarded systems*, SIAM J. Control Opt. **27** (1989), 407–431.

[28] Kailath, T., *Linear Systems*, Englewood–Cliffs, Prentice Hall, 1980.

[29] Kenney, C. and Laub, A. J., *Controllability and stability radii for companion form systems*, Math. Control Signal Systems **1** (1988), 239–256.

[30] Laub, A. J., *Numerical linear algebra aspects of control design computations*, IEEE Trans. Automat. Control **AC-30** (1985), 97–108.

[31] Lancaster, P. and Tismenetsky, M., *The theory of Matrices*, Academic Press, Orlando, 1985.

[32] Lasiecka, I. and Triggiani, R., *Differential and algebraic Riccati equations with application to boundary /point control problems: continuous theory and approximation theory*, Lecture Notes in Control and Information Sciences 164, Springer, Berlin, 1978.

[33] Lewkowicz, I., *When are the complex and the real stability radii equal*, IEEE Transactions on Automatic Control **AC-37** (1992), 880–883.

[34] Miminis, G., *Numerical algorithms for controllability and eigenvalue computation*, Masters thesis, School of Computer Science, McGill Univ., 1981.

[35] Oates, K., *A study of control system radii for approximations of infinite dimensional systems*, MS thesis, Department of Mathematics, VPI–SU, Blacksburg, VA, 1991, ICAM Report 91-08-01.

[36] Paige, C., *Properties of numerical algorithms related to computing controllability*, IEEE Trans. Automat. Control **AC-26** (1981), 130–139.

[37] Pandolfi, L., *Controllability properties of perturbed distributed parameter systems*, Lin. Alg. Appl. **122** (1989), 525–538.

[38] Peichl, G. H. and Wang, C.: *On the uniform stabilizability and the margin of stabilizability of the finite dimensional approximations of distributed parameter systems*, J. Math. Systems Estimation and Control **7** (1997), 277–304.

[39] Peichl, G. H. and Wang, C.: *Asymptotic Analysis of stabilizability of a control system for a discretized damped wave equation*, Numerical Functional Analysis **19** (1998), 91-113.

[40] Pritchard, A. J. and Townley, S., *A stability radius for infinite dimensional systems*, Distributed parameter systems, (F.Kappel, K.Kunisch, W.Schappacher, eds.), Lecture Notes in Control and Information Sciences 102, 1986, 272–291.

[41] Salamon, D., *Controllability and observability of time delay systems*, IEEE Trans. Automat. Control **AC-29** (1984), 432–439.

[42] Smith, G. D., *Numerical solution of partial differential equations*, Clarendon Press, Oxford, 1976.

[43] Wilkinson, J. H., *The Algebraic Eigenvalue Problem*, Clarendon Press, Oxford, 1965.

[44] Wicks, M. and DeCarlo, R., *Computing the distance to an uncontrollable system*, Proceedings of the 28th Conference on Decision and Control, Tampa, Fl, 1989, 2361–2366.

[45] Wicks, M. and DeCarlo, R., *Computing the distance to an uncontrollable system*, IEEE Transactions on Automatic Control **AC-36** (1991), 39–49.

3

Equilibrium Analysis for a Network Market Model *

Juan F. Escobar[1] and Alejandro Jofré[2]

[1] Department of Economics
Stanford University
jescobar@stanford.edu
[2] Centro de Modelamiento Matemático and Departamento de Ingenieria Matematica
Universidad de Chile
Blanco Encalada 2120 7o. piso
Santiago-Chile (56 2 6784454)
ajofre@dim.uchile.cl

Summary. In this paper we model and analyze a market equilibrium structure working on a network. The model is motivated by competition in electricity power generation markets, where consumers and producers are located in different nodes connected by power transmission lines. We analyze two different equilibrium concepts, namely, the Walrasian and the noncooperative Nash outcomes. By using concepts coming from Variational Analysis and Game Theory, we prove that both equilibria exist. Our existence proofs rely on fixed point theorems and epi-convergence stable approximations.

Key words: network market, Walrasian economic equilibrium, noncooperative economic equilibrium, variational analysis, Nash equilibrium, epi-convergence

Introduction

Motivated by the study of electricity power generation market models, in this paper we develop an equilibrium theory to analyze competition on a network market. The study of network market models started in the late seventies. However, it was only until the worldwide deregulation and privatization of some important industries (such as electricity, telecommunications, and water services) that the equilibrium analysis of network market gained especial attention by mathematicians, engineers and policy makers.

The model we offer is quite general. Indeed we consider a general connected graph that represents a network. There is a good (e.g. electricity, or water) that can be transported along the network's edges. Competing agents bid functions to a central authority. Each function represents how willing is the agent to pay or being paid for consumption or production of the good. Central agent's objective is to maximize the total surplus originated by trade of the good.

* This work was partially supported by ICM Complex Engineering Systems.

Two different equilibrium concepts are studied. On the one hand, we study a Walrasian equilibrium and assume that agents do not take into account the effects of these actions on network prices. This concept may not be interesting on its own when thinking about current markets. However, it provides a useful benchmark to more complex and long-sighted behaviors. On the other hand, we assume that agents perfectly foresight the impact of their bids on all relevant variables. This assumption is probably more suitable in practice. This setting prompts us to study the noncooperative solution of the game among agents and introduce the Nash equilibrium concept.

The difference between these two concepts is subtle, but its impact on the efficiency of the outcomes can be substantial. We do not deal with efficiency issues in this work. Simulation and computation can provide some guidance in this respect.

The main results of this work are existence theorems for Walrasian and noncooperative equilibria. The methods introduced in the proofs rely on standard fixed point tools. Our innovation comes from showing how stability analysis, more specifically the epi-convergence, can be combined with fixed point arguments to prove new economic equilibrium results. Recent works by Jofré et al. [14] go in this direction but the focus is different.

Our model particularized to a bid-based electricity pools can be deemed as an extension of the seminal work of Klemperer and Meyer [15], and the subsequent paper by Green and Newbery's [11] (among others). The existence results presented can also be considered as a formalization for models widely used in the engineering literature, see [13].

Variational inequalities tools have proven useful to study competition on networks. A good reference for this applications is Nagurney [17]. In the same vein the works by Wei and Smeers [21], Pang and Hobbs [18], and Hobbs et al. [12] provide spatial oligopolistic models and study existence and computability of equilibria. The market structures that can be study by employing variational inequalities is not connected to our study.

The structure of the paper is as follows. In Section 3.1 we develop the market equilibrium model. In Section 3.2 we introduce the two equilibrium notion (Walrasian and noncooperative). We discuss their interpretations and scope of applicability. In Section 3.3 we state and demonstrate the two main existence results. Both proofs are applications of fixed point arguments applied on suitable topological structures as to guarantee that payoff maps have continuity properties.

Before ending this introduction, we mention that this work is the first of several aiming to build a theoretical framework for competitive markets working on networks. Extensions of the existence result for noncooperative interaction equilibria are given by Escobar and Jofré [8]. In this work, it is also demonstrated that one can perturb slightly the fundamentals of the market (network topology, or payoffs), in such a way that equilibrium outcomes are not immensely altered by those perturbations. Some numerical algorithms for computation of equilibrium outcomes are introduced by Escobar et al. [9]. Some consequences from the point of view of standard economic theory are given by Escobar and Jofré [7].

3.1 The Market Equilibrium Model on a Network

Consider a network consisting of a set of nodes $\{1, \ldots, N\}$ and edges $\{1, \ldots, E\} \subseteq \{1, \ldots, N\} \times \{1, \ldots, N\}$. Each node hosts a (wholesale) consumer or a producer, but not both. Thus the set of nodes can be split in two sets

$$\{1, \ldots, N\} = \{1, \ldots, N_0\} \cup \{N_0 + 1, \ldots N\},$$

The set $\{1, \ldots, N_0\}$ corresponds to the set of nodes hosting (wholesale) consumers, while $\{N_0 + 1, \ldots N\}$ is the set of nodes hosting producers. A *production plan* is a vector $(q, f) \in \mathrm{R}^N \times \mathrm{R}^E$, meaning that the consumer (producer) at node n, henceforth consumer (producer) n, consumes (produces) q_n units of the good and that the flow of the good along edge e (henceforth path e) is f_e. We also consider a coordinating agent that can set production plans while respecting some network constraints.

Suppose that from the central agent's perspective, consumer $n \leq N_0$ has a (money-equivalent) utility function $x_n : \mathrm{R} \to \mathrm{R}$ and producer $n > N_0$ has a cost function $x_n : \mathrm{R} \to \mathrm{R}$. The central agent can manage the production plan and is interested in maximizing the total surplus defined by

$$\sum_{n=1}^{N_0} x_n(q_n) - \sum_{n=N_0+1}^{N} x_n(q_n)$$

subject to the technological and physical constraints that we specify below.

Production plans need to be feasible. Firstly, a production plan must satisfy a *market-clearing condition* at each node. The market-clearing condition is satisfied if and only if for all $n \leq N_0$

$$q_n + \sum_{e \in K_n} \mathrm{sign}(e, n) f_e = 0, \tag{3.1}$$

and for $n > N_0$

$$\sum_{e \in K_n} \mathrm{sign}(e, n) f_e = q_n, \tag{3.2}$$

where K_n is the set of edges connecting agent n, and $\mathrm{sign}(e, n) = -\mathrm{sign}(e, n')$ if $e = (n, n')$. Note that the signs of flows should be consistent with the sign conventions as usual in network flow problems. The interpretation of the equalities is as follows. For a consumption node, $n \leq N_0$, effective flows to the node, $\sum_{e \in K_n} \mathrm{sign}(e, n) f_e$, equal demand at the node, q_n, and for a production node, $n > N_0$, the sum of effective flows, $\sum_{e \in K_n} \mathrm{sign}(e, n) f_e$, equals the local production, q_n. One could consider these constraints with inequalities. That is not a difficulty.

On the other hand, we define the set of feasible production plans as

$$QF(\omega) = \{(q, f) \in \mathrm{R}^N \times \mathrm{R}^E \mid (q, f) \text{ satisfies } (3.1), (3.2) \ (q, f) \in F(\omega)\},$$

where ω, called a *network failure*, is a parameter that represents a sudden failure in the network (for example, a sudden path failure) and $F(\omega) \subseteq \mathrm{R}^N \times \mathrm{R}^E$ is compact-valued and measurable on ω. The set $F(\omega)$ represents network constraints other than the market-clearing conditions (3.1-3.2).

Thus, knowing the functions $(x_n)_{n=1}^N$, and the network failure ω, the central agent is interested in solving the following optimization problem

$$\max \left\{ \sum_{n=1}^{N_0} x_n(q_n) - \sum_{n=N_0+1}^{N} x_n(q_n) \mid (q, f) \in QF(\omega) \right\}. \tag{3.3}$$

We call this problem the *dispatch problem* and assume that its solutions exists. Define the set $Q(x, \omega) \subseteq \mathrm{R}^N$ of solutions to this program (3.3) (properly projected on R^N). Just for simplicity, we assume that $Q(x, \omega)$ is a singleton.

Let us now describe how prices are set. Following Schweppe et al. [20], we consider a shadow-value pricing rule defined from the dispatch program (3.3). That is, the price paid or received by agent (consumer or producer) at a node n is the Lagrange multiplier related to the corresponding market clear condition (3.1) or (3.2). We define $\Lambda(x, \omega) \subseteq \mathrm{R}^N$ as the set of described Lagrange multipliers and assume that it is nonempty.

Conditions on the functions x_n and $QF(x, \omega)$ can be given as to guarantee the existence of optimal solutions and Lagrange multipliers. We, however, are mainly interested in equilibrium analysis and take those conditions for granted.

Suppose that if consumer n consumes q_n and pays a price p_n per each unit consumed, its payoff is $\pi_n(p_n, q_n) = \bar{U}_n(q_n) - p_n q_n$. Analogously, producer n's payoff function takes the form $\pi_n(p_n, q_n) = p_n q_n - \bar{c}_n(q_n)$ where q_n is its production and p_n is the price received per unit produced. Functions $\bar{U}_n, \bar{c}_n$ are exogenous variables in our model that are not affected by any of the actions agents in the model take.

Consumers and producers have an active role in this market. Each agent bids a function x_n. The tuple $x = (x_n)_{n=1}^N$ becomes an input to the dispatch program (3.3) the central agent runs.

Agents *simultaneously* bid the functions $(x_n)_{n=1}^N$. That is, the control variable for consumer (producer) n is the utility (cost) function x_n (c_n) that becomes an input for the dispatch program (3.3). When bidding this function, agent n does not know the parameter ω.

The bid of agent n may or may not reflect its actual function $\bar{U}_n$ or $\bar{c}_n$. Indeed, if an agent n finds profitable lying in respect to its actual function $\bar{U}_n$ or $\bar{c}_n$, then it will do it. We, however, restrict the set of feasible decisions to agent n. Agent n must choose a function belonging to a nonempty set of functions X_n that is exogenously defined.

When choosing its bid, each agent thinks of the market failure ω as randomly distributed according to the probability distribution P. Just for simplicity, we assume that all agents agree in the distribution of ω. Define the support of P as Ω.

Summing up, the following is the basic setup of how the market works: (1) Agents simultaneously bid functions $(x_n)_{n=1}^N$; (2) The network failure ω is known; (3) The central agent takes $(x_n)_{n=1}^N$, ω as inputs to the optimization problem 3.3; (4) Agents must behave as mandated by the central agent.

We define $X = \Pi_{n=1}^N X_n$ and endow it with the (product) topology of the pointwise convergence. We further consider the obvious vectorial structure over the set of real-valued functions on R. Throughout the paper we assume the following conditions.

(A0) X is compact. For all n, X_n is a convex set.

(A1) For all $x \in X$, the function $q \in \mathrm{R}^N \mapsto \sum_{n=1}^{N_0} x_n(q_n) - \sum_{n=N_0+1}^{N} x_n(q_n)$ is concave.

(A2) For all $x \in X$ and $\omega \in \Omega$, $Q(x, \omega)$ and $\Lambda(x, \omega)$ are uniformly bounded.

(A3) For all n, π_n is continuous.

3.2 Walrasian and Strategic Noncooperative Equilibria

Different behavioral assumptions are plausible in the model we described in the previous section. Agents can form different expectations regarding how their actions impact the market outcome and how the rest of the agents will react in response to its own actions. We distinguish two kinds of behavior: the Walrasian and the strategic non-cooperative.

3.2.1 Walrasian Equilibrium

Typically, a Walrasian behavior is referred to a situation where agents participating in the market do not take into account the effect of their own actions on the market prices. This is the standard competitive paradigm analyzed by Arrow and Debreu [1] in the general equilibrium framework.

In our model agents' actions are to bid functions that are an input for the dispatch program the market coordinator run. The natural extension of the Walrasian behavior in this setting is to assume that each agent does not realize that its bid is affecting nodal prices throughout the network. However, each agent does realize that its bid determines the quantity it ends up consuming or producing.

Definition 1. *A Walrasian equilibrium is a set of functions* $(\bar{x}_n)_{n=1}^N$, *and a measurable selections* $\lambda(\omega)$ *from* $\Lambda(\bar{x}, \omega)$ *such that for all* n

$$\bar{x}_n \in \arg\max\left\{ \int \pi_n(\lambda_n(\omega), q_n((x_n, \bar{x}_{-n}), \omega))d\mathrm{P}(\omega) : x_n \in X_n \right\}.$$

Note that in the definition each agent maximizes its expected payoff taking the price selection $\lambda(\omega)$ as not being affected by its bid $\bar{x}_n$. Probably, this assumption is hardly a good assumption for modelling many actual markets. However, it may be attractive because it makes the model numerically tractable and provides a useful benchmark to more sophisticated interactions. Consult Motto et al. [16] for applications of the concept to power networks.

3.2.2 A Noncooperative Equilibrium

Now we introduce a more realistic behavioral assumption, where agents fully realize of the consequences of their actions on prices and quantities. The natural framework for this analysis is game theory and the analysis of the Nash equilibrium of the game among agents.

We will allow each agent to randomize over its set of feasible functions X_n. Therefore, the control variable each agent has is a probability measure m_n over X_n. Given how the other agents randomize, agent n chooses $\bar{m}_n$ optimally.

We need to expand the feasible set of each agent to allow randomization because the equilibrium where agents do not randomize may fail to exist. Technically, there are non-convexities prompted by the intrinsic non-concavity of the Lagrange multipliers as maps on $x \in X$, that make the standard arguments not applicable. The kind of randomization we consider is know as mixed strategies and is an usual concept in the economic literature.

Definition 2. *A* noncooperative equilibrium *is a measure* $\bar{m} = (\bar{m}_n)_{n=1}^N$ *defined over the point-wise (Borel) σ-field on X and a measurable selection $\lambda(x,\omega)$ from $\Lambda(x,\omega)$ such that for all n*

$$\bar{m}_n \in \arg\max \left\{ \int \Pi_n d(m_n \times \bar{m}_n)(c) \mid \right.$$

$$\left. m_n \text{ is defined over the point-wise } \sigma\text{-field on } X_n \right\},$$

where

$$\Pi_n(c) = \int \pi_n(\lambda_n(x,\omega), q_n(x,\omega)) \mathrm{P}(d\omega) \tag{3.4}$$

is agent n's expected payoff given the vector of bids $c = (c_n)$.

Remark 1. Note the fundamental difference between Definition 1 and Definition 2. Only in the first definition the selection of Lagrange multipliers is for fixed bids $\bar{x}_n$.

3.3 Existence Results

In this section we shall prove the existence of Walrasian and noncooperative equilibria. The proofs of these results rely crucially on epi-convergence and hypo-convergence. These convergence notions are concepts of Variational Analysis associated with the convergence of solutions of optimization problems when we approximate the problems. We refer to chapter 7 in Rockafellar and Wets [19] and references therein for details.

The following theorem guarantees the existence of a Walrasian equilibrium.

Theorem 1. *Suppose that*

1. *For all $\omega \in \Omega$, for all n, and for all $x \in X$, $Q_n(x,\omega)$ is a concave function of x_n; and*
2. *$\pi_n(p_n, q_n)$ is a concave and increasing function in q_n.*

Then, a Walrasian equilibrium exists.

Proof. Consider the following $(N + 1)$-person game. Agent 0, the coordinating agent, chooses a function $\lambda : \Omega \to \mathbb{R}^N$. Agent $n \in \{1, \ldots, N\}$ chooses a function $x_n \in X_n$. Agent 0's payoff is the following Lagrangian-like function

$$- \int \left(\sum_{n=1}^{N_0} x_n(Q_n(x, \omega)) - \sum_{n=N_0+1}^{N} x_n(q_n(x, \omega)) + \langle \lambda(\omega), \zeta(x) \rangle \right) dP(\omega),$$

where $\zeta(x) \in \mathbb{R}^N$ is a vector containing the N constraints (3.1)-(3.2) evaluated at the optimal solution of the dispatch program given x and ω. Agent $n \in \{1, \ldots, N\}$ has a payoff given by

$$\int \pi_n(\lambda_n(\omega), Q_n(x, \omega)) dP(\omega).$$

It is clear that a Nash equilibrium $(\bar{x}, \lambda)$ of this game will be a Walrasian equilibrium of the network market. To prove that $\lambda(\omega)$ will be a Lagrange multiplier just note that, from Theorem 14.60 in Rockafellar and Wets [19], we can interchange integration and maximization. That is, we can think of the problem agent 0 solves in the Nash equilibrium of the game as the dual problem related to (3.3) being solved for each $\omega \in \Omega$ given $x \in X$.

Let us demonstrate that the described game possesses a Nash equilibrium. To prove that, we will employ Glicksberg's theorem and verify that

(i) Each player has a compact strategy set contained in a metric space;
(ii) Payoff functions are continuous on all the choices variables;
(iii) Payoffs are concave in the own decision variable.

For references about Glicksberg's theorem we refer to Fudenberg and Tirole [10].

Note first that (iii) follows from the two concavity assumptions 1. and 2. Let us prove (i). For agent $n > 0$, its strategy set X_n is pointwise compact by assumption. For agent 0, by virtue of (A2), we can assume that Lagrange multipliers are bounded. So, without loss of generality, agent 0 chooses a function from the set of all functions bounded by a sufficiently big constant κ. It is clear that we can take this set to be closed for the pointwise convergence. Therefore, agent 0 picks its strategy from a compact set.

To prove (ii), we argue first that for each $\omega \in \Omega$, the function $x \in X \mapsto Q(x, \omega)$ is continuous. Take any sequence x^ν pointwise converging to x on X. Consider the convex function $(q, f) \in QF(\omega) \mapsto J^\nu(q, f) = \sum_{n=1}^{N_0} x_n^\nu(q_n) - \sum_{n=N_0+1}^{N} x_n^\nu(q_n)$. This is a sequence of concave functions that converges pointwise to $(q, f) \in QF(\omega) \mapsto J(q, f) = \sum_{n=1}^{N_0} x_n(q_n) - \sum_{n=N_0+1}^{N} x_n(q_n)$. We can therefore apply Theorem 7.17 in Rockafellar and Wets [19] to deduce that $(J^\nu)_\nu$ hypo-converges to $J(x)$. Then, applying the fundamental hypo-convergence property detailed in Proposition 7.30 in Rockafellar and Wets [19], we can deduce that any accumulation point of the sequence $Q(x^\nu, \omega)$ must be a solution for

$$\max\{J(q, f) \mid (q, f) \in QF(\omega)\}.$$

But by definition, the only solution to this problem is $Q(x, \omega)$. By (A2), Q is uniformly bounded. We then deduce that $Q(x^\nu, \omega) \to Q(x, \omega)$. Finally, (ii) follows by virtue of Lebesgue's dominated convergence theorem once the domination condition for each of the payoffs has been verified. $\qquad\square$

Now we turn to the study of a noncooperative solution to the network market model. The proof of existence of a noncooperative equilibrium relies once again in stability analysis. The proof is, in a sense, constructive.

Theorem 2. *Suppose that* P *is atomless. Then, a noncooperative equilibrium exists.*

Proof. We will prove this result in a number of steps. In Step 1 we prove that solutions and Lagrange multipliers to the dispatch program (3.3) are smooth maps of x. In Step 2, we then prove that Step 1 implies that the payoffs map (of payoffs that can be supported by optimal solutions and Lagrange multipliers from the dispatch program (3.3)) is smooth enough and convex-valued. We finally argue in Step 3 that given that the payoff map is well behaved, we can approximate it by smooth games. Each of these games has an equilibrium. In turn, this equilibrium sequence will converge to a point that will be an equilibrium for the game we are interested in.

Step 1. For each ω, $Q(\cdot, \omega) \colon X \rightrightarrows \mathrm{R}^N$ is continuous on X, and $\Lambda(\cdot, \omega) \colon X \rightrightarrows \mathrm{R}^N$ is outer semi-continuous on X.

It suffices to show that the maps have closed graphs.

The continuity of $Q(\cdot, \omega)$ was already shown in Theorem 1. To prove that $\Lambda(\cdot, \omega)$ has a closed graph, consider $x^\nu \to x$ on X and $\lambda^\nu \in \Lambda(x^\nu, \omega)$ such that $\lambda^\nu \to \lambda$. We will prove that $\lambda \in \Lambda(x, \omega)$. Write the dual problem associated to the dispatch program 3.3 when we consider the inputs (x^ν, ω) as

$$\min_\lambda T^\nu(\lambda),$$

where $T^\nu(\lambda)$ is defined as the value of the maximization problem we considered to prove the continuity of $Q(\cdot, \omega)$ in Theorem 1, but adding the dual term related to restrictions (3.1)-(3.2). By definition, the solution set for this maximization problem is $\Lambda(x^\nu, \omega)$. We can refine the argument we gave there, and prove that T^ν converges continuously to T, where T is defined analogously. Once again, this follows by proving the hypo-convergence of the objective function. We then deduce that T^ν epi-converges to T. It then follows that any limit point of solutions of the minimization problem above is a solution to the limit maximization problem. In other words, the sequence $\lambda^\nu \in \Lambda(x^\nu, \omega)$ has to converge to a point in $\Lambda(x, \omega)$. This proves the Step 1.

Step 2. Given $x \in X$, define the payoffs map by

$$E\alpha(x) = \left\{ \int t(d)d\mathrm{P}(\omega) \mid t \text{ an } L_1\text{-selection from } \alpha(x, \cdot) \right\}, \tag{3.5}$$

where $\alpha(x, \omega) = \{(\pi_n(p_n, Q_n(x, \omega))_{n=1}^N \in \mathrm{R}^N \mid p \in \Lambda(x, \omega)\}$. Then, $E\alpha$ is nonempty-convex-valued, has a bounded range, and is outer semi-continuous. It further can be interpreted as the union of all possible agents' payoffs that can be supported by selections $\lambda(x, \omega) \in \Lambda(x, \omega)$ for x fixed.

The existence of measurable selections follows from Castaing's theorem, see Theorem 14.5 in Rockafellar and Wets [19]. So, $E\alpha(x)$ is a nonempty set. Since P is atomless, we can apply Auman's theorem [2] to deduce that $E\alpha(x)$ is convex-valued. $E\alpha$ is bounded because of assumptions (A2) and (A3).

Finally, let us prove that $E\alpha$ has a closed graph. Let x^ν converge to x, and $u^\nu \in E\alpha(x^\nu)$ converging to $u \in R^G$. Take t^ν an L_1-selection from $\alpha(x^\nu, \cdot)$ such that $u^\nu = \int t^\nu dP$. Therefore

$$\lim_{\nu \to +\infty} \int t^\nu dP = u.$$

Because (t^ν) is a bounded sequence, from Tychonoff's theorem, there exists (t^{ν_k}) pointwise converging to t and, due to the Dominated Convergence Theorem, we can assume that $\lim_{k \to +\infty} \int t^{\nu_k} dP(\omega) = \int t dP(\omega)$. Thus $u = \int t dP$. Then, from Step 1 and the continuity assumption (A3), it follows that t is an L_1-selection from $alpha(x, \cdot)$ completing this step.

Step 3. A non-cooperative equilibrium exists. Further, we can obtain it from Balder's construction [3; 4].

Let us consider the following scheme. For each $\nu \in N$ we can find $V^\nu : X \to R^N$ which is a continuous $1/\nu$-selection from $E\alpha$. Step 3 together with the continuous selection theorem allows us to do that. Consider, for each ν, the N-person strategic-form game $G^\nu = (X, V^\nu)$, where X denotes the set of feasible strategies and V^ν denotes the payoff vector. From Glicksberg's theorem, G^ν possesses a mixed-strategy equilibrium measure m^ν, that is defined over the product point-wise σ-field on X. Furthermore, there exists a measure $\bar{m} = (\bar{m})_{n=1}^N$ which is a weak cluster point for the sequence $(\bar{m}^\nu)_{\nu \in N}$. (For measure related convergence results, consult Billinsgley [5]). From Balder's result [3; 4], we can deduce that $\bar{m}$ is an equilibrium measure, for some selection λ from Λ.   □

References

[1] Arrow, K.J., G. Debreu. 1954. Existence of an equilibrium for a competitive economy. *Econometrica*, 22, 265-290.

[2] Aumann, R.J. 1965. Integrals of set-valued functions. *Journal of Mathematical Analysis and Aplications*, 12, 1-12.

[3] Balder, E. 2001. 2001. On equilibria for discontinuous games: Nash approximation schemes. Preprint.

[4] Balder, E. 2004. An equilibrium existence result for games with incomplete information and indeterminate outcomes. *Journal of Mathematical Economics*, 40:297-320.

[5] Billingsley, P (1968): *Convergence of Probability Measures*. New York, Wiley.

[6] Day, C., B. Hobbs, J.-S. Pang. 2002. Oligopolistic competition in power markets: A conjectured supply function approach. *IEEE Transactions on Power Systems*, 17:597-607.

[7] Escobar, J.F., A. Jofré. 2004. Electricity wholesale markets: An oligopoly-network framework. Submitted to *Review of Economic Studies*.

[8] Escobar, J.F., A. Jofré. 2005. A Variational convergence approach for equilibrium analysis on network markets. Mimeo

[9] Escobar, J.F., A. Jofré, R. Palma. 2005. A bid-based pool power market model: Theory and computation. Mimeo

[10] Fudenberg D., and J. Tirole (1991): *Game Theory*. Cambridge, MIT Press.

[11] Green, R.J., D. Newbery. 1992. Competition in the British electricity spot market. *Journal of Political Economy*, 100:929-953.

[12] Hobbs, B.F., C.B. Metzler, J.-S. Pang. 2000. Strategic gaming analysis for electric power networks: An MPEC approach. *IEEE Transactions on Power Systems*, 15:638-645.

[13] IEEE PES Tutorial. 1999. *Game Theory Applications in Electric Power Markets*. IEEE.

[14] Jofré, A., R.T. Rockafellar, R.J-B Wets. 2005. A variational inequality scheme for determining economic equilibrium *Variational Analysis and Applications*. F. Giannessi, A. Maugeri, eds. Kluwer, Boston.

[15] Klemperer P., M. Meyer. 1989. Supply function equilibria in oligopoly under uncertainty. *Econometrica*, 57:1243-1277.

[16] Motto, A., F. Galiana, A. Conejo, M. Haneault. 2003. On Walrasian equilibrium for pool-based electricity markets. *IEEE Transactions on Power Systems*, 17:774-781.

[17] Nagurney, A. 1999. *Network Economics: A Variational Inequality Approach*. Kluwer, Boston.

[18] Pang, J.-S., B. Hobbs. 2002. Spatial oligopolistic equilibria with arbitrage, shared resources, and price functions conjectures. To appear *Mathematical Programming*.

[19] Rockafellar, R.T., R.J-B Wets. 1998. *Variational Analysis*. Springer, New York.

[20] Schweppe, F., M. Caramanis, R. Tabors, R. Bohn. 1988. *Spot Pricing of Electricity*. Kluwer, Boston.

[21] Wei, J.-Y., Y. Smeers. 1999. Spatial oligopolistic models with Cournot firms and regulated transmission prices. *Operations Research*, 47:102-112.

4

Distributed Solution of Optimal Control Problems Governed by Parabolic Equations [*]

Matthias Heinkenschloss[1] and Michael Herty[2]

[1] Department of Computational and Applied Mathematics
   MS-134, Rice University
   6100 Main Street
   Houston, TX 77005-1892
   `heinken@rice.edu`
[2] Fachbereich Mathematik
   Technische Universität Kaiserslautern
   Postfach 30 49, D-67653 Kaiserslautern, Germany
   `herty@mathematik.uni-kl.de`

Summary. We present a spatial domain decomposition (DD) method for the solution of discretized parabolic linear–quadratic optimal control problems. Our DD preconditioners are extensions of Neumann-Neumann DD methods, which have been successfully applied to the solution of single elliptic partial differential equations and of linear–quadratic optimal control problems governed by elliptic equations.

We use a decomposition of the spatial domain into non-overlapping subdomains. The optimality conditions for the parabolic linear–quadratic optimal control problem are split into smaller problems restricted to spatial subdomain-time cylinders. These subproblems correspond to parabolic linear–quadratic optimal control problems on subdomains with Dirichlet data on interfaces. The coupling of these subdomain problems leads to a Schur complement system in which the unknowns are the state and adjoint variables on the subdomain interfaces in space and time.

The Schur complement system is solved using a preconditioned Krylov subspace method. The preconditioner is obtained from the solution of appropriate subdomain parabolic linear–quadratic optimal control problems. The dependence of the performance of these preconditioners on mesh size and subdomain size is studied numerically. Our tests indicate that their dependence on mesh size and subdomain size is similar to that of its counterpart applied to elliptic equations only. Our tests also suggest that the preconditioners are insensitive to the size of the control regularization parameter.

Key words: optimal control, parabolic equations, domain decomposition, Neumann-Neumann methods

[*] This work was supported in part by NSF Grants ACI-0121360 and CNS-0435425 and DAAD Grant D/04/23833.

4.1 Introduction

This paper presents a spatial domain decomposition method for the solution of linear-quadratic parabolic optimal control problems. Such problems arise directly in many applications, but also as subproblems in Newton or sequential quadratic programming methods for the solution of nonlinear parabolic optimal control problems. The motivation for this work is threefold. First, our approach attempts to address the storage issue that arises in the numerical solution of parabolic optimal control problems out of the strong coupling in space and time of state (PDE solution), the adjoint, and the control. Secondly, our spatial domain decomposition method introduces parallelism at the optimization level. The last motivation arises from the availability of sensor networks that offer in-network computing capabilities, allow neighbor-to-neighbor communication, but for which global communication requires large amounts of resources because of communication bandwidth and battery power limitations. Our domain decomposition method offers the possibility for in-network computing, in which the global problem is solved using spatially distributed processors that communicate with their neighbors.

Domain decomposition methods have been applied previously to linear-quadratic time dependent optimal control problems. They split into time domain decomposition methods [4; 10; 13; 14; 17] and spatial domain decomposition methods [1; 2; 3]. Like [1; 2; 3], the approach in this paper is also based on a decomposition of the spatial domain. The resulting subproblems are smaller linear-quadratic parabolic optimal control problems posed on a spatial subdomain-time cylinder. The difference between the approaches [1; 2; 3] and our approach lies in the way the subdomain problems are coupled and in the solution method for the coupled subdomain problems.

Our spatial domain decomposition method for linear-quadratic parabolic optimal control problems is based on the so-called Neumann-Neumann domain decomposition methods. Of the domain decomposition method for elliptic partial differential equations (PDEs), Neumann-Neumann methods are among the most successful ones. Their derivation and discussions of their convergence properties can be found in the books [18; 20; 21] and the references given therein. Recently, Neumann-Neumann methods were generalized to solve linear-quadratic elliptic optimal control problems. The results in [11; 12] have shown that their performance on linear-quadratic elliptic optimal control model problems is comparable to their good performance for single elliptic PDEs. This paper extends Neumann-Neumann methods to the solution of linear-quadratic parabolic optimal control problems.

To illustrate our ideas, we consider the example problem

$$
\begin{aligned}
\min \quad & \frac{\alpha_1}{2} \int_0^T \int_\Omega (y(x,t) - \hat{y}(x,t))^2 dx dt \\
& + \frac{\alpha_2}{2} \int_\Omega (y(x,T) - \hat{y}_T(x))^2 dx \\
& + \frac{\alpha_3}{2} \int_0^T \int_\Omega u^2(x,t) dx dt
\end{aligned}
\tag{4.1a}
$$

s.t. $\partial_t y(x,t) - \Delta y(x,t) = f(x,t) + u(x,t)$ in $\Omega \times (0,T)$,   (4.1b)

$\quad\ y(x,t) = 0$ on $\partial\Omega \times (0,T)$,   (4.1c)

$\quad\ y(x,0) = y_0(x)$ in Ω,   (4.1d)

where $\hat{y}, \hat{y}_T, f$ are given functions and $\alpha_1, \alpha_2 \geq 0, \alpha_3 > 0$ are given parameters. The problem (4.1) has to be solved for y and u. Detailed model problem assumptions will be introduced in the next section.

Section 4.3 introduces the domain decomposition formulation for a discretization of (4.1). We decompose Ω into nonoverlapping domains $\Omega_i, i = 1,\ldots, s$. Essentially, we decompose the system of optimality conditions for (4.1) by expressing states, controls, and adjoints inside the subdomains $\Omega_i \times (0,T), i = 1,\ldots, s$, as functions of the states and adjoints on the subdomain interfaces $(\partial\Omega_i \setminus \partial\Omega) \times (0,T)$. Transmission conditions that couple the subdomain problems are then viewed as operator equations in states and adjoints restricted to the subdomain interfaces $(\partial\Omega_i \setminus \partial\Omega) \times (0,T)$. As we have mentioned before, our domain decomposition method extends Neumann-Neumann domain decomposition methods well known for the solution of elliptic PDEs to the solution of linear-quadratic parabolic optimal control problems. Often the discretization of the system of optimality conditions for (4.1) is not equal to the system of optimality conditions for the discretization of (4.1). Therefore, we formulate our domain decomposition method not for the infinite dimensional problem, but for a discretization of (4.1). This ensures that the problem solved by our domain decomposition method is the discretization of (4.1), i.e. the problem one would typically solve and not some perturbation of it.

Section 4.4 reports on some initial numerical results.

4.2 The Example Problem

In this section, we define the setting for the model problem (4.1), recall a result on the existence and uniqueness of its solution, and review the well-known necessary and sufficient optimality conditions. Furthermore, we introduce a discretization of (4.1). The results in this section are well known, but are recalled here to serve as the background for our domain decomposition formulation.

4.2.1 The Infinite Dimensional Problem

Let $\Omega \subset \mathbb{R}^d, d = 1, 2, 3$, be an open, bounded set with Lipschitz boundary (if $d = 2$ or 3). We consider the state space

$$\mathcal{Y} = W(0,T) = \left\{ y \ : \ y \in L^2(0,T; H^1_0(\Omega)), y' \in L^2(0,T; H^{-1}(\Omega)) \right\},$$

and the control space

$$\mathcal{U} = L^2(0,T; L^2(\Omega)).$$

We assume that $y_0, \hat{y}_T \in L^2(\Omega), \hat{y} \in L^2(0,T; L^2(\Omega)), f \in L^2(0,T; L^2(\Omega))$ are given functions, and that $\alpha_1, \alpha_2 \geq 0, \alpha_3 > 0$ are given parameters. We define the bilinear forms

$$a : H_0^1(\Omega) \times H_0^1(\Omega) \to \mathbb{R}, \qquad a(y, \psi) = \int_\Omega \nabla y(x) \nabla \psi(x) dx,$$

$$b : L^2(\Omega) \times H_0^1(\Omega) \to \mathbb{R}, \qquad b(u, \psi) = -\int_\Omega u(x) \psi(x) dx,$$

and we use $\langle \cdot, \cdot \rangle_{L^2(\Omega)}$ and $\| \cdot \|_{L^2(\Omega)}$ to denote the inner product and the norm in $L^2(\Omega)$.

We are interested in the solution $y \in Y$, $u \in U$ of the optimal control problem

$$\min \; \frac{\alpha_1}{2} \int_0^T \|y(t) - \hat{y}(t)\|_{L^2(\Omega)}^2 dt$$

$$+ \frac{\alpha_2}{2} \|y(T) - \hat{y}_T\|_{L^2(\Omega)}^2$$

$$+ \frac{\alpha_3}{2} \int_0^T \|u(t)\|_{L^2(\Omega)}^2 dt \tag{4.2a}$$

$$\text{s.t.} \; \langle y'(t), \phi \rangle_{L^2(\Omega)} + a(y(t), \phi) + b(u(t), \phi)$$

$$= \langle f(t), \phi \rangle_{L^2(\Omega)}, \quad \forall \phi \in H_0^1(\Omega), \tag{4.2b}$$

$$y(0) = y_0. \tag{4.2c}$$

Theorem 1. *The optimal control problem (4.2) has a unique solution $(u_*, y_*) \in \mathcal{U} \times \mathcal{Y}$, which, together with the adjoint variable $p_* \in \mathcal{Y}$, is characterized by the necessary and sufficient optimality conditions*

$$-\langle p'(t), \psi \rangle_{L^2(\Omega)} + a(\psi, p(t)) = -\alpha_1 \langle y(t) - \hat{y}(t), \psi \rangle_{L^2(\Omega)}, \tag{4.3a}$$

$$p(T) = -\alpha_2(y(T) - \hat{y}_T), \tag{4.3b}$$

$$\alpha_3 \langle u(t), \mu \rangle_{L^2(\Omega)} + b(\mu, p(t)) = 0, \tag{4.3c}$$

$$\langle y'(t), \phi \rangle_{L^2(\Omega)} + a(y(t), \phi) + b(u(t), \phi) = \langle f(t), \phi \rangle_{L^2(\Omega)}, \tag{4.3d}$$

$$y(0) = y_0. \tag{4.3e}$$

for all $\psi, \phi \in H_0^1(\Omega)$, $\mu \in L^2(\Omega)$.

Proof. The assertion of the theorem is well known and a proof can be found, e.g., in [16, p. 114,116]. □

4.2.2 Discretization

Let $V^h \subset H_0^1(\Omega)$, $U^h \subset L^2(\Omega)$ be finite dimensional subspaces with bases $\phi_1, \ldots, \phi_m$ and $\mu_1, \ldots, \mu_n$, respectively. We approximate the states and controls by $y_h \in H^1(0, T; V^h)$, $u_h \in L^2(0, T; U^h)$ defined as

$$y_h(t) = \sum_{l=1}^m y_l(t) \phi_l, \quad u_h(t) = \sum_{l=1}^n u_l(t) \mu_l. \tag{4.4}$$

We define $\mathbf{A} \in \mathbb{R}^{m \times m}$, $\mathbf{B} \in \mathbb{R}^{m \times n}$, $\mathbf{M} \in \mathbb{R}^{m \times m}$, $\mathbf{Q} \in \mathbb{R}^{n \times n}$, $\mathbf{f} \in L^2(0, T; \mathbb{R}^m)$, $\mathbf{c} \in L^2(0, T; \mathbb{R}^m)$ and $\mathbf{d} \in \mathbb{R}^m$ as follows:

$$\mathbf{A}_{jl} = a(\phi_l, \phi_j), \quad \mathbf{M}_{jl} = \langle \phi_l, \phi_j \rangle_{L^2(\Omega)},$$

$$\mathbf{c}_j(t) = -\alpha_1 \langle \hat{y}(t), \phi_j \rangle_{L^2(\Omega)},$$
$$\mathbf{f}_j(t) = \langle f(t), \phi_j \rangle_{L^2(\Omega)},$$
$$\mathbf{d}_j = -\alpha_2 \langle \hat{y}_T, \phi_j \rangle_{L^2(\Omega)},$$

for $j, l = 1, \dots, m$, and

$$\mathbf{B}_{jl} = b(\mu_l, \phi_j), \quad \mathbf{Q}_{jl} = \langle \mu_l, \mu_j \rangle_{L^2(\Omega)},$$

for $j, l = 1, \dots, n$. We set $\mathbf{y}(t) = (y_1(t), \dots, y_m(t))^T$ and $\mathbf{u}(t) = (u_1(t), \dots, u_n(t))^T$ where y_i, u_i, p_i are the functions in (4.4).

We now replace y, u by y_h, u_h defined in (4.4) and require (4.2b) to hold for $\phi = \phi_l$, $l = 1, \dots, m$. This finite element semi-discretization of the optimal control problem (4.2) leads to a large-scale linear–quadratic problem of the form

$$\min \quad \int_0^T \frac{\alpha_1}{2} \mathbf{y}(t)^T \mathbf{M} \mathbf{y}(t) + \mathbf{c}(t)^T \mathbf{y}(t) dt$$

$$+ \frac{\alpha_2}{2} \mathbf{y}(T)^T \mathbf{M} \mathbf{y}(T) + \mathbf{d}^T \mathbf{y}(T)$$

$$+ \int_0^T \frac{\alpha_3}{2} \mathbf{u}(t)^T \mathbf{Q} \mathbf{u}(t) dt \tag{4.5a}$$

$$\text{s.t.} \ \mathbf{M} \mathbf{y}'(t) + \mathbf{A} \mathbf{y}(t) + \mathbf{B} \mathbf{u}(t) = \mathbf{f}(t), \quad t \in (0, T), \tag{4.5b}$$

$$\mathbf{y}(0) = \mathbf{y}_0. \tag{4.5c}$$

To fully discretize (4.5), we use the backward Euler method on an equidistant time grid

$$t_k = k\Delta t, \quad k = 0, \dots, K$$

with time step size $\Delta t = T/K$. We use $\mathbf{y}_k$ and $\mathbf{u}_k$ to denote the approximate semi-discretized state $\mathbf{y}$ and control $\mathbf{u}$ at t_k. Our discretization of (4.5) is given by

$$\min \quad \sum_{k=1}^{K} \frac{\alpha_1 \Delta t}{2} \mathbf{y}_k^T \mathbf{M} \mathbf{y}_k + \Delta t \mathbf{c}(t_k)^T \mathbf{y}_k$$

$$+ \frac{\alpha_2}{2} \mathbf{y}_K^T \mathbf{M} \mathbf{y}_K + \mathbf{d}^T \mathbf{y}_K + \sum_{k=1}^{K} \frac{\alpha_3 \Delta t}{2} \mathbf{u}_k^T \mathbf{Q} \mathbf{u}_k \tag{4.6a}$$

$$\text{s.t.} \ \mathbf{M} \frac{\mathbf{y}_{k+1} - \mathbf{y}_k}{\Delta t} + \mathbf{A} \mathbf{y}_{k+1} + \mathbf{B} \mathbf{u}_{k+1} = \mathbf{f}(t_{k+1}), \tag{4.6b}$$

$$k = 0, \dots, K - 1.$$

In (4.6), $\mathbf{y}_0$ is the given vector of initial data.

Theorem 2. *The vectors* $\mathbf{y}_1, \ldots, \mathbf{y}_K, \mathbf{u}_1, \ldots, \mathbf{u}_K$ *solve* (4.6) *if and only if there exist* $\mathbf{p}_1, \ldots, \mathbf{p}_K$ *such that*

$$-\mathbf{M}\frac{\mathbf{p}_{k+1} - \mathbf{p}_k}{\Delta t} + \mathbf{A}^T \mathbf{p}_k + \alpha_1 \mathbf{M} \mathbf{y}_k = -\mathbf{c}(t_k), \tag{4.7a}$$

$$k = 1, \ldots, K - 1,$$

$$\left(\frac{1}{\Delta t}\mathbf{M} + \mathbf{A}^T\right)\mathbf{p}_K + \left(\alpha_1 + \frac{\alpha_2}{\Delta t}\right)\mathbf{M}\mathbf{y}_K = -\mathbf{c}(t_K) - \frac{1}{\Delta t}\mathbf{d}, \tag{4.7b}$$

$$\alpha_3 \mathbf{Q} \mathbf{u}_k + \mathbf{B}^T \mathbf{p}_k = 0, \tag{4.7c}$$

$$k = 1, \ldots, K,$$

$$\mathbf{M}\frac{\mathbf{y}_k - \mathbf{y}_{k-1}}{\Delta t} + \mathbf{A}\mathbf{y}_k + \mathbf{B}\mathbf{u}_k = \mathbf{f}(t_k), \tag{4.7d}$$

$$k = 1, \ldots, K.$$

Proof. The Lagrangian for (4.6) is given by

$$
\begin{aligned}
L(\mathbf{y}_1, \ldots, &\mathbf{y}_K, \mathbf{u}_1, \ldots, \mathbf{u}_K, \mathbf{p}_1, \ldots, \mathbf{p}_K) \\
&= \sum_{k=1}^{K} \frac{\alpha_1 \Delta t}{2} \mathbf{y}_k^T \mathbf{M} \mathbf{y}_k + \Delta t \mathbf{c}(t_k)^T \mathbf{y}_k \\
&\quad + \frac{\alpha_2}{2} \mathbf{y}_K^T \mathbf{M} \mathbf{y}_K + \mathbf{d}^T \mathbf{y}_K + \sum_{k=1}^{K} \frac{\alpha_3 \Delta t}{2} \mathbf{u}_k^T \mathbf{Q} \mathbf{u}_k, \\
&\quad + \sum_{k=0}^{K-1} \mathbf{p}_{k+1}^T \Delta t \left(\mathbf{M}\frac{\mathbf{y}_{k+1} - \mathbf{y}_k}{\Delta t} + \mathbf{A}\mathbf{y}_{k+1} + \mathbf{B}\mathbf{u}_{k+1} - \mathbf{f}(t_{k+1}) \right).
\end{aligned}
$$

In the definition of the Lagrangian we use the weighted Euclidean inner product between the Lagrange multipliers $\mathbf{p}_{k+1}$ and the constraints, with weight given by Δt, since it corresponds to an integral for the semi-discrete problem (4.5). Of course, this weighting is equivalent to a scaling of the Lagrange multipliers.

The problem (4.6) is convex, linear-quadratic. The necessary and sufficient optimality conditions for (4.6) are obtained by setting the derivatives of the Lagrangian to zero, which gives (4.7).   □

4.3 Domain Decomposition Formulation of the Example Problem

In this section we introduce our spatial domain decomposition algorithm for the fully discretized problem. This study is useful because the discretization of the adjoint equations (4.3a) is usually not equal to the adjoint equations of the discretization of the optimal control problem (4.6). In this section we apply the spatial domain decomposition algorithm to the optimality conditions for the discretization of the optimal control problem (4.6). Hence, the solution generated by our domain decomposition algorithm is the solution of the full discretization of (4.2).

4.3.1 Domain Decomposition in Space

We discretize (4.2) using conforming linear finite elements. Thus, given a triangulation $\{T_l\}$ of Ω, the space V^h used in the discretization of the states is given by

$$V^h = \left\{ v \in H_0^1(\Omega) \ : \ v|_{T_l} \in P^1(T_l) \text{ for all } l \right\}.$$

We divide Ω into nonoverlapping subdomains Ω_i, $i = 1, \ldots, s$, such that each T_l belongs to exactly one $\overline{\Omega}_i$. We define $\Gamma_i = \partial\Omega_i \setminus \partial\Omega$ and $\Gamma = \cup_{i=1}^s \Gamma_i$.

Let $\{x_j\}_{j=1}^m$ be the set of vertices of $\{T_l\}$ that lie inside Ω_i and let $\{\phi_j\}_{j=1}^m$ be the piecewise linear nodal basis for V^h. Let m_I^i be the number of vertices in Ω_i, let m_Γ^i be the number of vertices on $\Gamma \cap \partial\Omega_i$, and let $m_\Gamma = \sum_{i=1}^s m_\Gamma^i$ be the number of vertices on the subdomain interfaces Γ. Hence the number of discretized state variables for a given time t is given by $m = m_\Gamma + \sum_{i=1}^s m_I^i$.

For the semidiscretization of the control, we use functions that are continuous on each Ω_i, $i = 1, \ldots, s$, and linear on each $\Omega_i \cap T_l$, but that are not assumed to be continuous at $\partial\Omega_i \cap \partial\Omega_j$, $i \neq j$. In particular, for each point $x_l \in \partial\Omega_i \cap \partial\Omega_j$, $i \neq j$, there are two discrete controls $u_h(x_{k_i}, t)$, $u_h(x_{k_j}, t)$ belonging to subdomains Ω_i and Ω_j, respectively (see the right plot in Figure 4.1). Hence, our control discretization depends on the partition $\{\Omega_i\}_{i=1}^d$ of the domain Ω. See [11; 12] for more discussion. We define the discrete spaces

$$U_i^h = \left\{ u \in C^0(\Omega_i) \ : \ u \text{ is linear on } \Omega_i \cap T_l \text{ for all } T_l \subset \overline{\Omega}_i \right\}.$$

We identify U_i^h with a subspace of $L^2(\Omega)$ by extending functions $u_i \in U_i^h$ by zero onto Ω. We define

$$U^h = \cup_{i=1}^s U_i^h \subset L^2(\Omega).$$

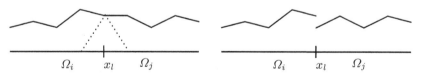

Fig. 4.1. Sketch of the Control Discretization in Space for the Case $\Omega \subset \mathbb{R}$.

Let $\{\mu_j^i\}_{j=1}^{n^i}$ be the piecewise linear nodal basis for U_i^h, where n^i is the number of vertices in $\overline{\Omega}_i$. We identify μ_j^i with a function in $L^2(\Omega)$ by extending μ_j^i by zero outside $\overline{\Omega}_i$. We set

$$\mu_1 = \mu_1^1, \ldots, \mu_{n^i} = \mu_{n^1}^1, \ \mu_{n^1+1} = \mu_1^2, \ldots, \mu_{n^1+n^2} = \mu_{n^2}^2, \ldots.$$

The number of discretized control variables for a given time t is given by $n = \sum_{i=1}^s n^i$.

4.3.2 Decomposition of the Discretized Example Problem

We can use the decomposition of Ω to decompose the matrices $\mathbf{A}$, etc. For $i = 1, \ldots, s$ we define

$$a_i : H^1(\Omega_i) \times H^1(\Omega_i) \to \mathbb{R}, \qquad a_i(y_i, \psi_i) = \int_{\Omega_i} \nabla y_i(x) \nabla \psi_i(x) dx,$$

$$b_i : L^2(\Omega_i) \times H^1(\Omega_i) \to \mathbb{R}, \qquad b_i(u, \psi) = -\int_{\Omega_i} u_i(x) \psi_i(x) dx.$$

For $i = 1, \ldots, s$, we define the submatrices

$$\begin{aligned}
(\mathbf{A}_{II}^i)_{jl} &= a_i(\phi_l, \phi_j), \quad x_j, x_l \in \Omega_i, \\
(\mathbf{A}_{I\Gamma}^i)_{jl} &= a_i(\phi_l, \phi_j), \quad x_j \in \Omega_i, \, x_l \in \partial\Omega_i \setminus \partial\Omega, \\
(\mathbf{A}_{\Gamma I}^i)_{jl} &= a_i(\phi_l, \phi_j), \quad x_j \in \partial\Omega_i \setminus \partial\Omega, \, x_l \in \Omega_i, \\
(\mathbf{A}_{\Gamma\Gamma}^i)_{jl} &= a_i(\phi_l, \phi_j), \quad x_j, x_l \in \partial\Omega_i \setminus \partial\Omega,
\end{aligned}$$

and $\mathbf{A}_{\Gamma\Gamma} = \sum_{i=1}^s \mathbf{A}_{\Gamma\Gamma}^i$. After a suitable reordering of rows and columns, the stiffness matrix can be written as

$$\mathbf{A} = \begin{pmatrix} \mathbf{A}_{II}^1 & & & \mathbf{A}_{I\Gamma}^1 \\ & \ddots & & \vdots \\ & & \mathbf{A}_{II}^s & \mathbf{A}_{I\Gamma}^s \\ \mathbf{A}_{\Gamma I}^1 & \cdots & \mathbf{A}_{\Gamma I}^s & \mathbf{A}_{\Gamma\Gamma} \end{pmatrix}.$$

Similar decompositions can be introduced for $\mathbf{M}$ and for the vectors $\mathbf{c}(t_k)$, $\mathbf{f}(t_k)$, $\mathbf{d}$. For example, for $i = 1, \ldots, s$, we define

$$\begin{aligned}
(\mathbf{d}_I^i)_j &= -\alpha_2 \langle \hat{y}_T, \phi_j \rangle_{L^2(\Omega_i)}, \quad x_j \in \Omega_i, \\
(\mathbf{d}_\Gamma^i)_j &= -\alpha_2 \langle \hat{y}_T, \phi_j \rangle_{L^2(\Omega_i)}, \quad x_j \in \partial\Omega_i \setminus \partial\Omega,
\end{aligned}$$

and $\mathbf{d}_\Gamma = \sum_{i=1}^s \mathbf{d}_\Gamma^i$. After a suitable reordering, the vector $\mathbf{d}$ can be written as

$$\mathbf{d} = \begin{pmatrix} \mathbf{d}_I^1 \\ \vdots \\ \mathbf{d}_I^s \\ \mathbf{d}_\Gamma \end{pmatrix}.$$

The vectors $\mathbf{y}_k, \mathbf{p}_k$ can be partitioned accordingly. For example, $(\mathbf{y}_I^i)_k$ denotes the subvector of $\mathbf{y}_k$ with indices l such that $x_l \in \Omega_i$, $(\mathbf{y}_\Gamma)_k$ denotes the subvector of $\mathbf{y}_k$ with indices l such that $x_l \in \Gamma$, and $(\mathbf{y}_\Gamma^i)_k$ denotes the subvector of $\mathbf{y}_k$ with indices l such that $x_l \in \Gamma \cap \partial\Omega_i$.

For $i = 1, \ldots, s$, we define the submatrices

$$\begin{aligned}
(\mathbf{B}_{II}^i)_{jl} &= b_i(\mu_l, \phi_j), \quad x_j \in \Omega_i, x_l \in \overline{\Omega}_i, \\
(\mathbf{B}_{\Gamma I}^i)_{jl} &= b_i(\mu_l, \phi_j), \quad x_j \in \partial\Omega_i \setminus \partial\Omega, x_l \in \overline{\Omega}_i, \\
(\mathbf{Q}_{II}^i)_{jl} &= \langle \mu_l, \phi_j \rangle_{L^2(\Omega_i)}, \quad x_j \in \Omega_i, x_l \in \overline{\Omega}_i.
\end{aligned}$$

After a suitable reordering of rows and columns, the matrix $\mathbf{B}$ can be written as

$$\mathbf{B} = \begin{pmatrix} \mathbf{B}_{II}^1 & & \\ & \ddots & \\ & & \mathbf{B}_{II}^s \\ \mathbf{B}_{\Gamma I}^1 & \cdots & \mathbf{B}_{\Gamma I}^s \end{pmatrix}, \quad \mathbf{Q} = \begin{pmatrix} \mathbf{Q}_{II}^1 & & \\ & \ddots & \\ & & \mathbf{Q}_{II}^s \end{pmatrix}.$$

Note that in our particular control discretization, all basis functions μ_l^i for the discretized control u_h have support in only one subdomain Ω_i (see the right plot in Figure 4.1). Consequently, there is no $\mathbf{B}_{\Gamma\Gamma}^i$. The vectors $\mathbf{u}_k$ can be partitioned into

$$\mathbf{u}_k = \begin{pmatrix} (\mathbf{u}_I^1)_k \\ \vdots \\ (\mathbf{u}_I^s)_k \end{pmatrix}.$$

Due to our control discretization in space, there is no subvector $(\mathbf{u}_\Gamma)_k$ corresponding to interface nodes.

Let

$$I_\Gamma^i \in \mathbb{R}^{m_\Gamma^i \times m_\Gamma} \tag{4.8}$$

be the matrix with zero or one entries that extracts out of a vector $\mathbf{v}_\Gamma \in \mathbb{R}^{m_\Gamma}$ the subvector $\mathbf{v}_\Gamma^i \in \mathbb{R}^{m_\Gamma^i}$ whose components correspond to vertices $x_l \in \Gamma \cap \partial\Omega_i$, i.e., $\mathbf{v}_\Gamma^i = I_\Gamma^i \mathbf{v}_\Gamma$.

The optimality conditions (4.7) can now be decomposed into the systems

$$-\mathbf{M}_{II}^i \frac{(\mathbf{p}_I^i)_{k+1} - (\mathbf{p}_I^i)_k}{\Delta t} + (\mathbf{A}_{II}^i)^T (\mathbf{p}_I^i)_k + \alpha_1 \mathbf{M}_{II}^i (\mathbf{y}_I^i)_k$$
$$-\mathbf{M}_{I\Gamma}^i \frac{(\mathbf{p}_\Gamma^i)_{k+1} - (\mathbf{p}_\Gamma^i)_k}{\Delta t} + (\mathbf{A}_{\Gamma I}^i)^T (\mathbf{p}_\Gamma^i)_k + \alpha_1 \mathbf{M}_{I\Gamma}^i (\mathbf{y}_\Gamma^i)_k$$
$$= -\mathbf{c}_I^i(t_k), \tag{4.9a}$$
$$k = 1, \ldots, K-1,$$

$$\left(\tfrac{1}{\Delta t}\mathbf{M}_{II}^i + (\mathbf{A}_{II}^i)^T\right)(\mathbf{p}_I^i)_K + (\alpha_1 + \tfrac{\alpha_2}{\Delta t})\mathbf{M}_{II}^i(\mathbf{y}_I^i)_K$$
$$+ \left(\tfrac{1}{\Delta t}\mathbf{M}_{I\Gamma}^i + (\mathbf{A}_{\Gamma I}^i)^T\right)(\mathbf{p}_\Gamma^i)_K + (\alpha_1 + \tfrac{\alpha_2}{\Delta t})\mathbf{M}_{I\Gamma}^i(\mathbf{y}_\Gamma^i)_K$$
$$= -\mathbf{c}_I^i(t_K) - \tfrac{1}{\Delta t}\mathbf{d}_I^i, \tag{4.9b}$$

$$\alpha_3 \mathbf{Q}_{II}(\mathbf{u}_I^i)_k + (\mathbf{B}_{II}^i)^T(\mathbf{p}_I^i)_k + (\mathbf{B}_{\Gamma I}^i)^T(\mathbf{p}_\Gamma^i)_k = 0, \tag{4.9c}$$
$$k = 1, \ldots, K,$$

$$\mathbf{M}_{II}^i \frac{(\mathbf{y}_I^i)_k - (\mathbf{y}_I^i)_{k-1}}{\Delta t} + \mathbf{A}_{II}^i(\mathbf{y}_I^i)_k + \mathbf{B}_{II}^i(\mathbf{u}_I^i)_k$$
$$+ \mathbf{M}_{I\Gamma}^i \frac{(\mathbf{y}_\Gamma^i)_k - (\mathbf{y}_\Gamma^i)_{k-1}}{\Delta t} + \mathbf{A}_{I\Gamma}^i(\mathbf{y}_\Gamma^i)_k = \mathbf{f}_I^i(t_k), \tag{4.9d}$$
$$k = 1, \ldots, K,$$

for $i = 1, \ldots, s$, where $(\mathbf{y}_I^i)_0$ is the subvector of the given vector of initial data corresponding to nodes in the interior of Ω_i, and into the interface coupling condition

$$\sum_{i=1}^{s}(I_{\Gamma}^{i})^{T}\Big(-\mathbf{M}_{\Gamma I}^{i}\frac{(\mathbf{p}_{I}^{i})_{k+1}-(\mathbf{p}_{I}^{i})_{k}}{\Delta t}+(\mathbf{A}_{I\Gamma}^{i})^{T}(\mathbf{p}_{I}^{i})_{k}$$

$$-\mathbf{M}_{\Gamma\Gamma}^{i}\frac{(\mathbf{p}_{\Gamma}^{i})_{k+1}-(\mathbf{p}_{\Gamma}^{i})_{k}}{\Delta t}+(\mathbf{A}_{\Gamma\Gamma}^{i})^{T}(\mathbf{p}_{\Gamma}^{i})_{k}$$

$$+\alpha_{1}\mathbf{M}_{\Gamma I}^{i}(\mathbf{y}_{I}^{i})_{k}+\alpha_{1}\mathbf{M}_{\Gamma\Gamma}^{i}(\mathbf{y}_{\Gamma}^{i})_{k}\Big)=-\mathbf{c}_{\Gamma}(t_{k}), \qquad (4.10a)$$

$$k=1,\ldots,K-1,$$

$$\sum_{i=1}^{s}(I_{\Gamma}^{i})^{T}\Big(\big(\tfrac{1}{\Delta t}\mathbf{M}_{\Gamma I}^{i}+(\mathbf{A}_{I\Gamma}^{i})^{T}\big)(\mathbf{p}_{I}^{i})_{K}+(\alpha_{1}+\tfrac{\alpha_{2}}{\Delta t})\mathbf{M}_{\Gamma I}^{i}(\mathbf{y}_{I}^{i})_{K}$$

$$+\big(\tfrac{1}{\Delta t}\mathbf{M}_{\Gamma\Gamma}^{i}+(\mathbf{A}_{\Gamma\Gamma}^{i})^{T}\big)(\mathbf{p}_{\Gamma}^{i})_{K}+(\alpha_{1}+\tfrac{\alpha_{2}}{\Delta t})\mathbf{M}_{\Gamma\Gamma}^{i}(\mathbf{y}_{\Gamma}^{i})_{K}\Big)$$

$$=-\mathbf{c}_{\Gamma}(t_{K})-\tfrac{1}{\Delta t}\mathbf{d}_{\Gamma}, \qquad (4.10b)$$

$$\sum_{i=1}^{s}(I_{\Gamma}^{i})^{T}A\Big(\mathbf{M}_{\Gamma I}^{i}\frac{(\mathbf{y}_{I}^{i})_{k}-(\mathbf{y}_{I}^{i})_{k-1}}{\Delta t}+\mathbf{A}_{\Gamma I}^{i}(\mathbf{y}_{I}^{i})_{k}+\mathbf{B}_{\Gamma I}^{i}(\mathbf{u}_{I}^{i})_{k}$$

$$+\mathbf{M}_{\Gamma\Gamma}^{i}\frac{(\mathbf{y}_{\Gamma}^{i})_{k}-(\mathbf{y}_{\Gamma}^{i})_{k-1}}{\Delta t}+\mathbf{A}_{\Gamma\Gamma}^{i}(\mathbf{y}_{\Gamma}^{i})_{k}\Big)=\mathbf{f}_{\Gamma}(t_{k}), \qquad (4.10c)$$

$$k=1,\ldots,K,$$

where I_{Γ}^{i} is the matrix defined in (4.8) and where $(\mathbf{y}_{\Gamma}^{i})_{0}$ is the subvector of the given vector of initial data corresponding to nodes on $\partial\Omega_{i}\setminus\partial\Omega$.

We now view the solution $(\mathbf{y}_{I}^{i})_{k}$, $(\mathbf{u}_{I}^{i})_{k}$, $(\mathbf{p}_{I}^{i})_{k}$, $k=1,\ldots,K$, of (4.9) as a function of $(\mathbf{y}_{\Gamma})_{1},\ldots,(\mathbf{y}_{\Gamma})_{K}$, $(\mathbf{p}_{\Gamma})_{1},\ldots,(\mathbf{p}_{\Gamma})_{K}$. Then (4.10) represents a system of equations in $(\mathbf{y}_{\Gamma})_{1},\ldots,(\mathbf{y}_{\Gamma})_{K}$, $(\mathbf{p}_{\Gamma})_{1},\ldots,(\mathbf{p}_{\Gamma})_{K}$. Before we give a precise statement of this system, we give an interpretation of the system (4.9), $i=1,\ldots,s$, as the optimality conditions of an optimal control problem.

Theorem 3. *Given* $(\mathbf{y}_{\Gamma}^{i})_{1},\ldots,(\mathbf{y}_{\Gamma}^{i})_{K}$, $(\mathbf{p}_{\Gamma}^{i})_{1},\ldots,(\mathbf{p}_{\Gamma}^{i})_{K}$, *the system* (4.9), $i=1,\ldots,s$, *can be interpreted as the necessary and sufficient optimality conditions for the following discrete subdomain optimal control problems in the unknowns* $(\mathbf{y}_{I}^{i})_{1},\ldots,(\mathbf{y}_{I}^{i})_{K}$, $(\mathbf{u}_{I}^{i})_{1},\ldots,(\mathbf{u}_{I}^{i})_{K}$.

$$\min \quad \sum_{k=1}^{K}\frac{\alpha_{1}\Delta t}{2}(\mathbf{y}_{I}^{i})_{k}^{T}\mathbf{M}_{II}^{i}(\mathbf{y}_{I}^{i})_{k}$$

$$+\Delta t\Big(\mathbf{c}_{I}^{i}(t_{k})-\mathbf{M}_{I\Gamma}^{i}\frac{(\mathbf{p}_{\Gamma}^{i})_{k+1}-(\mathbf{p}_{\Gamma}^{i})_{k}}{\Delta t}$$

$$+(\mathbf{A}_{\Gamma I}^{i})^{T}(\mathbf{p}_{\Gamma}^{i})_{k}+\alpha_{1}\mathbf{M}_{I\Gamma}^{i}(\mathbf{y}_{\Gamma}^{i})_{k}\Big)^{T}(\mathbf{y}_{I}^{i})_{k}$$

$$+\frac{\alpha_{2}}{2}(\mathbf{y}_{I}^{i})_{K}^{T}\mathbf{M}_{II}^{i}(\mathbf{y}_{I}^{i})_{K}$$

$$+\big(\mathbf{d}_{I}^{i}+\mathbf{M}_{I\Gamma}^{i}(\mathbf{p}_{\Gamma}^{i})_{K+1}+\alpha_{2}\mathbf{M}_{I\Gamma}^{i}(\mathbf{y}_{\Gamma}^{i})_{K}\big)^{T}(\mathbf{y}_{I}^{i})_{K}$$

$$+\sum_{k=1}^{K}\frac{\alpha_{3}\Delta t}{2}(\mathbf{u}_{I}^{i})_{k}^{T}\mathbf{Q}_{II}^{i}(\mathbf{u}_{I}^{i})_{k}+\Delta t(\mathbf{p}_{\Gamma}^{i})_{k}^{T}\mathbf{B}_{\Gamma I}^{i}(\mathbf{u}_{I}^{i})_{k} \qquad (4.11a)$$

s.t. $\mathbf{M}_{II}^i \dfrac{(\mathbf{y}_I^i)_k - (\mathbf{y}_I^i)_{k-1}}{\Delta t} + \mathbf{A}_{II}^i(\mathbf{y}_I^i)_k + \mathbf{B}_{II}^i(\mathbf{u}_I^i)_k$

$$= \mathbf{f}_I^i(t_k) - \mathbf{M}_{I\Gamma}^i \dfrac{(\mathbf{y}_\Gamma^i)_k - (\mathbf{y}_\Gamma^i)_{k-1}}{\Delta t} - \mathbf{A}_{I\Gamma}^i(\mathbf{y}_\Gamma^i)_k, \qquad (4.11b)$$

$$k = 1, \ldots, K.$$

Proof. The Lagrangian for (4.11) is given by

$$L_I^i((\mathbf{y}_I^i)_1, \ldots, (\mathbf{y}_I^i)_K, (\mathbf{u}_I^i)_1, \ldots, (\mathbf{u}_I^i)_K, (\mathbf{p}_I^i)_1, \ldots, (\mathbf{p}_I^i)_K)$$

$$= \sum_{k=1}^{K} \frac{\alpha_1 \Delta t}{2}(\mathbf{y}_I^i)_k^T \mathbf{M}_{II}^i(\mathbf{y}_I^i)_k$$

$$+ \Delta t \left(\mathbf{c}_I^i(t_k) - \mathbf{M}_{I\Gamma}^i \frac{(\mathbf{p}_\Gamma^i)_{k+1} - (\mathbf{p}_\Gamma^i)_k}{\Delta t} \right.$$

$$\left. + (\mathbf{A}_{\Gamma I}^i)^T (\mathbf{p}_\Gamma^i)_k + \alpha_1 \mathbf{M}_{I\Gamma}^i(\mathbf{y}_\Gamma^i)_k \right)^T (\mathbf{y}_I^i)_k$$

$$+ \frac{\alpha_2}{2}(\mathbf{y}_I^i)_K^T \mathbf{M}_{II}^i(\mathbf{y}_I^i)_K$$

$$+ \left(\mathbf{d}_I^i + \mathbf{M}_{I\Gamma}^i(\mathbf{p}_\Gamma^i)_{K+1} + \alpha_2 \mathbf{M}_{I\Gamma}^i(\mathbf{y}_\Gamma^i)_K \right)^T (\mathbf{y}_I^i)_K$$

$$+ \sum_{k=1}^{K} \frac{\alpha_3 \Delta t}{2}(\mathbf{u}_I^i)_k^T \mathbf{Q}_{II}^i(\mathbf{u}_I^i)_k + \Delta t (\mathbf{p}_\Gamma^i)_k^T \mathbf{B}_{\Gamma I}^i(\mathbf{u}_I^i)_k$$

$$+ \sum_{k=1}^{K} (\mathbf{p}_I^i)_k^T \Delta t \left(\mathbf{M}_{II}^i \frac{(\mathbf{y}_I^i)_k - (\mathbf{y}_I^i)_{k-1}}{\Delta t} + \mathbf{A}_{II}^i(\mathbf{y}_I^i)_k + \mathbf{B}_{II}^i(\mathbf{u}_I^i)_k \right.$$

$$\left. - \mathbf{f}_I^i(t_k) + \mathbf{M}_{I\Gamma}^i \frac{(\mathbf{y}_\Gamma^i)_k - (\mathbf{y}_\Gamma^i)_{k-1}}{\Delta t} + \mathbf{A}_{I\Gamma}^i(\mathbf{y}_\Gamma^i)_k \right).$$

The problem (4.11) is convex, linear-quadratic. The necessary and sufficient optimality conditions for (4.11) are obtained by setting the derivatives of the Lagrangian to zero and are given by (4.9). □

As we have stated before, we view the solutions $(\mathbf{y}_I^i)_1, \ldots, (\mathbf{y}_I^i)_K$, $(\mathbf{p}_I^i)_1, \ldots, (\mathbf{p}_I^i)_K$, $(\mathbf{u}_I^i)_1, \ldots, (\mathbf{u}_I^i)_K$ of (4.9), $i = 1, \ldots, s$, as affine linear maps of $(\mathbf{y}_\Gamma^i)_1, \ldots, (\mathbf{y}_\Gamma^i)_K$, $(\mathbf{p}_\Gamma^i)_1, \ldots, (\mathbf{p}_\Gamma^i)_K$, and the view (4.10) as a system of linear equations in $(\mathbf{y}_\Gamma)_1, \ldots, (\mathbf{y}_\Gamma)_K$, $(\mathbf{p}_\Gamma)_1, \ldots, (\mathbf{p}_\Gamma)_K$. This leads to the linear map

$$\mathbf{S}_i^{\Delta t} : \left(\mathbb{R}^{m_\Gamma^i} \right)^{2K} \to \left(\mathbb{R}^{m_\Gamma^i} \right)^{2K} \qquad (4.12a)$$

84   Matthias Heinkenschloss and Michael Herty

defined by

$$
\mathbf{S}_i^{\Delta t}\Big((\mathbf{y}_\Gamma^i)_1,\ldots,(\mathbf{y}_\Gamma^i)_K,(\mathbf{p}_\Gamma^i)_1,\ldots,(\mathbf{p}_\Gamma^i)_K\Big)
$$

$$
=
\begin{pmatrix}
-\mathbf{M}_{\Gamma I}^i \frac{(\mathbf{p}_I^i)_2-(\mathbf{p}_I^i)_1}{\Delta t}+(\mathbf{A}_{I\Gamma}^i)^T(\mathbf{p}_I^i)_1\ldots \\
\ldots-\mathbf{M}_{\Gamma\Gamma}^i \frac{(\mathbf{p}_\Gamma^i)_2-(\mathbf{p}_\Gamma^i)_1}{\Delta t}+(\mathbf{A}_{\Gamma\Gamma}^i)^T(\mathbf{p}_\Gamma^i)_1 \\
+\alpha_1\mathbf{M}_{\Gamma I}^i(\mathbf{y}_I^i)_1+\alpha_1\mathbf{M}_{\Gamma\Gamma}^i(\mathbf{y}_\Gamma^i)_1 \\
\vdots \\
-\mathbf{M}_{\Gamma I}^i \frac{(\mathbf{p}_I^i)_K-(\mathbf{p}_I^i)_{K-1}}{\Delta t}+(\mathbf{A}_{I\Gamma}^i)^T(\mathbf{p}_I^i)_{K-1}\ldots \\
\ldots-\mathbf{M}_{\Gamma\Gamma}^i \frac{(\mathbf{p}_\Gamma^i)_K-(\mathbf{p}_\Gamma^i)_{K-1}}{\Delta t}+(\mathbf{A}_{\Gamma\Gamma}^i)^T(\mathbf{p}_\Gamma^i)_{K-1} \\
+\alpha_1\mathbf{M}_{\Gamma I}^i(\mathbf{y}_I^i)_{K-1}+\alpha_1\mathbf{M}_{\Gamma\Gamma}^i(\mathbf{y}_\Gamma^i)_{K-1} \\
(\frac{1}{\Delta t}\mathbf{M}_{\Gamma I}^i+(\mathbf{A}_{I\Gamma}^i)^T)(\mathbf{p}_I^i)_K+(\alpha_1+\frac{\alpha_2}{\Delta t})\mathbf{M}_{\Gamma I}^i(\mathbf{y}_I^i)_K\ldots \\
\ldots+(\frac{1}{\Delta t}\mathbf{M}_{\Gamma\Gamma}^i+(\mathbf{A}_{\Gamma\Gamma}^i)^T)(\mathbf{p}_\Gamma^i)_K+(\alpha_1+\frac{\alpha_2}{\Delta t})\mathbf{M}_{\Gamma\Gamma}^i(\mathbf{y}_\Gamma^i)_K \\
\mathbf{M}_{\Gamma I}^i \frac{(\mathbf{y}_I^i)_1}{\Delta t}+\mathbf{A}_{\Gamma I}^i(\mathbf{y}_I^i)_1+\mathbf{B}_{\Gamma I}^i(\mathbf{u}_I^i)_1\ldots \\
\ldots+\mathbf{M}_{\Gamma\Gamma}^i \frac{(\mathbf{y}_\Gamma^i)_1}{\Delta t}+\mathbf{A}_{\Gamma\Gamma}^i(\mathbf{y}_\Gamma^i)_1 \\
\vdots \\
\mathbf{M}_{\Gamma I}^i \frac{(\mathbf{y}_I^i)_K-(\mathbf{y}_I^i)_{K-1}}{\Delta t}+\mathbf{A}_{\Gamma I}^i(\mathbf{y}_I^i)_K+\mathbf{B}_{\Gamma I}^i(\mathbf{u}_I^i)_K\ldots \\
\ldots+\mathbf{M}_{\Gamma\Gamma}^i \frac{(\mathbf{y}_\Gamma^i)_K-(\mathbf{y}_\Gamma^i)_{K-1}}{\Delta t}+\mathbf{A}_{\Gamma\Gamma}^i(\mathbf{y}_\Gamma^i)_K
\end{pmatrix}
\tag{4.12b}
$$

where $(\mathbf{y}_I^i)_1,\ldots,(\mathbf{y}_I^i)_K,(\mathbf{p}_I^i)_1,\ldots,(\mathbf{p}_I^i)_K,(\mathbf{u}_I^i)_1,\ldots,(\mathbf{u}_I^i)_K$ is the solution of (4.9) (or, equivalently, of (4.11)) with $(\mathbf{y}_0)_I^i=\mathbf{0}$, $\mathbf{f}_I^i=\mathbf{0}$, $\mathbf{c}_I^i=\mathbf{0}$, $\mathbf{d}_I^i=\mathbf{0}$. Furthermore, we define

$$
\mathbf{r}_i^{\Delta t}\in\left(\mathbb{R}^{m_\Gamma^i}\right)^{2K}
\tag{4.13a}
$$

by

$$
\mathbf{r}_i^{\Delta t} =
\begin{pmatrix}
-\mathbf{c}_\Gamma^i(t_1) + \mathbf{M}_{\Gamma I}^i \frac{(\mathbf{p}_I^i)_2 - (\mathbf{p}_I^i)_1}{\Delta t} \\
\quad -(\mathbf{A}_{I\Gamma}^i)^T (\mathbf{p}_I^i)_1 - \alpha_1 \mathbf{M}_{\Gamma I}^i (\mathbf{y}_I^i)_1 \\
\\
\vdots \\
\\
-\mathbf{c}_\Gamma^i(t_{K-1}) + \mathbf{M}_{\Gamma I}^i \frac{(\mathbf{p}_I^i)_K - (\mathbf{p}_I^i)_{K-1}}{\Delta t} \\
\quad -(\mathbf{A}_{I\Gamma}^i)^T (\mathbf{p}_I^i)_{K-1} - \alpha_1 \mathbf{M}_{\Gamma I}^i (\mathbf{y}_I^i)_{K-1} \\
\\
-\mathbf{c}_\Gamma^i(t_K) - \frac{1}{\Delta t}\mathbf{d}_\Gamma^i - (\frac{1}{\Delta t}\mathbf{M}_{\Gamma I}^i + (\mathbf{A}_{I\Gamma}^i)^T)(\mathbf{p}_I^i)_K \\
\quad -(\alpha_1 + \frac{\alpha_2}{\Delta t})\mathbf{M}_{\Gamma I}^i (\mathbf{y}_I^i)_K \\
\\
\mathbf{f}_\Gamma^i(t_1) - \mathbf{M}_{\Gamma I}^i \frac{(\mathbf{y}_I^i)_1 - (\mathbf{y}_I^i)_0}{\Delta t} - \mathbf{A}_{\Gamma I}^i(\mathbf{y}_I^i)_1 + \mathbf{B}_{\Gamma I}^i(\mathbf{u}_I^i)_1 \\
\quad + \mathbf{M}_{\Gamma\Gamma}^i \frac{(\mathbf{y}_\Gamma^i)_0}{\Delta t} \\
\\
\vdots \\
\\
\mathbf{f}_\Gamma^i(t_K) - \mathbf{M}_{\Gamma I}^i \frac{(\mathbf{y}_I^i)_K - (\mathbf{y}_I^i)_{K-1}}{\Delta t} \\
\quad -\mathbf{A}_{\Gamma I}^i(\mathbf{y}_I^i)_K + \mathbf{B}_{\Gamma I}^i(\mathbf{u}_I^i)_K
\end{pmatrix}
\tag{4.13b}
$$

where $(\mathbf{y}_I^i)_1, \ldots, (\mathbf{y}_I^i)_K, (\mathbf{p}_I^i)_1, \ldots, (\mathbf{p}_I^i)_K, (\mathbf{u}_I^i)_1, \ldots, (\mathbf{u}_I^i)_K$ is the solution of (4.9) (or, equivalently, of (4.11)) with $(\mathbf{y}_\Gamma^i)_1 = \ldots = (\mathbf{y}_\Gamma^i)_K = (\mathbf{p}_\Gamma^i)_1 = \ldots = (\mathbf{p}_\Gamma^i)_K = \mathbf{0}$.

Let $\mathbf{I}_\Gamma^{i,\Delta t} = \mathrm{diag}(I_\Gamma^i, \ldots, I_\Gamma^i) \in \mathbb{R}^{2Km_\Gamma^i \times 2Km_\Gamma^i}$, where I_Γ^i is defined in (4.8). The system (4.9), (4.10a) can now be written as an operator equation

$$
\sum_{i=1}^s (\mathbf{I}_\Gamma^{i,\Delta t})^T \mathbf{S}_i^{\Delta t} \mathbf{I}_\Gamma^{i,\Delta t} \Big((\mathbf{y}_\Gamma)_1, \ldots, (\mathbf{y}_\Gamma)_K, (\mathbf{p}_\Gamma)_1, \ldots, (\mathbf{p}_\Gamma)_K \Big)
$$
$$
= \sum_{i=1}^s (\mathbf{I}_\Gamma^{i,\Delta t})^T \mathbf{r}_i
\tag{4.14}
$$

in the unknowns $(\mathbf{y}_\Gamma)_1, \ldots, (\mathbf{y}_\Gamma)_K, (\mathbf{p}_\Gamma)_1, \ldots, (\mathbf{p}_\Gamma)_K \in \mathbb{R}^{m_\Gamma}$. If the solution $(\mathbf{y}_\Gamma)_1, \ldots, (\mathbf{y}_\Gamma)_K, (\mathbf{p}_\Gamma)_1, \ldots, (\mathbf{p}_\Gamma)_K$ of (4.14) is computed, then the remaining components $(\mathbf{y}_I^i)_1, \ldots, (\mathbf{y}_I^i)_K, (\mathbf{p}_I^i)_1, \ldots, (\mathbf{p}_I^i)_K, (\mathbf{u}_I^i)_1, \ldots, (\mathbf{u}_I^i)_K, i = 1, \ldots, s$, can be computed by solving (4.9) (or, equivalently, (4.11)).

In the next theorem, we will show how to apply the inverse of the subdomain operator $\mathbf{S}_i, i = 1, \ldots, s$. For this result it is useful to introduce the notation

$$
\mathbf{A}^i = \begin{pmatrix} \mathbf{A}_{II}^i & \mathbf{A}_{I\Gamma}^i \\ \mathbf{A}_{\Gamma I}^i & \mathbf{A}_{\Gamma\Gamma}^i \end{pmatrix}, \quad
\mathbf{M}^i = \begin{pmatrix} \mathbf{M}_{II}^i & \mathbf{M}_{I\Gamma}^i \\ \mathbf{M}_{\Gamma I}^i & \mathbf{M}_{\Gamma\Gamma}^i \end{pmatrix}, \quad
\mathbf{B}^i = \begin{pmatrix} \mathbf{B}_{II}^i \\ \mathbf{B}_{\Gamma I}^i \end{pmatrix}
\tag{4.15a}
$$

and

$$
\mathbf{y}_k^i = \begin{pmatrix} (\mathbf{y}_I^i)_k \\ (\mathbf{y}_\Gamma^i)_k \end{pmatrix}, \quad
\mathbf{p}_k^i = \begin{pmatrix} (\mathbf{p}_I^i)_k \\ (\mathbf{p}_\Gamma^i)_k \end{pmatrix}.
\tag{4.15b}
$$

Furthermore, let

$$I^i \in \mathbb{R}^{m_{\Gamma}^i \times m^i} \tag{4.16}$$

be the matrix with zero or one entries that extracts out of a vector $\mathbf{v}^i \in \mathbb{R}^{m^i}$ the subvector $\mathbf{v}_{\Gamma}^i \in \mathbb{R}^{m_{\Gamma}^i}$ whose components correspond to vertices $x_l \in \Gamma \cap \partial\Omega_i$.

Theorem 4. Let $\mathbf{r}^i = (\mathbf{r}_1^i, \ldots, \mathbf{r}_{2K}^i) \in (\mathbb{R}^{m_{\Gamma}^i})^{2K}$ be given. The solution $(\mathbf{y}_{\Gamma}^i)_1, \ldots, (\mathbf{y}_{\Gamma}^i)_K, (\mathbf{p}_{\Gamma}^i)_1, \ldots, (\mathbf{p}_{\Gamma}^i)_K \in \mathbb{R}^{m_{\Gamma}}$ of

$$\mathbf{S}_i^{\Delta t}\Big((\mathbf{y}_{\Gamma}^i)_1, \ldots, (\mathbf{y}_{\Gamma}^i)_K, (\mathbf{p}_{\Gamma}^i)_1, \ldots, (\mathbf{p}_{\Gamma}^i)_K \Big) = \mathbf{r}_i$$

is given by

$$(\mathbf{y}_{\Gamma}^i)_k = I^i \mathbf{y}_k^i, \quad k = 1, \ldots, K, \quad (\mathbf{p}_{\Gamma}^i)_k = I^i \mathbf{p}_k^i, \quad k = 1, \ldots, K,$$

where I^i is the matrix defined in (4.16) and where $\mathbf{y}_k^i, \mathbf{u}_k^i, \mathbf{p}_k^i, k = 1, \ldots, K$, solve

$$-\mathbf{M}^i \frac{\mathbf{p}_{k+1}^i - \mathbf{p}_k^i}{\Delta t} + (\mathbf{A}^i)^T \mathbf{p}_k^i + \alpha_1 \mathbf{M}^i \mathbf{y}_k^i = \begin{pmatrix} \mathbf{0} \\ \mathbf{r}_k^i \end{pmatrix}, \tag{4.17a}$$

$$k = 1, \ldots, K-1,$$

$$\left(\tfrac{1}{\Delta t}\mathbf{M}^i + (\mathbf{A}^i)^T \right) \mathbf{p}_K^i + (\alpha_1 + \tfrac{\alpha_2}{\Delta t})\mathbf{M}^i \mathbf{y}_K^i = \begin{pmatrix} \mathbf{0} \\ \mathbf{r}_K^i \end{pmatrix}, \tag{4.17b}$$

$$\alpha_3 \mathbf{Q}_{II}(\mathbf{u}_I^i)_k + (\mathbf{B}^i)^T \mathbf{p}_k^i = \mathbf{0}, \tag{4.17c}$$

$$k = 1, \ldots, K,$$

$$\mathbf{M}^i \frac{\mathbf{y}_k^i - \mathbf{y}_{k-1}^i}{\Delta t} + \mathbf{A}^i \mathbf{y}_k^i + \mathbf{B}^i \mathbf{u}_k^i = \begin{pmatrix} \mathbf{0} \\ \mathbf{r}_{K+k}^i \end{pmatrix}, \tag{4.17d}$$

$$k = 1, \ldots, K,$$

$$\mathbf{y}_0^i = \mathbf{0}. \tag{4.17e}$$

The equations (4.17) are the system of necessary and sufficient optimality conditions for the optimal control problem

$$\min \sum_{k=1}^{K} \frac{\alpha_1 \Delta t}{2}(\mathbf{y}_k^i)^T \mathbf{M}^i \mathbf{y}_k^i - \Delta t(\mathbf{0}^T, (\mathbf{r}_k^i)^T)\mathbf{y}_k^i$$

$$+ \frac{\alpha_2}{2}(\mathbf{y}_K^i)^T \mathbf{M}^i \mathbf{y}_K^i + \sum_{k=1}^{K} \frac{\alpha_3 \Delta t}{2}(\mathbf{u}_I^i)_k^T \mathbf{Q}_{II}^i(\mathbf{u}_I^i)_k \tag{4.18a}$$

$$\text{s.t. } \mathbf{M}^i \frac{\mathbf{y}_k^i - \mathbf{y}_{k-1}^i}{\Delta t} + \mathbf{A}^i \mathbf{y}_k^i + \mathbf{B}^i \mathbf{u}_k^i = \begin{pmatrix} \mathbf{0} \\ \mathbf{r}_{K+k}^i \end{pmatrix}, \tag{4.18b}$$

$$k = 1, \ldots, K,$$

$$\mathbf{y}_0^i = \mathbf{0}. \tag{4.18c}$$

Proof. By definition (4.12) of $\mathbf{S}_i$ the equation

$$\mathbf{S}_i^{\Delta t}\left((\mathbf{y}_\Gamma^i)_1, \ldots, (\mathbf{y}_\Gamma^i)_K, (\mathbf{p}_\Gamma^i)_1, \ldots, (\mathbf{p}_\Gamma^i)_K\right) = \mathbf{r}^i$$

is equivalent to

$$
-\mathbf{M}_{II}^i \frac{(\mathbf{p}_I^i)_{k+1}-(\mathbf{p}_I^i)_k}{\Delta t} + (\mathbf{A}_{II}^i)^T(\mathbf{p}_I^i)_k + \alpha_1 \mathbf{M}_{II}^i(\mathbf{y}_I^i)_k
$$
$$
-\mathbf{M}_{I\Gamma}^i \frac{(\mathbf{p}_\Gamma^i)_{k+1}-(\mathbf{p}_\Gamma^i)_k}{\Delta t} + (\mathbf{A}_{\Gamma I}^i)^T(\mathbf{p}_\Gamma^i)_k
$$
$$
+\alpha_1 \mathbf{M}_{I\Gamma}^i(\mathbf{y}_\Gamma^i)_k = \mathbf{0}, \tag{4.19a}
$$
$$
k = 1, \ldots, K-1,
$$

$$
\left(\tfrac{1}{\Delta t}\mathbf{M}_{II}^i + (\mathbf{A}_{II}^i)^T\right)(\mathbf{p}_I^i)_K + (\alpha_1 + \tfrac{\alpha_2}{\Delta t})\mathbf{M}_{II}^i(\mathbf{y}_I^i)_K
$$
$$
+ \left(\tfrac{1}{\Delta t}\mathbf{M}_{I\Gamma}^i + (\mathbf{A}_{\Gamma I}^i)^T\right)(\mathbf{p}_\Gamma^i)_K
$$
$$
+(\alpha_1 + \tfrac{\alpha_2}{\Delta t})\mathbf{M}_{I\Gamma}^i(\mathbf{y}_\Gamma^i)_K = \mathbf{0}, \tag{4.19b}
$$

$$
\alpha_3 \mathbf{Q}_{II}(\mathbf{u}_I^i)_k + (\mathbf{B}_{II}^i)^T(\mathbf{p}_I^i)_k + (\mathbf{B}_{\Gamma I}^i)^T(\mathbf{p}_\Gamma^i)_k = \mathbf{0}, \tag{4.19c}
$$
$$
k = 1, \ldots, K,
$$

$$
\mathbf{M}_{II}^i \frac{(\mathbf{y}_I^i)_k-(\mathbf{y}_I^i)_{k-1}}{\Delta t} + \mathbf{A}_{II}^i(\mathbf{y}_I^i)_k + \mathbf{B}_{II}^i(\mathbf{u}_I^i)_k
$$
$$
+\mathbf{M}_{I\Gamma}^i \frac{(\mathbf{y}_\Gamma^i)_k-(\mathbf{y}_\Gamma^i)_{k-1}}{\Delta t} + \mathbf{A}_{I\Gamma}^i(\mathbf{y}_\Gamma^i)_k = \mathbf{0}, \tag{4.19d}
$$
$$
k = 1, \ldots, K,
$$

$$(\mathbf{y}_I^i)_0 = \mathbf{0}, \tag{4.19e}$$

and

$$
-\mathbf{M}_{\Gamma I}^i \frac{(\mathbf{p}_I^i)_{k+1}-(\mathbf{p}_I^i)_k}{\Delta t} + (\mathbf{A}_{I\Gamma}^i)^T(\mathbf{p}_I^i)_k
$$
$$
-\mathbf{M}_{\Gamma\Gamma}^i \frac{(\mathbf{p}_\Gamma^i)_{k+1}-(\mathbf{p}_\Gamma^i)_k}{\Delta t} + (\mathbf{A}_{\Gamma\Gamma}^i)^T(\mathbf{p}_\Gamma^i)_k
$$
$$
+\alpha_1 \mathbf{M}_{\Gamma I}^i(\mathbf{y}_I^i)_k + \alpha_1 \mathbf{M}_{\Gamma\Gamma}^i(\mathbf{y}_\Gamma^i)_k = \mathbf{r}_k^i, \tag{4.19f}
$$
$$
k = 1, \ldots, K-1,
$$

$$
\left(\tfrac{1}{\Delta t}\mathbf{M}_{\Gamma I}^i + (\mathbf{A}_{I\Gamma}^i)^T\right)(\mathbf{p}_I^i)_K + (\alpha_1 + \tfrac{\alpha_2}{\Delta t})\mathbf{M}_{\Gamma I}^i(\mathbf{y}_I^i)_K
$$
$$
+ \left(\tfrac{1}{\Delta t}\mathbf{M}_{\Gamma\Gamma}^i + (\mathbf{A}_{\Gamma\Gamma}^i)^T\right)(\mathbf{p}_\Gamma^i)_K
$$
$$
+(\alpha_1 + \tfrac{\alpha_2}{\Delta t})\mathbf{M}_{\Gamma\Gamma}^i(\mathbf{y}_\Gamma^i)_K = \mathbf{r}_K^i, \tag{4.19g}
$$

$$\mathbf{M}^i_{\Gamma I}\frac{(\mathbf{y}^i_I)_k-(\mathbf{y}^i_I)_{k-1}}{\Delta t}+\mathbf{A}^i_{\Gamma I}(\mathbf{y}^i_I)_k+\mathbf{B}^i_{\Gamma I}(\mathbf{u}^i_I)_k$$
$$+\mathbf{M}^i_{\Gamma\Gamma}\frac{(\mathbf{y}^i_\Gamma)_k-(\mathbf{y}^i_\Gamma)_{k-1}}{\Delta t}+\mathbf{A}^i_{\Gamma\Gamma}(\mathbf{y}^i_\Gamma)_k=\mathbf{r}^i_{K+k},\qquad(4.19h)$$
$$k=1,\dots,K,$$

$$(\mathbf{y}^i_\Gamma)_0=\mathbf{0}.\qquad(4.19i)$$

Using the notation (4.15), the system (4.19) can be written in the compact form (4.17).

The proof that (4.17) are the necessary and sufficient optimality conditions for (4.18) can be carried out analogously the proofs of Theorem 2 or Theorem 3. We omit the details. □

Theorem 5. *The matrices* $\mathbf{S}^{\Delta t}_i$, $i=1,\dots,s$, *defined in (4.12) are symmetric.*

Proof. We define

$$\mathbb{A}^i_{II}=\begin{pmatrix}\frac{1}{\Delta t}\mathbf{M}^i_{II}+\mathbf{A}^i_{II}\\-\frac{1}{\Delta t}\mathbf{M}^i_{II}\ \frac{1}{\Delta t}\mathbf{M}^i_{II}+\mathbf{A}^i_{II}\\&\ddots&\ddots\\&&-\frac{1}{\Delta t}\mathbf{M}^i_{II}\ \frac{1}{\Delta t}\mathbf{M}^i_{II}+\mathbf{A}^i_{II}\end{pmatrix},$$

$$\mathbb{B}^i_{II}=\begin{pmatrix}\mathbf{B}^i_{II}\\&\ddots\\&&\mathbf{B}^i_{II}\end{pmatrix},$$

$$\mathbb{M}^i_{II}=\begin{pmatrix}\alpha_1\mathbf{M}^i_{II}\\&\ddots\\&&\alpha_1\mathbf{M}^i_{II}\\&&&(\alpha_1+\frac{\alpha_2}{\Delta t})\mathbf{M}^i_{II}\end{pmatrix},$$

$$\mathbb{Q}^i_{II}=\begin{pmatrix}\mathbf{Q}^i_{II}\\&\ddots\\&&\mathbf{Q}^i_{II}\end{pmatrix}.$$

The matrices $\mathbb{A}^i_{\Gamma I},\mathbb{B}^i_{\Gamma I}$, etc., are defined analogously. We set

$$\mathbf{x}^i_I=\left((\mathbf{y}^i_I)^T_1,\dots,(\mathbf{y}^i_I)^T_K,(\mathbf{u}^i_I)^T_1,\dots,(\mathbf{u}^i_I)^T_K,(\mathbf{p}^i_I)^T_1,\dots,(\mathbf{p}^i_I)^T_K\right)^T$$

and

$$\mathbf{x}^i_\Gamma=\left((\mathbf{y}^i_\Gamma)^T_1,\dots,(\mathbf{y}^i_\Gamma)^T_K,(\mathbf{p}^i_\Gamma)^T_1,\dots,(\mathbf{p}^i_\Gamma)^T_K\right)^T.$$

With this notation, the solution $(\mathbf{y}^i_I)_1,\dots,(\mathbf{y}^i_I)_K$, $(\mathbf{p}^i_I)_1,\dots,(\mathbf{p}^i_I)_K$, $(\mathbf{u}^i_I)_1,\dots,(\mathbf{u}^i_I)_K$ of (4.9) with $(\mathbf{y}_0)^i_I=\mathbf{0}$, $\mathbf{f}^i_I=\mathbf{0}$, $\mathbf{c}^i_I=\mathbf{0}$, $\mathbf{d}^i_I=\mathbf{0}$ is given by

$$\mathbf{x}^i_I=-\begin{pmatrix}\mathbb{M}^i_{II}&&(\mathbb{A}^i_{II})^T\\&\mathbb{Q}^i_{II}&(\mathbb{B}^i_{II})^T\\\mathbb{A}^i_{II}&\mathbb{B}^i_{II}\end{pmatrix}^{-1}\begin{pmatrix}\mathbb{M}^i_{I\Gamma}&(\mathbb{A}^i_{\Gamma I})^T\\&(\mathbb{B}^i_{\Gamma I})^T\\\mathbb{A}^i_{I\Gamma}\end{pmatrix}\mathbf{x}^i_\Gamma.$$

Furthermore,

$$
\begin{aligned}
\mathbf{S}_i^{\Delta t}(\mathbf{x}_\Gamma^i) \\
&= \begin{pmatrix} \mathbb{M}_{\Gamma\Gamma}^i & (\mathbb{A}_{\Gamma\Gamma}^i)^T \\ \mathbb{A}_{\Gamma\Gamma}^i & \end{pmatrix} \mathbf{x}_\Gamma^i + \begin{pmatrix} \mathbb{M}_{\Gamma I}^i & (\mathbb{A}_{I\Gamma}^i)^T \\ \mathbb{A}_{\Gamma I}^i & \mathbb{B}_{\Gamma I}^i \end{pmatrix} \mathbf{x}_I^i, \\
&= \Bigg[\begin{pmatrix} \mathbb{M}_{\Gamma\Gamma}^i & (\mathbb{A}_{\Gamma\Gamma}^i)^T \\ \mathbb{A}_{\Gamma\Gamma}^i & \end{pmatrix} \\
&\quad - \begin{pmatrix} \mathbb{M}_{\Gamma I}^i & (\mathbb{A}_{I\Gamma}^i)^T \\ \mathbb{A}_{\Gamma I}^i & \mathbb{B}_{\Gamma I}^i \end{pmatrix} \begin{pmatrix} \mathbb{M}_{II}^i & (\mathbb{A}_{II}^i)^T \\ \mathbb{Q}_{II}^i & (\mathbb{B}_{II}^i)^T \\ \mathbb{A}_{II}^i & \mathbb{B}_{II}^i \end{pmatrix}^{-1} \begin{pmatrix} \mathbb{M}_{I\Gamma}^i & (\mathbb{A}_{\Gamma I}^i)^T \\ & (\mathbb{B}_{\Gamma I}^i)^T \\ \mathbb{A}_{I\Gamma}^i & \end{pmatrix} \Bigg] \mathbf{x}_\Gamma^i.
\end{aligned}
$$

The last identity reveals the symmetry of $\mathbf{S}_i^{\Delta t}$.   ◻

4.3.3 Solution Algorithm

In the previous section we have shown that the discrete optimal control problem (4.6) is equivalent to the linear operator equation (4.14). We solve (4.14) using preconditioned GMRES [9; 19; 22] or the symmetric QMR (sQMR) [7; 8]. The inverse of the system operator $\sum_{i=1}^{s} (\mathbf{I}_\Gamma^{i,\Delta t})^T \mathbf{S}_i^{\Delta t} \mathbf{I}_\Gamma^{i,\Delta t}$ is approximated by a weighted sum of inverses of the subdomain operators $\mathbf{S}_i^{\Delta t}$. This choice is motivated by Neumann-Neumann domain decomposition preconditioners that have been used successfully for the solution of elliptic PDEs (see [18; 20; 21] and the references given therein) as well as of elliptic linear-quadratic optimal control problems [11; 12]. We let $D_\Gamma^i \in \mathbb{R}^{m_\Gamma^i \times m_\Gamma^i}$ be positive definite diagonal matrices such that

$$
\sum_{i=1}^{s} (\mathbf{I}_\Gamma^i)^T D_\Gamma^i = \mathbf{I}.
$$

In our case the entry $(D_\Gamma^i)_{ll}$ is equal to one over the number of subdomains containing the interface node x_l. We set

$$
\mathbf{D}_\Gamma^{i,\Delta t} = \mathrm{diag}(D_\Gamma^i, \dots, D_\Gamma^i) \in \mathbb{R}^{2Km_\Gamma^i \times 2Km_\Gamma^i}. \tag{4.20}
$$

The preconditioner for $\sum_{i=1}^{s} (\mathbf{I}_\Gamma^{i,\Delta t})^T \mathbf{S}_i^{\Delta t} \mathbf{I}_\Gamma^{i,\Delta t}$ is now given by

$$
\mathbf{P}^{\Delta t} = \sum_{i=1}^{s} (\mathbf{I}_\Gamma^{i,\Delta t})^T \mathbf{D}_\Gamma^{i,\Delta t} (\mathbf{S}_i^{\Delta t})^{-1} \mathbf{D}_\Gamma^{i,\Delta t} \mathbf{I}_\Gamma^{i,\Delta t}. \tag{4.21}
$$

The unknowns $(\mathbf{y}_\Gamma)_1, \dots, (\mathbf{y}_\Gamma)_K, (\mathbf{p}_\Gamma)_1, \dots, (\mathbf{p}_\Gamma)_K \in \mathbb{R}^{m_\Gamma}$ to be determined via GMRES are the discretizations of the states and the adjoints on the subdomain interfaces $\Gamma \times (0, T)$. The evaluation of a matrix vector product

$$
\sum_{i=1}^{s} (\mathbf{I}_\Gamma^{i,\Delta t})^T \mathbf{S}_i^{\Delta t} \mathbf{I}_\Gamma^{i,\Delta t} \Big((\mathbf{y}_\Gamma)_1, \dots, (\mathbf{y}_\Gamma)_K, (\mathbf{p}_\Gamma)_1, \dots, (\mathbf{p}_\Gamma)_K \Big)
$$

can be done in parallel on s processors. Each processor has to evaluate

$$\mathbf{S}_i^{\Delta t} \mathbf{I}_\Gamma^{i,\Delta t} \Big((\mathbf{y}_\Gamma)_1, \ldots, (\mathbf{y}_\Gamma)_K, (\mathbf{p}_\Gamma)_1, \ldots, (\mathbf{p}_\Gamma)_K \Big),$$

which corresponds to the solution of a subdomain optimal control problem (4.11) (cf. Theorems 3). These subdomain optimal control problems can be solved using standard techniques and their solution only involves discretizations of states, controls and adjoints on the smaller domains $\Omega_i \times (0, T)$. Similarly, the application of the preconditioner (4.21) to a vector $(\mathbf{v}_\Gamma)_1, \ldots, (\mathbf{v}_\Gamma)_K, (\mathbf{q}_\Gamma)_1, \ldots, (\mathbf{q}_\Gamma)_{K+1} \in \mathbb{R}^{m_\Gamma}$ can be done in parallel on s processors. Each processor has to evaluate

$$(\mathbf{S}_i^{\Delta t})^{-1} \mathbf{D}_\Gamma^{i,\Delta t} \mathbf{I}_\Gamma^{i,\Delta t} \Big((\mathbf{v}_\Gamma)_1, \ldots, (\mathbf{v}_\Gamma)_K, (\mathbf{q}_\Gamma)_1, \ldots, (\mathbf{q}_\Gamma)_K \Big),$$

which corresponds to the solution of a subdomain optimal control problem (4.18) (cf. Theorem 4). Again, these subdomain optimal control problems can be solved using standard techniques and their solution only involves discretizations of states, controls and adjoints on the smaller domains $\Omega_i \times (0, T)$.

4.4 Numerical Results

We consider (4.1) with $\Omega = (0, 1)$, $f = 0$ and $y_0(x) = \sin(2\pi x)$. The desired states $\hat{y}$ and $\hat{y}_T$ are given by the hat functions $\hat{y}(x, t) = \min\{2x, 2(1 - x)\}$ and $\hat{y}_T(x) = \min\{2x, 2(1 - x)\}$, respectively. For the spatial discretization of the problem we use piecewise linear finite elements on an equidistant grid with mesh size $\Delta x = 1/K$ and for the time discretization, we use the backward Euler method with step size $\Delta t = 1/K$. The domain $\Omega = (0, 1)$ is subdivided into equidistant subdomains $\Omega_i = ((i - 1)H, iH)$, $i = 1, \ldots, s$, $H = 1/s$. For $\alpha_1 = \alpha_2 = 10^3$, $\alpha_3 = 1$, and $s = 4$, the computed optimal control and corresponding state are shown in Figure 4.2.

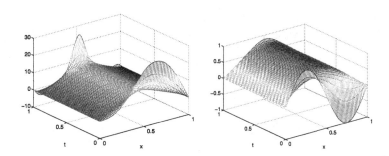

Fig. 4.2. Computed optimal control (left) and corresponding state (right) for the example problem with $\alpha_1 = \alpha_2 = 10^3$, $\alpha_3 = 1$ and $s = 4$ subdomains.

The evaluation of matrix-vector products of the form $\mathbf{S}_i \mathbf{v}_\Gamma^i$ and $(\mathbf{S}_i)^{-1} \mathbf{r}_\Gamma^i$, is equivalent to solving subdomain optimal control problems (4.11) and (4.18), respectively, (cf. Theorems 3 and 4). Matrix-vector products $\mathbf{S}_i \mathbf{v}_\Gamma^i$ and $(\mathbf{S}_i)^{-1} \mathbf{r}_\Gamma^i$ are computed by solving the reduced forms of the optimal control problems (4.11) and (4.18) using the conjugate gradient method. The reduced form of the optimal control problems (4.11) or (4.18) is the one in which the state is viewed as function of the control and the optimal control problem is posed as a minimization problem in the controls only. The conjugate gradient method is stopped when the norm of the gradient is less than 10^{-10}. The preconditioned GMRES/sQMR applied to linear operator equation (4.14) is stopped if the preconditioned residual is less than 10^{-4}.

Table 4.1. Number of preconditioned sQMR/GMRES iterations needed for the solution of (4.14) depending on the number of subdomains s and on the discretization size $\Delta x = \Delta t = 1/K$.

sQMR Iterations $\alpha_1 = \alpha_2 = 10, \alpha_3 = 1$				sQMR Iterations $\alpha_1 = \alpha_2 = 10^3, \alpha_3 = 1$		
$s \backslash K$	64	128	256	$s \backslash K$ 64	128	256
2	1	1	1	2 1	1	1
4	15	16	16	4 8	8	8
8	68	68	77	8 32	32	32
16	514	> 1000	545	16 124	140	121

GMRES Iterations $\alpha_1 = \alpha_2 = 10, \alpha_3 = 1$				GMRES Iterations $\alpha_1 = \alpha_2 = 10^3, \alpha_3 = 1$		
$s \backslash K$	64	128	256	$s \backslash K$ 64	128	256
2	1	1	1	2 1	1	1
4	15	16	16	4 7	7	7
8	54	55	60	8 27	28	29
16	166	177	184	16 95	99	103

Table 4.1 shows that for this example, preconditioned GMRES and preconditioned sQMR behave similarly. sQMR uses a three term recurrence and therefore the amount of memory required is independent of the number of iterations, whereas in GMRES it depends linearly on the number of iterations. The number of preconditioned GMRES/SQMR iterations is roughly proportional to the square of the number of subdomains. The deterioration of the performance of the Neumann-Neumann preconditioner with increasing number of subdomains is well known [5; 6; 15; 20; 21]. For elliptic PDEs and for linear-quadratic elliptic optimal control problems a coarse space is introduced to avoid this deterioration (see the references above). For linear-quadratic parabolic optimal control problems this is still under investigation. Table 4.1 also shows that the number of preconditioned GMRES/sQMR iterations is insensitive to the weighting parameters α_1, α_2. The conditioning of the optimal control problem (4.1) grows as α_1/α_3 and as α_2/α_3. For larger α_1/α_3 and α_2/α_3 the problem (4.1) becomes more difficult to solve numerically. The insensitivity of the number of preconditioned GMRES iterations again matches the observations made in [11; 12] for Neumann-Neumann methods applied to linear-quadratic elliptic optimal control

problems. The fact that preconditioned GMRES/sQMR performs better on the more difficult problem is likely due to the particular structure of our model problem, but not true in general, see, e.g., [11; 12].

4.5 Conclusions

We have presented a spatial domain decomposition (DD) preconditioner for the solution of discretized parabolic linear–quadratic optimal control problems. Our DD preconditioner is based on a decomposition of the spatial domain into non-overlapping subdomains. The optimality conditions for the parabolic linear–quadratic optimal control problem is split into smaller problems restricted to spatial subdomain-time cylinders. These subproblems correspond to parabolic linear–quadratic optimal control problems on subdomains with Dirichlet data on interfaces. The coupling of these subdomain problems leads to a Schur complement system in which the unknowns are the state and adjoint variables on the subdomain interfaces in space and time.

The Schur complement system is solved using a preconditioned GMRES or a preconditioned sQMR. The application of the Schur complement to a vector requires the (parallel) solution of smaller subdomain parabolic linear–quadratic optimal control problems. States, controls and adjoints for these smaller subdomain problems are only needed locally, but do not have to be communicated or stored globally. The preconditioner is obtained from the solution of appropriate subdomain parabolic linear–quadratic optimal control problems. The application of this preconditioner to a vector also requires the (parallel) solution of smaller subdomain parabolic linear–quadratic optimal control problems. Again, states, controls and adjoints for these smaller subdomain problems are only needed locally, but do not have to be communicated or stored globally.

Our numerical tests indicate that the dependence of the performance of our preconditioner on mesh size and subdomain size is similar to that of its counterpart applied to elliptic equations only and to that of its counterpart applied to elliptic linear-quadratic optimal control problems. In particular, the number of preconditioned GMRES/sQMR iterations is roughly proportional to the square of the number of subdomains. With the introduction of a coarse space one might be able to prevent this deterioration. This is the subject of future research. Our tests also show that the preconditioners are insensitive to the size of the control regularization parameter.

References

[1] J.-D. Benamou. A domain decomposition method for control problems. In P. Bjørstad, M. Espedal, and D. Keyes, editors, *DD9 Proceedings*, pages 266–273, Bergen, Norway, 1998. Domain Decomposition Press. Available electronically from http://www.ddm.org/DD9/index.html.
[2] J. D. Benamou. Domain decomposition, optimal control of systems governed by partial differential equations, and synthesis of feedback laws. *J. Optim. Theory Appl.*, 102(1):15–36, 1999.

[3] A. Bounaim. On the optimal control problem of the heat equation: New formulation of the problem using a non-overlapping domain decomposition technique. Technical report, Scientific Computing Group, Department of Informatics. University of Oslo, 2002.

[4] A. Comas. Time domain decomposition methods for second order linear quadratic optimal control problems. Master's thesis, Department of Computational and Applied Mathematics, Rice University, Houston, TX, 2004.

[5] M. Dryja, B. F. Smith, and O. B. Widlund. Schwarz analysis of iterative substructuring algorithms for elliptic problems in three dimensions. *SIAM J. Numer. Anal.*, 31(3):1662–1694, 1994.

[6] M. Dryja and O. B. Widlund. Schwarz methods of Neumann-Neumann type for three-dimensional elliptic finite element methods. *Comm. Pure Appl. Math.*, 48:121–155, 1995.

[7] R. W. Freund and N. M. Nachtigal. A new Krylov-subspace method for symmetric indefinite linear systems. In W. F. Ames, editor, *Proceedings of the 14th IMACS World Congress on Computational and Applied Mathematics*, pages 1253–1256. IMACS, 1994.

[8] R. W. Freund and N. M. Nachtigal. Software for simplified Lanczos and QMR algorithms. *Applied Numerical Mathematics*, 19:319–341, 1995.

[9] A. Greenbaum. *Iterative Methods for the Solution of Linear Systems*. SIAM, Philadelphia, 1997.

[10] M. Heinkenschloss. Time–domain decomposition iterative methods for the solution of distributed linear quadratic optimal control problems. *Journal of Computational and Applied Mathematics*, 173:169–198, 2005.

[11] M. Heinkenschloss and H. Nguyen. Balancing Neumann-Neumann methods for elliptic optimal control problems. In R. Kornhuber, R. H. W. Hoppe, J. Periaux, O. Pironneau, O. B. Widlund, and J. Xu, editors, *Domain Decomposition methods in Science and Engineering*, Lecture Notes in Computational Science and Engineering, Vol. 40, pages 589–596, Heidelberg, 2004. Springer-Verlag.

[12] M. Heinkenschloss and H. Nguyen. Neumann-Neumann domain decomposition preconditioners for linear–quadratic elliptic optimal control problems. Technical Report TR04–01, Department of Computational and Applied Mathematics, Rice University, 2004.

[13] J. E. Lagnese and G. Leugering. Time-domain decomposition of optimal control problems for the wave equation. *System and Control Letters*, 48:229–242, 2003.

[14] J. E. Lagnese and G. Leugering. *Domain Decomposition Methods in Optimal Control of Partial Differential Equations*, volume 148 of *International Series of Numerical Mathematics*. Birkhäuser Verlag, Basel, 2004.

[15] P. Le Tallec. Domain decomposition methods in computational mechanics. In J. T. Oden, editor, *Computational Mechanics Advances, Volume 1(2)*, pages 121–220, New York, 1994. North Holland.

[16] J.-L. Lions. *Optimal Control of Systems Governed by Partial Differential Equations*. Springer Verlag, Berlin, Heidelberg, New York, 1971.

[17] Y. Maday and G. Turinici. A parareal in time procedure for the control of partial differential equations. *Comptes Rendus de l'Académie des Sciences. Série I. Mathématique*, 335:387–392, 2002.

[18] A. Quarteroni and A. Valli. *Domain Decomposition Methods for Partial Differential Equations*. Oxford University Press, Oxford, 1999.

[19] Y. Saad. *Iterative Methods for Sparse Linear Systems*. SIAM, Philadelphia, 2nd edition, 2003.

[20] B. Smith, P. Bjørstad, and W. Gropp. *Domain Decomposition. Parallel Multilevel Methods for Elliptic Partial Differential Equations*. Cambridge University Press, Cambridge, 1996.

[21] A. Toselli and O. Widlund. *Domain Decomposition Methods - Algorithms and Theory*. Computational Mathematics, Vol. 34. Springer–Verlag, Berlin, 2004.

[22] H. A. van der Vorst. *Iterative Krylov Methods for Large Linear Systems*, volume 13 of *Cambridge Monographs on Applied and Computational Mathematics*. Cambridge University Press, Cambridge, 2003.

Modeling and Implementation of Risk-Averse Preferences in Stochastic Programs Using Risk Measures

Pavlo A. Krokhmal[1] and Robert Murphey[2]

[1] Department of Mechanical and Industrial Engineering
University of Iowa
3131 Seamans Center
Iowa City, IA 52242-1257
krokhmal@engineering.uiowa.edu
[2] Air Force Research Lab
Munitions Directorate
Eglin AFB, FL 32542
robert.murphey@eglin.af.mil

Summary. We consider modeling and implementation of risk-averse preferences in stochastic programming problems using axiomatically defined risk measures. We derive representations for several classes of risk measures (e.g., coherent risk measures, deviation measures) via solutions of specially formulated stochastic programming problems that facilitate incorporation of risk measures in multistage stochastic programming problems. As an illustration of the general approach, we consider a two-stage stochastic weapon-target assignment problem, where a coherent risk measure is used to capture the risk of the second-stage (recourse) action.

Key words: risk measures, stochastic programming, stochastic dominance, convex analysis

5.1 Introduction

Over the last several decades, modeling and implementation of risk-averse preferences in decision-making problems involving uncertainties have represented a major research area in the decision making science and, in particular, mathematical programming. One of the first and most significant contributions in this context has been made by von Neumann and Morgenstern [1] with the development of the utility theory. The von Neumann and Morgenstern's (vNM) argument is that any preference relation $\succeq$ defined on a set of random outcomes (risks) and satisfying certain assumptions (axioms), namely *completeness, transitivity, continuity,* and *independence,* can be equivalently represented by some function $U \colon \mathbb{R} \mapsto \mathbb{R}$, such that the inequality

$E[U(X)] \geq E[U(Y)]$ holds if and only if outcome X is preferred to outcome Y, $X \succeq Y$. Along with its numerous modifications and extensions, the vNM utility theory is now widely adopted as a model of rational choice, especially in economics and social sciences (see, among others, [2; 3; 4]). Practical applications of the vNM utility theory are, however, not as widespread as its scholarly and academic use. This is sometimes attributed to the difficulties in identifying the specific form of a utility function that reflects the preferences of the given decision-maker. Also, the vNM utility is known to be *ordinal*, i.e., it does not *measure* preferences, but rather *orders* them. Most engineering decision-making applications, including those encountered in financial engineering, typically require precisely quantified preferences.

An alternative approach, originated by the seminal Markowitz [5; 6] mean-variance (MV) model, is embodied by the *risk-reward optimization* paradigm, where the computational tractability of bi-criteria optimization of empirically chosen risk and reward functions is favored over the theoretical merits of vNM axiomatic approach. Since 1952, Markowitz's risk-reward optimization framework received extensive development in directions of both increasing the numerical efficiency and refinement of the models for risk measurement and estimation. For example, it was recognized that the symmetric attitude of the classical MV model does not always yield an adequate estimation of risks induced by the uncertainties, which initiated a substantial body of research work on development of *downside risk measures* and models. Among those, we mention the *lower standard semi-deviation* [6, see also 7; 8; 9]; *lower partial moment*, or *Expected Regret*, in stochastic programming also known as *integrated chance constraint* [10; 11; 12; 13], *Value-at-Risk* [14; 15; 16; 17], *Conditional Value-at-Risk* [18; 19; 20], *Conditional Drawdown-at-Risk* [21; 22], *Expected Shortfall* [23; 24], *Maximum Loss* [25; 22] etc.

At the same time, significant attention has been devoted to integration of the risk-reward framework with the principles of vNM approach. For instance, it has been shown that under certain conditions the Markowitz MV framework is consistent with the vNM utility theory [26]. Ogryczak and Ruszczyński [7; 8; 9] developed mean-semideviation models that are consistent with stochastic dominance concepts [27; 28; 29]. Optimization models featuring stochastic dominance constraints have been recently introduced by Ruszczyński and Dentcheva [30].

Many of the recent advances in risk theory and risk-averse decision-making are arguably associated with axiomatic approach to construction of risk measures pioneered by Artzner et al. [31; 32; 33] and Delbaen [34]. The authors have proposed an axiomatic definition of *coherent* risk measures, i.e., risk measures with a prescribed set of properties that would ensure their robust and consistent behavior in applications. Among the risk measures that satisfy the coherency properties, there are Conditional Value-at-Risk, Maximum Loss [35; 24], coherent risk measures based on one-sided moments [36], etc. Recently, Rockafellar et al. [37; 38] have extended the formal axiomatic theory of risk measures to the case of *deviation measures*, and demonstrated a close relationship between coherent risk measures and deviation measures. *Spectral measures of risk* have been proposed by Acerbi [39]; optimization of *convex risk functions*, similar in properties to coherent risk measures, has been considered by Ruszczyński and Shapiro [40].

The present endeavor essentially exploits this axiomatic approach in order to devise simple computational recipes for incorporation of risk-averse preferences into stochastic programming problems, which can be viewed as yet another evidence of the benefits of the axiomatic approach to risk analysis. Namely, we introduce representations for several classes of axiomatically defined risk measures via solutions of appropriately constructed stochastic programming problems. As an example of the general approach, we consider a two-stage weapon-target assignment problem, where the risk of the second-stage decision is controlled.

5.2 Risk Measures as Stochastic Programming Problems

In this section we develop representations for various classes of axiomatically defined risk measures that facilitate incorporation of these risk measures in (multistage) stochastic programming problems.

First, let us outline a generic definition of risk measure. Generally, risk measure $\mathscr{R}(X)$ of a random outcome X from some probability space $(\Omega, \mathscr{F}, \mu)$ may be defined as a mapping $\mathscr{R}: \mathcal{X} \mapsto \overline{\mathbb{R}}$, where $\mathcal{X}$ is a linear space of $\mathscr{F}$-measurable functions $X: \Omega \mapsto \overline{\mathbb{R}}$. In a more general setting one may assume $\mathcal{X}$ to be a separated locally convex space; for our purposes it suffices to consider $\mathcal{X} = \mathscr{L}^p(\Omega, \mathscr{F}, \mu), 1 \leq p \leq \infty$, where the particular value of p shall be clear from the context. Traditionally to convex analysis, we call function $f: \mathcal{X} \mapsto \overline{\mathbb{R}}$ *proper* if $f(X) > -\infty$ for all $X \in \mathcal{X}$ and dom $f \neq \varnothing$, i.e., exists $X \in \mathcal{X}$ such that $f(X) < +\infty$ (see, e.g., [41; 42]). In the remainder of the paper, we confine ourselves to risk measures that are proper and not identically equal to $+\infty$.

The axiomatic approach to construction of risk measures has been first presented in the landmark paper by Artzner et al. [32] (see also Delbaen [34]), where the authors have established a set of properties (axioms) that a "reasonable" measure of risk has to satisfy, and called these properties "coherence." Later, Rockafellar et al. [37; 38], Ruszczyński and Shapiro [43], Acerbi [39] have extended the axiomatic framework of risk analysis with introduction of deviation measures, convex risk functions, spectral risk measures, etc. As axiomatic definition does not provide an explicit form for the corresponding class of risk measures, Artzner et al. [32] have demonstrated that coherent risk measures admit a representation in the form

$$\mathscr{R}(X) = - \inf_{\mu \in \mathcal{A}} \mathsf{E}_\mu X,$$

where $\mathcal{A}$ is a family of probability measures on $(\Omega, \mathscr{F})$. Similar expressions have been obtained in [37; 38; 43] for deviation measures, convex risk functions, etc.

Below we introduce a different type of representation, where risk measure is obtained via solution of a stochastic programming problem, whose form is governed by the set of axioms that this risk measure has to satisfy.

5.2.1 Coherent Risk Measures

The first class of risk measures that we consider in this section are the coherent risk measures. According to [32] and [34], coherent risk measure can be defined as a mapping $\mathscr{R} \colon \mathcal{X} \mapsto \overline{\mathbb{R}}$ that satisfies the next four axioms:

(A1) *monotonicity:* $X \geq 0 \;\Rightarrow\; \mathscr{R}(X) \leq 0$ for all $X \in \mathcal{X}$,

(A2) *sub-additivity:* $\mathscr{R}(X+Y) \leq \mathscr{R}(X) + \mathscr{R}(Y)$ for all $X, Y \in \mathcal{X}$,

(A3) *positive homogeneity:* $\mathscr{R}(\lambda X) = \lambda \mathscr{R}(X)$ for all $X \in \mathcal{X}$, $\lambda > 0$,

(A4) *translation invariance:* $\mathscr{R}(X+a) = \mathscr{R}(X) - a$ for all $X \in \mathcal{X}$, $a \in \mathbb{R}$.

Since axioms (A2) and (A3) immediately yield convexity of coherent risk measures, little will be lost in the generality of the above definition if one replaces the sub-additivity requirement (A2) with the stronger requirement of convexity:

(A2') *convexity:* $\mathscr{R}(\lambda X + (1-\lambda)Y) \leq \lambda \mathscr{R}(X) + (1-\lambda)\mathscr{R}(Y)$, $X, Y \in \mathcal{X}$, $0 \leq \lambda \leq 1$.

From the axioms (A1)–(A4) one can easily derive the following useful properties of coherent risk measures (see, for example, 34):

(C1) $\mathscr{R}(0) = 0$ [A1+A2 or A3],

(C2) $X \leq Y \;\Rightarrow\; \mathscr{R}(X) \geq \mathscr{R}(Y)$ [A1+A2],

(C3) $\mathscr{R}(a) = -a, \; a \in \mathbb{R}$ [A4+C1],

(C4) $X \geq a \;\Rightarrow\; \mathscr{R}(X) \leq -a, \; a \in \mathbb{R}$ [C2+C3],

(C5) $\mathscr{R}(X + \mathscr{R}(X)) = 0$ [A4+C3],

(C6) $\mathscr{R}(X)$ is continuous in its effective domain,

where expressions in brackets indicate the axioms or properties that support the given statement. Also, throughout the paper the inequalities $X \geq a$, $X \leq Y$, etc., are assumed to hold almost surely.

From the definition of coherent risk measures it is easy to see that, for example, $\mathsf{E}(-X)$ and $\mathrm{ess.inf}(-X)$, where

$$\mathrm{ess.inf}\, X = \begin{cases} \max\{\eta \in \mathbb{R} \mid \eta \leq X\}, & \text{if } \{\eta \in \mathbb{R} \mid \eta \leq X\} \neq \varnothing, \\ -\infty, & \text{otherwise,} \end{cases}$$

are coherent risk measures; more examples can be found in [37]. Below we present simple computational formulas that aid in construction of coherent risk measures and may be seamlessly incorporated into stochastic programs. Namely, we execute the idea that one of the axioms (A3) or (A4) can be relaxed and then "reinforced" by solving an appropriately defined mathematical programming problem. In other words, one can construct a coherent risk measure via solution of a stochastic programming problem that involves a function $\phi \colon \mathcal{X} \mapsto \mathbb{R}$ satisfying only three of the four axioms (A1)–(A4). Our focus on axioms (A3) and (A4) is explained by the fact that they determine the behavior of a coherent risk measure with respect to scalar addition and multiplication.

First we introduce a representation for coherent risk measures that is based on relaxation of the translation invariance axiom (A4). The next theorem shows that if one selects a function $\phi \colon \mathcal{X} \mapsto \mathbb{R}$ satisfying axioms (A1)–(A3), but not (A4), then there exists a simple stochastic optimization problem involving ϕ whose optimal value would satisfy (A1)–(A4).

Theorem 1. *Let function $\phi \colon \mathcal{X} \mapsto \mathbb{R}$ satisfy axioms (A1)–(A3), but not (A4), and be a lsc function such that $\phi(\eta) > -\eta$ for all real $\eta \neq 0$. Then the optimal value of the stochastic programming problem*

$$\rho(X) = \inf_{\eta} \ \eta + \phi(X + \eta) \tag{5.1}$$

is a proper coherent risk measure, and the infimum is attained for all X, so $\inf_\eta$ in (5.1) may be replaced by $\min_{\eta \in \mathbb{R}}$.

Proof. Convexity, lower semicontinuity, and sublinearity of ϕ in $\mathcal{X}$ imply that function $f_X(\eta) = \eta + \phi(X + \eta)$ is also convex, lsc, and proper in $\mathbb{R}$ for any $X \in \mathcal{X}$. For the infimum of $f_X(\eta)$ to be achievable at finite η, its recession function has to be positive: $f_X 0^+(\pm 1) > 0$, which is equivalent to $f_X 0^+(\xi) > 0$, $\xi \neq 0$ due to the positive homogeneity of $f_X(\cdot)$. By definition of recession function [41; 42] and positive homogeneity of ϕ, we have that the last condition holds if $\phi(\xi) > -\xi$ for all $\xi \neq 0$:

$$f_X 0^+(\xi) = \lim_{\tau \to \infty} \frac{\eta + \tau\xi + \phi(X + \eta + \tau\xi) - \eta - \phi(X + \eta)}{\tau} = \xi + \phi(\xi).$$

Hence, $\rho(X)$ defined by (5.1) is a proper lsc function, and minimum in (5.1) is attained at finite η for all $X \in \mathcal{X}$. Below we verify that $\rho(X)$ satisfies axioms (A1)–(A4).

(A1) Let $X \geq 0$. Then $\phi(X) \leq 0$ as ϕ satisfies (A1), which implies

$$\min_{\eta \in \mathbb{R}} \ \eta + \phi(X + \eta) \ \leq \ 0 + \phi(X + 0) \ \leq \ 0.$$

(A2) For any $Z \in \mathcal{X}$ let $\eta_Z \in \arg\min_{\eta \in \mathbb{R}} \{\eta + \phi(Z - \eta)\} \subset \mathbb{R}$, then

$$\rho(X) + \rho(Y) = \eta_X + \phi(X - \eta_X) + \eta_Y + \phi(Y - \eta_Y)$$
$$\geq \eta_X + \eta_Y + \phi(X + Y - \eta_X - \eta_Y)$$
$$\geq \eta_{X+Y} + \phi(X + Y - \eta_{X+Y})$$
$$= \rho(X + Y).$$

(A3) For any fixed $\lambda > 0$ we have

$$\rho(\lambda X) = \min_{\eta \in \mathbb{R}} \left\{ \eta + \phi(\lambda X + \eta) \right\}$$
$$= \lambda \min_{\eta \in \mathbb{R}} \left\{ \eta/\lambda + \phi(X + \eta/\lambda) \right\}$$
$$= \lambda \rho(X). \tag{5.2}$$

(A4) Similarly, for any fixed $a \in \mathbb{R}$,

$$\rho(X + a) = \min_{\eta \in \mathbb{R}} \left\{ \eta + \phi(X + a + \eta) \right\}$$
$$= -a + \min_{\eta \in \mathbb{R}} \left\{ (\eta + a) + \phi(X + (\eta + a)) \right\}$$
$$= -a + \rho(X). \tag{5.3}$$

Thus, $\rho(X)$ defined as in (5.1) is a proper coherent risk measure.   □

It is all-important that the stochastic programming problem (5.1) is *convex*, due to the convexity of function ϕ.

Example 1. A famous special case of (5.1) is the optimization formula for Conditional Value-at-Risk measure [18; 19]:

$$\mathsf{CVaR}_\alpha(X) = \min_{\eta \in \mathbb{R}} \ \eta + \alpha^{-1} \mathsf{E}(X + \eta)^-, \quad 0 < \alpha \leq 1, \tag{5.4}$$

where $(X)^- = \max\{-X, 0\}$, and function $\phi(X) = \alpha^{-1}\mathsf{E}(X)^-$ evidently satisfies the conditions of Theorem 1. Space $\mathcal{X}$ in this case can be selected as $\mathscr{L}^2(\Omega, \mathscr{F}, \mathsf{P})$. One of the many appealing features of (5.4) is that it has a simple intuitive interpretation: if $-X$ represents loss/unsatisfaction, then $\mathsf{CVaR}_\alpha(X)$, is, roughly speaking, the conditional expectation of losses that may occur in $\alpha \cdot 100\%$ of the worst cases. In the case of a continuously distributed X, this rendition is exact: $\mathsf{CVaR}_\alpha(X) = -\mathsf{E}\big[X \mid X \leq -\mathsf{VaR}_\alpha(X)\big]$, where $\mathsf{VaR}_\alpha(X)$ is defined as $\mathsf{VaR}_\alpha(X) = -\inf\big\{ \varsigma \mid \mathsf{P}[X \leq \varsigma] > \alpha \big\}$, i.e., the negative of the α-quantile of X. In the general case, the formal definition of $\mathsf{CVaR}_\alpha(X)$ becomes more intricate [19], but the representation (5.4) still applies.

A generalization of (5.4) can be constructed as

$$\mathscr{R}_{\alpha,\beta}(X) = \min_{\eta \in \mathbb{R}} \ \eta + \alpha \, \mathsf{E}(X + \eta)^- - \beta \, \mathsf{E}(X + \eta)^+, \tag{5.5}$$

where, in accordance with the conditions of Theorem 1, one has to put $\alpha > 1$ and $0 \leq \beta < 1$. Naturally, $(X)^+ = \max\{X, 0\}$ in (5.5).

Example 2. Let $\mathcal{X} = \mathcal{L}^p(\Omega, \mathcal{F}, \mu)$, and consider $\phi(X) = \alpha \|(X)^-\|_p$, $\alpha > 1$, where $\|X\|_p = (E|X|^p)^{1/p}$. Clearly, ϕ satisfies the conditions of Theorem (1), therefore the optimal value of

$$\rho(X) = \min_{\eta \in \mathbb{R}} \eta + \alpha \|(X + \eta)^-\|_p, \qquad (5.6)$$

is a proper coherent risk measure. It is easy to see that $\rho(X)$ as defined above is majorized by the corresponding coherent risk measure of semi-$\mathcal{L}_p$ type [38]:

$$\rho(X) \le -EX + \alpha \|(X - EX)^-\|_p. \qquad (5.7)$$

Example 3. If the requirement of finiteness of ϕ in (5.1) is relaxed, i.e., the image of ϕ is $(-\infty, +\infty]$, then optimal value of (5.1) still defines a coherent risk measure, however, the infimum may not be achievable. An example is served by the so-called MaxLoss measure,

$$\mathrm{MaxLoss}(X) = -\mathrm{ess.inf}\ X = \inf_{\eta} \eta + \phi^*(X + \eta),$$

where

$$\phi^*(X) = \begin{cases} +\infty, & X < 0, \\ 0, & X \ge 0. \end{cases}$$

It is easy to see that ϕ^* is positive homogeneous convex, non-increasing, lsc, and satisfies $\phi^*(\eta) > -\eta$ for all $\eta \ne 0$, but is not finite.

Formula (5.1) readily extends to the case of multiple functions ϕ_i, $i = 1, \ldots, n$, that are cumulatively used in measuring the risk of element $X \in \mathcal{X}$ and conform to the conditions of Theorem 1. Namely, one has that

$$\rho_n(X) = \min_{\eta_i \in \mathbb{R},\ i=1,\ldots,n} \sum_{i=1}^{n} \Big(\eta_i + \phi_i(X + \eta_i)\Big), \qquad (5.8)$$

is a proper coherent risk measure.

It can be shown that the optimal solution $\eta(X)$ in (5.1) also possesses some interesting properties as a function of X. In establishing of these properties the following notation is convenient. Assuming that the set $\arg\min_{x \in \mathbb{R}} f(x)$ is closed for some function $f: \mathbb{R} \mapsto \mathbb{R}$, we denote its left endpoint as

$$\mathrm{Arg}\min_{x \in \mathbb{R}} f(x) = \min \{y \mid y \in \arg\min_{x \in \mathbb{R}} f(x)\}.$$

Theorem 2. *Let function $\phi \colon \mathcal{X} \mapsto \mathbb{R}$ satisfy the conditions of Theorem 1. Then function*

$$\eta(X) = \mathrm{Arg}\min_{\eta \in \mathbb{R}} \eta + \phi(X + \eta) \qquad (5.9)$$

exists and satisfies properties (A3) and (A4). If, additionally, $\phi(X) = 0$ for every $X \ge 0$, then $\eta(X)$ satisfies (A1), along with inequality $\eta(X) \le \rho(X)$, where $\rho(X)$ is the optimal value of (5.1).

Proof. Conditions on function ϕ ensure that the set of optimal solutions of problem (5.1) is closed and finite, whence follows the existence of $\eta(X)$ in (5.9).

(A3) From (5.2) we have for any $\lambda > 0$

$$\eta(\lambda X) = \underset{\eta \in \mathbb{R}}{\operatorname{Arg\,min}} \left\{ \eta + \phi(\lambda X + \eta) \right\} \qquad (5.10)$$

$$= \underset{\eta \in \mathbb{R}}{\operatorname{Arg\,min}} \left\{ \eta/\lambda + \phi(X + \eta/\lambda) \right\},$$

from which follows $\eta(\lambda X) = \lambda \eta(X)$.

(A4) Similarly, (5.3) yields

$$\eta(X + a) = \underset{\eta \in \mathbb{R}}{\operatorname{Arg\,min}} \left\{ \eta + \phi(X + a + \eta) \right\}$$

$$= \underset{\eta \in \mathbb{R}}{\operatorname{Arg\,min}} \left\{ (\eta + a) + \phi(X + (\eta + a)) \right\},$$

which, in turn, leads to the sought relation $\eta(X + a) = \eta(X) - a$.

(A1) Let ϕ be such that $\phi(X) = 0$ for every $X \geq 0$. Then, (C2) immediately yields $\phi(X) \geq 0$ for all $X \in \mathcal{X}$, which proves

$$\eta(X) \leq \eta(X) + \phi(X + \eta(X)) = \rho(X).$$

By the definition of $\eta(X)$, we have for all $X \geq 0$

$$\eta(X) + \phi(X + \eta(X)) \leq 0 + \phi(X + 0) = 0,$$

or

$$\eta(X) \leq -\phi(X + \eta(X)). \qquad (5.11)$$

Assume that $\eta(X) > 0$ which implies $\phi(\eta(X)) = 0$. From (A2) it follows that $\phi(X + \eta(X)) \leq \phi(X) + \phi(\eta(X)) = 0$, leading to $\phi(X + \eta(X)) = 0$, and, consequently, to $\eta(X) \leq 0$ by (5.11). The contradiction furnishes the statement of the theorem. $\square$

Note that if ϕ satisfies all conditions of Theorem 2, the optimal solution $\eta(X)$ of the stochastic optimization problem (5.1) complies with all axioms for the coherent risk measures, except (A2), thereby failing to be convex.

Example 4. A well-known example of two risk measures obtained by solving a stochastic programming problem of type (5.1) is again provided by formula (5.4) due to Rockafellar and Uryasev [18; 19], and its counterpart

$$\mathsf{VaR}_\alpha(X) = \underset{\eta \in \mathbb{R}}{\operatorname{Arg\,min}} \ \eta + \alpha^{-1} \mathsf{E}(X + \eta)^+.$$

The Value-at-Risk measure $\mathsf{VaR}_\alpha(X)$ (14; 15; 16, etc), despite being adopted as the *de facto* standard for measurement of risk in finance and banking industry, is notorious for its poor behavior in risk estimation and control. In fact, the early developments in the modern risk theory can be viewed as attempts of constructing risk management concepts that overcome the handicaps of the Value-at-Risk measure [44].

Now, observe that formula (5.1) in Theorem 1 is analogous to the operation of *infimal convolution*, well-known in convex analysis:

$$(f \,\square\, g)(x) = \inf_y \ f(x-y) + g(y).$$

Continuing the analogy between representation (5.1) and the operation of infimal convolution, consider the operation of right scalar multiplication

$$(\phi\eta)(X) = \eta\,\phi(\eta^{-1}X), \quad \eta \geq 0,$$

where for $\eta = 0$ we set $(\phi 0)(X) = (\phi 0^+)(X)$. If ϕ is proper and convex, then it is known that $(\phi\eta)(X)$ is a convex proper function in $\eta \geq 0$ for any $X \in \mathrm{dom}\,\phi$ (see, for example, [41]). Interestingly enough, this fact can be pressed into service to formally define coherent risk measure as the optimal value of stochastic programming problem

$$\underline{\rho}(X) = \inf_{\eta \geq 0} \ \eta\,\phi(\eta^{-1}X), \tag{5.12}$$

if function ϕ, along with some technical conditions similar to those of Theorem 1, satisfies axioms (A1), (A2′), and (A4), but not (A3). Note that excluding the positive homogeneity (A3) from the list of properties of ϕ denies also its convexity, thus one must replace (A2) with (A2′) to ensure that (5.12) is a convex programming problem. In the terminology of convex analysis the function $\rho(X)$ defined by (5.12) is known as the *positively homogeneous convex function generated by* ϕ.

Likewise, by direct verification of conditions (A1)–(A4) it can be demonstrated that

$$\overline{\rho}(X) = \sup_{\eta > 0} \ \eta\,\phi(\eta^{-1}X), \tag{5.13}$$

is a proper coherent risk measure, provided that $\phi(X)$ satisfies (A1), (A2), and (A4). By (C1), axioms (A1) and (A2) imply that $\phi(0) = 0$, which allows one to rewrite (5.13) as

$$\overline{\rho}(X) = \sup_{\eta > 0} \frac{\phi(\eta X + 0) - \phi(0)}{\eta} = \phi 0^+(X), \tag{5.14}$$

where the last inequality in (5.14) comes from the definition of the recession function [41; 42]. Note that quantities defined by (5.12) and (5.13) coincide when function ϕ is positive homogeneous, i.e., it is a coherent risk measure itself:

$$\inf_{\eta > 0} \ \eta\,\phi(\eta^{-1}X) = \sup_{\eta > 0} \ \eta\,\phi(\eta^{-1}X) = \phi(X). \tag{5.15}$$

Optimization of functions $\phi(X)$ that satisfy axioms (A1), (A2'), and $(A4)$ has been recently considered by Ruszczyński and Shapiro [40] (the authors called them *convex risk functions*). Hence, (5.12) and (5.13) imply that the positively homogeneous convex function generated by a convex risk function, and the recession function of a convex risk function are also coherent risk measures, certain conditions applied.

Yet, the practical usefulness of representations (5.12) or (5.13) for coherent risk measures seems rather questionable, as (5.12)–(5.13) would generally lead to non-convex programming problems, should X parametrically depend on a decision vector $x \in \mathbb{R}^n$. On the contrary, representation (5.1) yields a nice convex optimization problem, provided that X is convex in x.

5.2.2 Deviation Measures

Since being introduced in Artzner et al. [32], the axiomatic approach to construction of risk measures has been repeatedly employed by many authors for development of other types of risk measures, tailored to specific preferences and applications (see [37; 38; 39; 40]). In this subsection we consider *deviation measures* as introduced by Rockafellar et al. [37]. Namely, deviation measure is a mapping $\mathscr{D} \colon \mathcal{X} \mapsto [0, +\infty]$ that satisfies

(D1) $\mathscr{D}(X) > 0$ for any non-constant $X \in \mathcal{X}$, whereas $\mathscr{D}(X) = 0$ for constant X,

(D2) $\mathscr{D}(X + Y) \le \mathscr{D}(X) + \mathscr{D}(Y)$ for all $X, Y \in \mathcal{X}$, $0 \le \lambda \le 1$,

(D3) $\mathscr{D}(\lambda X) = \lambda \mathscr{D}(X)$ for all $X \in \mathcal{X}$, $\lambda > 0$,

(D4) $\mathscr{D}(X + a) = \mathscr{D}(X)$ for all $X \in \mathcal{X}$, $a \in \mathbb{R}$.

Again, from axioms (D1) and (D2) follows the convexity of $\mathscr{D}(X)$. In [37] it was shown that deviation measures that further satisfy

(D5) $\mathscr{D}(X) \le \mathsf{E}X - \text{ess.inf}\, X$ for all $X \in \mathcal{X}$,

are characterized by the one-to-one correspondence

$$\mathscr{D}(X) = \mathscr{R}(X - \mathsf{E}X) \tag{5.16}$$

with *expectation-bounded* coherent risk measures, i.e., risk measures that satisfy (A1)–(A4) and an additional requirement

(A5) $\mathscr{R}(X) > \mathsf{E}(-X)$ for all non-constant $X \in \mathcal{X}$, whereas $\mathscr{R}(X) = \mathsf{E}(-X)$ for all constant X.

Using this result, it is easy to provide analog of formula (5.1) for deviation measures.

Theorem 3. *Let function* $\phi \colon \mathcal{X} \mapsto \mathbb{R}$ *satisfy axioms (A1)–(A3), and be a lsc function such that* $\phi(X) > -\mathsf{E}X$ *for all* $X \ne 0$. *Then the optimal value of the stochastic programming problem*

$$\mathscr{D}(X) = \mathsf{E}X + \inf_{\eta} \{\eta + \phi(X + \eta)\} \qquad (5.17)$$

is a deviation measure, and the infimum is attained for all X, so $\inf_{\eta}$ in (5.17) may be replaced by $\min_{\eta \in \mathbb{R}}$.

Proof. Since formula (5.17) differs from (5.1) by the constant summand $\mathsf{E}X$, we only have to verify that $\mathscr{R}(X) = \inf_{\eta} \{\eta + \phi(X + \eta)\}$ satisfies (A5). As $\phi(X) > -\mathsf{E}X$ for all $X \neq 0$, we have that $\phi(X + \eta_X) > -\mathsf{E}(X + \eta_X)$ for all non-constant $X \in \mathcal{X}$, where $\eta_X \in \arg\min_{\eta}\{\eta + \phi(X + \eta)\}$. From the last inequality it follows that $\eta_X + \phi(X + \eta_X) > -\mathsf{E}X$, or $\mathscr{R}(X) > \mathsf{E}(-X)$ for all non-constant $X \in \mathcal{X}$. Thus, $\mathscr{D}(X) > 0$ for all non-constant X. For $a \in \mathbb{R}$, $\inf_{\eta} \{\eta + \phi(a + \eta)\} = -a$, whence $\mathscr{D}(a) = 0$. $\qquad\qquad \square$

Given the close relation between deviation measures and coherent risk measures, it is straightforward to apply the results developed in the previous subsection to deviation measures.

5.2.3 A Class of SSD-consistent Risk Measures

As it has been mentioned in the Introduction, significant attention has been devoted in the literature to development of risk models and measures that are consistent with the utility theory of von Neumann and Morgenstern [1], which represents one of the cornerstones of the decision-making science.

The vNM theory argues that when the preference relation $\succeq$ of the decision-maker satisfy certain axioms (completeness, transitivity, continuity, and independence), there exists a function $u \colon \mathbb{R} \mapsto \mathbb{R}$, called the Bernoulli utility index, such that an outcome X is preferred to outcome Y ("$X \succeq Y$") if and only if $\mathsf{E}[u(X)] \geq \mathsf{E}[u(Y)]$. If function u is increasing and concave, the corresponding preference is said to be risk averse. Rothschild and Stiglitz [28] have bridged the vNM utility theory with the concept of second-order stochastic dominance, by showing that X dominating Y by the second-order stochastic dominance, $X \succeq_{SSD} Y$, is equivalent to relation $\mathsf{E}[u(X)] \geq \mathsf{E}[u(Y)]$ holding true for all concave increasing functions u, where the inequality is strict for at least one such u. Recall that random outcome X dominates outcome Y by the second order stochastic dominance if

$$\int_{-\infty}^{z} \mathsf{P}[X \leq t]\, dt \leq \int_{-\infty}^{z} \mathsf{P}[Y \leq t]\, dt \quad \text{for all } z \in \mathbb{R}.$$

Since coherent risk measures are generally inconsistent with the second-order stochastic dominance (see an explicit example in 45), it is of interest to introduce risk measures that comply with this property. To this end, we replace the monotonicity axiom (A1) in the definition of coherent risk measures by the requirement of *SSD isotonicity* [35]:

$$X \succeq_{SSD} Y \quad \Rightarrow \quad \mathscr{R}(X) \leq \mathscr{R}(Y).$$

Namely, we consider risk measures $\mathscr{R}: \mathcal{X} \mapsto \overline{\mathbb{R}}$ that satisfy the following set of axioms:

(A1′) *SSD isotonicity:* $X \succeq_{SSD} Y \;\Rightarrow\; \mathscr{R}(X) \le \mathscr{R}(Y)$ for $X, Y \in \mathcal{X}$,

(A2′) *convexity:*

$$\mathscr{R}(\lambda X + (1-\lambda)Y) \le \lambda\mathscr{R}(X) + (1-\lambda)\mathscr{R}(Y) \quad X, Y \in \mathcal{X}, \; 0 \le \lambda \le 1,$$

(A3) *positive homogeneity:* $\mathscr{R}(\lambda X) = \lambda\mathscr{R}(X), \quad X \in \mathcal{X}, \; \lambda > 0,$

(A4) *translation invariance:* $\mathscr{R}(X + a) = \mathscr{R}(X) - a, \quad X \in \mathcal{X}, \; a \in \mathbb{R}.$

Again, it is possible to develop an analog of formula (5.1), which would allow for construction of risk measures with the above properties using functions that comply with (A1′), (A2′), and (A3).

Theorem 4. *Let function* $\phi: \mathcal{X} \mapsto \mathbb{R}$ *satisfy axioms (A1′), (A2′), and (A3), and be a lsc function such that* $\phi(\eta) > -\eta$ *for all real* $\eta \ne 0$. *Then the optimal value of the stochastic programming problem*

$$\rho(X) = \min_{\eta \in \mathbb{R}} \; \eta + \phi(X + \eta) \tag{5.18}$$

exists and is a proper function that satisfies (A1′), (A2′), (A3), and (A4).

Proof. The proof of existence and all properties except (A1′) is identical to that of Theorem 1. Property (A1′) follows elementarily: if $X \succeq_{SSD} Y$, then $X + c \succeq_{SSD} Y + c$, and consequently, $\phi(X + c) \le \phi(Y + c)$ for $c \in \mathbb{R}$, whence

$$\begin{aligned}\rho(X) &= \eta_X + \phi(X + \eta_X) \\ &\le \eta_Y + \phi(X + \eta_Y) \\ &\le \eta_Y + \phi(Y + \eta_Y) \\ &= \rho(Y),\end{aligned}$$

where, as usual, $\eta_Z \in \arg\min_\eta \eta + \phi(Z + \eta) \subset \mathbb{R}$. ☐

Functions ϕ that satisfy the conditions of Theorem 4 can be easily constructed in the scope of the presented approach. Let $\phi(X) = \mathsf{E}[u(X)]$, where $u: \mathbb{R} \mapsto \mathbb{R}$ is a convex, positively homogeneous, decreasing function such that $u(\eta) > -\eta$ for all $\eta \ne 0$. Obviously, function $\phi(X)$ defined in this way satisfies the conditions of Theorem 4. Since $u(\cdot)$ is convex and decreasing, one has that $-\mathsf{E}[u(X)] \ge -\mathsf{E}[u(Y)]$, and, consequently, $\phi(X) \le \phi(Y)$ whenever $X \succeq_{SSD} Y$. It is easy to see that such a function ϕ has the form

$$\phi(X) = \alpha\mathsf{E}(X)^- - \beta\mathsf{E}(X)^+, \quad \alpha \in (1, +\infty), \; \beta \in [0, 1).$$

Thus, in accordance to Theorems 1 and 4, the coherent risk measure $\mathscr{R}_{\alpha,\beta}$ (5.5) is also consistent with the second-order stochastic dominance. In other words, by employing

$\mathcal{R}_{\alpha,\beta}$ in stochastic programming problem one introduces risk preferences that are consistent with both concepts of coherence and second-order stochastic dominance. A special case of (5.5) is the Conditional Value-at-Risk, which is known to be consistent with the second-order stochastic dominance [35].

5.2.4 Incorporation of Risk Measures in Multistage Stochastic Programming Problems

We summarize the foregoing discussion by commenting on the use of the developed representations for risk measures in multistage stochastic programming problems. As an illustration, consider a standard two-stage linear stochastic programming problem (see, e.g., Birge and Louveaux [46]):

$$\min\quad c^T x + \mathsf{E}\left[\min_y\; q(w)^T y(w)\right] \qquad (5.19)$$
$$\text{s.t.}\quad Ax = b,$$
$$T(w)\, x + W(w)\, y(w) = h(w),\quad w \in \Omega,$$
$$x \geq 0,\quad y(w) \geq 0,$$

where the second constraint is assumed to hold almost surely. In formulation (5.19), the first-stage decision x is generated *before* the realization of stochastic parameter $\xi(w)$, comprising the elements of $q(w)$, $T(w)$, $W(w)$, and $h(w)$, can be observed. Hence, such an x must take into account the expected cost of the second-stage decision, or recourse action $y(w)$. The second-stage decision $y(w)$, generated *after* observing the realization of the random parameter $\xi(w)$, depends, in turn, on both the first-stage decision x and realization of $\xi(w)$.

Now, consider the case when one is interested in a first-stage decision x that hedges against the *risk* of the second-stage action, not its expected cost. How can one implement risk-averse preferences in a two-stage stochastic programming problem, if the risk of the second-stage[3] decision is of interest? Formally, we can write a two-stage stochastic linear programming problem with risk recourse in the deterministic equivalent form as

$$\min\left\{c^T x + \mathscr{Q}(x) \mid Ax = b,\; x \geq 0\right\},\; \text{where}\; \mathscr{Q}(x) = \mathscr{R}\Big[Q\big(x, \xi(w)\big)\Big], \quad (5.20)$$

and

$$Q\big(x, \xi(w)\big)$$
$$= \min\left\{q(w)^T y(w) \mid T(w)\, x + W(w)\, y(w) = h(w),\; y(w) \geq 0\right\}, \qquad (5.21)$$

where $\mathscr{R}$ is a risk measure. The difficulty here, comparing to static stochastic programming problems, is that the risk measure $\mathscr{R}$ would have as an argument the

---

[3] Since the second-stage decision is a function of the first-stage decision, one may equally conceive that the risk is associated with the first-stage decision.

second-stage value function $Q(x, \xi(\omega))$, a random quantity that itself is a solution of optimization problem (5.21). The developed above representations of the form (5.1) circumvent this difficulty by exploiting the specific "nested" structure of multistage stochastic programming problems. For instance, the introduced above two-stage linear stochastic programming problem with risk recourse (5.20)–(5.21) can be cased using representation (5.1) for risk measure $\mathscr{R}$ as

$$\min \quad c^T x + \eta + \phi\Big(q(\omega)^T y(\omega) + \eta\Big) \tag{5.22}$$

$$\text{s.t.} \quad Ax = b,$$
$$T(\omega)\, x + W(\omega)\, y(\omega) = h(\omega), \quad \omega \in \Omega,$$
$$x \geq 0, \quad y(\omega) \geq 0, \quad \eta \in \mathbb{R}.$$

Due to convexity and positive homogeneity of the considered classes of risk measures, the risk recourse function $\mathscr{Q}(x)$ (5.20) is also convex and positive homogeneous. Evidently, function ϕ in the objective of (5.22) is also convex and positively homogeneous, which opens possibilities for adapting the existing solution algorithms (e.g., the L-shaped method) for solving stochastic programming problems of form (5.22). Observe also that variable η in (5.22) can be regarded as a first-stage decision variable; its properties as a solution of (5.22) are discussed in Theorem 2.

5.3 An example: Two-stage Stochastic Weapon-target Assignment Problem

In this section we present an example of two-stage decision-making problem where risk of the second-stage decision (*recourse* action) has to be taken into account while generating the first-stage decision. Namely, we consider stochastic version of a non-linear knapsack problem with a convex separable objective

$$\min \quad \sum_{k=1}^{K} V_k q_k^{x_k} \tag{5.23}$$

$$\text{s.t.} \quad \sum_{k=1}^{K} n_k x_k \leq M,$$
$$x_k \in \mathbb{Z}_+, \quad k = 1, \dots, K.$$

The knapsack-type formulations, linear and non-linear, routinely arise in various resource allocation problems and applications, including production planning, inventory management, capital budgeting, etc. (for a review of methods and applications for non-linear knapsack problems, see, e.g., [47; 48] and references therein). Similarly to the classical linear knapsack instances, non-linear knapsack problems of type (5.23) admit polynomial ε-approximation schemes [48].

The considered here particular form (5.23) of the non-linear knapsack problem is inherent to the context of the deterministic weapon-target assignment (WTA) problem [49; 50; 51]. In this framework, M stands for the number of (uniform) weapons to be allocated against targets in K categories, with the objective to maximize damage to these targets. Each category $k = 1, \ldots, K$, contains n_k targets with value V_k and persistence q_k (in other words, the probability to neutralize a target in category k using one weapon is equal to $1 - q_k$). The decision variable x_k denotes the number of weapons assigned to a target in category $k = 1, \ldots, K$.

In many situations it suffices to determine the total number of weapons $n_k x_k$ allocated to each category k of targets. If this is the case, solution of integer programming (IP) problem (5.23) can be approximated by solution of a linear programming (LP) problem. Note that the objective of (5.23) is separable, i.e., it is a linear combination of univariate functions that are furthermore convex, hence one can replace functions $\varphi_k(x_k) = (q_k)^{x_k}$ in the objective of (5.23) by piecewise-linear functions $\ell_k(x_k)$ such that $\ell_k(i) = \varphi_k(i)$, $i = 0, \ldots, M$:

$$\ell_k(x_k) = \max_{m=0,\ldots,M-1} \left\{ q_k^m \left[(1 - q_k)(m - x_k) + 1 \right] \right\}. \tag{5.24}$$

The resulting LP problem reads as

$$\min \quad \sum_{k=1}^{K} V_k u_k \tag{5.25a}$$

$$\text{s.t.} \quad \sum_{k=1}^{K} n_k x_k \leq M, \tag{5.25b}$$

$$u_k \geq q_k^m \left[(1 - q_k)(m - x_k) + 1 \right], \tag{5.25c}$$

$$k = 1, \ldots, K, \quad m = 0, \ldots, M - 1,$$

$$x_k \geq 0, \quad u_k \geq 0, \quad k = 1, \ldots, K. \tag{5.25d}$$

Proposition 1. *There exists an optimal solution* $\{u_k^*, x_k^*\} \in \mathbb{R}^{2K}$ *of* (5.25) *where at most one variable* x_k^* *is non-integer-valued. If, at optimality, all variables* x_k^* *are integer-valued, then* $\{x_k^*\} \in \mathbb{Z}_+^K$ *is also optimal for* (5.23).

Proof. According to the basics of linear programming, an optimal solution of LP problem with bounded feasible set

$$\min_{z \in \mathbb{R}^n} \left\{ c^T z \mid A z \leq 0 \right\},$$

is achieved in extreme points (vertices) of the feasible region $A z \leq 0$. To be a vertex of the polyhedral $A z \leq 0$, point $z^* \in \mathbb{R}^n$ has to satisfy n linearly independent equations of system $A z = 0$ (indeed, a point in $\mathbb{R}^n$ can be defined as the intersection of n different (linearly independent) hyperplanes).

Without loss of generality, we assume that the feasible set of (5.25) is bounded (we may impose constraints $u_k \leq C$ for some large $C > 0$ without affecting the set of optimal solutions). Then, consider a feasible point $z^* = (u_1, \ldots, u_K, x_1, \ldots, x_K) \in \mathbb{R}^{2K}$ that has J non-integer components $x_{k_1}, \ldots, x_{k_J}$, where $2 \leq J \leq K$. This point may satisfy at most $2K - J + 1$ linearly independent equalities that define the boundary of the feasible set (5.25b)–(5.25d). Indeed, each of $K - J$ integer-valued components $x_{i_0} = m_0 \in \{1, 2, \ldots, M - 1\}$, and the corresponding u_{i_0} may satisfy two equalities (5.25c)

$$u_{i_0} = q_{i_0}^{m_0} \left[(1 - q_{i_0})(m_0 - x_{i_0}) + 1 \right]$$

and

$$u_{i_0} = q_{i_0}^{m_0 - 1} \left[(1 - q_{i_0})(m_0 - 1 - x_{i_0}) + 1 \right],$$

(if $m_0 = 0$, then x_{i_0} and u_{i_0} satisfy one equality $x_{i_0} = 0$ from set (5.25d) and one equality $u_{i_0} = 1$ from set (5.25c); the case $m_0 = M$ is treated similarly). Each non-integer x_{k_j} and the corresponding u_{k_j} may satisfy at most one equality (5.25c)

$$u_{k_j} = q_{k_j}^m \left[(1 - q_{k_j})(m - x_{k_j}) + 1 \right].$$

Additionally, point z^* may satisfy equality

$$\sum_{k=1}^{K} n_k x_k = M. \tag{5.26}$$

Thus, a feasible point with $2 \leq J \leq K$ non-integer components $x_{k_1}, \ldots, x_{k_J}$ may satisfy at most $2(K - J) + J + 1 = 2K - J + 1 < 2K$ different equalities that define the boundary of the feasible region (5.25b)–(5.25d). Hence, such a point cannot be a vertex of the feasible set of (5.25), therefore problem (5.25) has an optimal solution with at most one non-integer-valued variable x_k^*. Observe that if all x_k^* are integer, then along with the corresponding variables u_k^*, they satisfy $2K$ equalities (5.25c), and, possibly, (5.25d). Then, if equality $\sum_{k=1}^{K} n_k x_k^* = M$ holds, it is a linear combination of K equalities $x_k^* = m_k$, where $m_k \in \mathbb{Z}_+$. Hence, point $z_k^* \in \mathbb{R}^{2K}$ with all integer-valued components x_k^* satisfies exactly $2K$ different equalities that define the boundary of the feasible region of (5.25).

If all x_k are integer-valued at optimality, they also deliver optimum to (5.23). Indeed, by construction of (5.25), for all $\{x_k\}$ feasible to (5.25) one has

$$\sum_{k=1}^{K} V_k u_k^* = \sum_{k=1}^{K} V_k (q_k)^{x_k^*} \leq \sum_{k=1}^{K} V_k \ell_k (x_k).$$

Obviously, the feasible set of (5.23) is contained in that of (5.25). Hence, the last inequality holds for all integer-valued x_k' feasible to (5.23), and, by definition (5.24) of functions $\ell_k(\cdot)$,

$$\sum_{k=1}^{K} V_k \ell_k(x_k') = \sum_{k=1}^{K} V_k (q_k)^{x_k'},$$

which proves that $\{x_k^*\} \in \mathbb{Z}_+^K$ is optimal for (5.23). $\qquad\qquad\qquad\square$

Corollary 1. Problem (5.25) has an optimal solution that allocates integer number $n_k x_k^*$ of weapons to each category k of targets, $k = 1, \ldots, K$.

Depending on the application, any of the parameters in the non-linear knapsack problem (5.23) may be deemed uncertain, yet the most important source of uncertainty is associated with parameters n_k, since it affects the feasibility of the knapsack constraint. In our case, this translates into the uncertainty in the number of targets to allocate the resources against. Below we present a formulation of a two-stage stochastic non-linear knapsack problem corresponding to a weapon-target assignment where the number of targets is unknown beforehand.

Adopting the discussed deterministic setup (5.23), we assume that each category $k = 1, \ldots, K$, contains n_k of known, or detected targets, and an uncertain number $\xi_k(\omega)$ of targets that may be detected at some time in future (here ω is a random element from some probability space). To allocate the munitions to both the detected targets and targets that may be detected in future, we define decision variables x_k and $y_k(\omega)$ as, respectively, the number of weapons assigned to each of n_k detected targets and $\xi_k(\omega)$ undetected targets in category k. In the stochastic programming terminology, x_k and $y_k(\omega)$ represent the first- and second-stage decision variables. The first-stage decision is made *before* the realization of uncertainties (in our case, the number of second-stage targets $\xi_k(\omega)$) is observed, whereas the second-stage decision, or the recourse action, is made *after* observation of the realization of uncertainties. Ordinarily, this would lead to a classical two-stage stochastic programming formulation of the form (5.19), where the first-stage decision takes into account the expected cost of the recourse action. In the described context, however, it may be more desirable to have a first-stage decision that hedges against the *risk* of the second-stage decision, not its expected value:

$$\min \quad \sum_{k=1}^{K} V_k q_k^{x_k} + \mathscr{R}\left[\sum_{k=1}^{K} V_k q_k^{y_k(\omega)}\right] \qquad\qquad (5.27)$$

$$\text{s.t.} \quad \sum_{k=1}^{K} n_k x_k \le M,$$

$$\sum_{k=1}^{K}\left(n_k x_k + \xi_k(\omega) y_k(\omega)\right) \le M, \quad \omega \in \Omega,$$

$$x_k, \ y_k(\omega) \in \mathbb{Z}_+, \quad k = 1, \ldots, K.$$

In other words, we are interested in an assignment where the first-stage decision is weighted against a conservative estimation of the cost of the recourse action rendered by the risk measure $\mathscr{R}$.

In general, the choice of a risk measure must be tailored to the application at hand. In our case, the Conditional Value-at-Risk measure, which associates the risk with the average dissatisfaction across several worst-case scenarios, represents a reasonable selection (see section 5.2.1). Observe that higher values of the term in brackets in (5.27) are less desirable, and hence constitute more risk. To account for this fact, we recast the expression for Conditional Value-at-Risk in the form:

$$\mathscr{R}_\alpha(X) = \min_{\eta \in \mathbb{R}} \eta + (1 - \alpha)^{-1} E(X - \eta)^+, \qquad (5.28)$$

where the parameter α allows for adjusting the degree of risk-aversity. When $\alpha = 0$, $\mathscr{R}_\alpha(X)$ in (5.28) reduces to the expectation operator, $\mathscr{R}_0(X) = EX$, yielding risk-neutral preferences; $\alpha \to 1$ corresponds to hedging against the worst-case outcome, i.e., it represents extremely risk-averse preferences: $\mathscr{R}_{\alpha \to 1}(X) = \text{ess.sup } X$.

Assuming that the uncertainty in the number $\xi_k(\omega)$ of the second-stage targets can be described by a set of S scenarios ξ_{ks}, $s = 1, \ldots, S$, and employing Proposition 1 we can write a formulation for the two-stage stochastic programming problem with risk recourse (5.27) similar to (5.25):

$$\min \quad \sum_{k=1}^{K} V_k u_k + \eta + (1 - \alpha)^{-1} \sum_{s=1}^{S} p_s w_s \qquad (5.29)$$

$$\text{s.t.} \quad \sum_{k=1}^{K} n_k x_k \le M',$$

$$u_k \ge q_k^m \big[(1 - q_k)(m - x_k) + 1\big], \quad \forall\, k, \;\; m = 0, \ldots, M - 1,$$

$$\sum_{k=1}^{K} \xi_{ks} y_{ks} \le M - M', \quad s = 1, \ldots, S,$$

$$v_{ks} \ge q_k^m \big[(1 - q_k)(m - y_{ks}) + 1\big], \quad \forall\, k, \; s, \;\; m = 0, \ldots, M - 1,$$

$$w_s \ge \sum_{k=1}^{K} V_k v_{ks} - \eta, \quad s = 1, \ldots, S,$$

$$x_k, u_k, y_{ks}, v_{ks}, w_s \ge 0, \quad k = 1, \ldots, K, \quad s = 1, \ldots, S, \quad M' \in \mathbb{Z}_+.$$

Because (5.29) has only one integer variable, it is solvable in polynomial time, therefore (5.29) may be regarded as essentially LP problem. Similarly to (5.25) it can be shown that (5.29) has an optimal solution with at most one non-integer-valued variable x_{k_0} and one non-integer-valued variable $y_{k_s s}$ for each $s = 1, \ldots, S$. This also entails that the cumulative assignments $n_k x_k$, $\xi_{ks} y_{ks}$, $s = 1, \ldots, S$ of weapons to targets in category k are integer-valued.

For illustrative purposes, problem (5.29) has been solved in the following setup. There are $M = 24$ weapons to be allocated to targets in two categories ($K = 2$). The first category contains low-priority targets that are also easier to destroy ($V_1 = 1$, $q_1 = 0.20$), whereas the second category contains harder targets of higher priority:

$V_2 = 3$, $q_2 = 0.35$. There are $n_1 = 3$ and $n_2 = 1$ of detected targets in each category; the uncertain numbers $\xi_{1,2}(\omega)$ of undetected targets in both categories vary from 1 to 5 and are described by $S = 10$ scenarios. Problem (5.29) was solved for values of parameter $\alpha = 0, 0.1, 0.2, \ldots, 0.9$, and the optimal objective value is shown in Figure 5.1. The numbers in boxes indicate the first-stage solution, e.g., (6,5) means $n_1x_1 = 6$, $n_2x_2 = 5$ (the optimal solution for $\alpha = 0$). Observe that as the value of α increases (i.e., the preferences introduced by the CVaR risk measure become more risk-averse), the optimal objective value (5.29) also increases. Intuitively, the more risk-averse one is with respect to the cost of the second-stage assignment, the fewer weapons one should assign against the first-stage targets, thereby reducing the second-stage cost.

This trend is clearly manifested by the optimal solution of (5.29). Indeed, the number of munitions allocated to categories 1 and 2 at the first stage is $n_1x_1 = 6$, $n_2x_2 = 5$, i.e., 2 weapons per category 1 target, and 5 weapons for the detected target in category 2. For $\alpha = 0.1, \ldots, 0.7$, the optimal assignment decreases the number of weapons allocated to targets allocated to the first-stage target in category 1 to $n_2x_2 = 4$. As α increases to 0.8 and 0.9, the first-stage assignment reduces further to $n_1x_1 = 5$, $n_2x_2 = 4$ (see Figure 5.1).

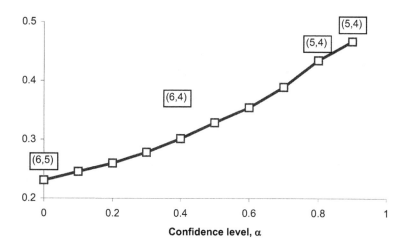

Fig. 5.1. Optimal solution of the two-stage stochastic WTA problem with risk recourse (5.29).

Finally, we comment on how closely the optimal solution of (5.29) approximates the solution of the original IP problem (5.27). By solving the IP problem (5.27) with the above parameters, we observed that the optimal solution of (5.29) coincided with that of (5.27) for $\alpha = 0, 0.1, \ldots, 0.7$. When $\alpha = 0.8$ and 0.9, 3 weapons are assigned to category 1 (one weapon per target in the category), and 4 weapons are assigned to the target in category 2 at the first stage.

Conclusions

In this paper, we have considered the issue of modeling of risk-averse preferences in (multi-stage) stochastic programming problems using risk measures. We utilized the axiomatic approach to construction of coherent risk measures and deviation measures in order to develop simple representations for these risk measures via solutions of specially designed stochastic programming problems. It has been shown that these types of representations allow for construction of risk coherent measures that are consistent with the second order stochastic dominance. As an example, we considered a two-stage stochastic Weapon-Target Assignment problem, where a coherent risk measure is used to determine the risk of the second-stage (recourse) action.

References

[1] von Neumann, J. and Morgenstern, O., *Theory of Games and Economic Behavior*, Princeton University Press, Princeton, NJ, 1953rd ed., 1944.

[2] Fishburn, P. C., *Utility Theory for Decision-Making*, Wiley, New York, 1970.

[3] Fishburn, P. C., *Non-Linear Preference and Utility Theory*, Johns Hopkins University Press, Baltimore, 1988.

[4] Karni, E. and Schmeidler, D., "Utility Theory with Uncertainty," *Handbook of Mathematical Economics*, edited by Hildenbrand and Sonnenschein, Vol. IV, North-Holland, Amsterdam, 1991.

[5] Markowitz, H. M., "Portfolio Selection," *Journal of Finance*, Vol. 7, No. 1, 1952, pp. 77–91.

[6] Markowitz, H. M., *Portfolio Selection*, Wiley and Sons, New York, 1st ed., 1959.

[7] Ogryczak, W. and Ruszczyński, A., "From stochastic dominance to mean-risk models: Semideviations as risk measures," *European Journal of Operational Research*, Vol. 116, 1999, pp. 33–50.

[8] Ogryczak, W. and Ruszczyński, A., "On consistency of stochastic dominance and mean-semideviation models," *Mathematical Programming*, Vol. 89, 2001, pp. 217—232.

[9] Ogryczak, W. and Ruszczyński, A., "Dual stochastic dominance and related mean-risk models," *SIAM Journal on Optimization*, Vol. 13, No. 1, 2002, pp. 60–78.

[10] Bawa, V. S., "Optimal Rules For Ordering Uncertain Prospects," *Review of Financial Studies*, Vol. 2, No. 1, 1975, pp. 95–121.

[11] Dembo, R. S. and Rosen, D., "The Practice of Portfolio Replication: A Practical Overview of Forward and Inverse Problems," *Annals of Operations Research*, Vol. 85, 1999, pp. 267–284.

[12] Testuri, C. and Uryasev, S., "On Relation Between Expected Regret and Conditional Value-at-Risk," 2000, Research Report 2000-9. ISE Dept., Univ. of Florida.

[13] van der Vlerk, M. H., "Integrated Chance Constraints in an ALM Model for Pension Funds," 2003, Working paper.

[14] JP Morgan, *Riskmetrics*, JP Morgan, New York, 1994.

[15] Jorion, P., *Value at Risk: The New Benchmark for Controlling Market Risk*, McGraw-Hill, 1997.

[16] Duffie, D. and Pan, J., "An Overview of Value-at-Risk," *Journal of Derivatives*, Vol. 4, 1997, pp. 7–49.

[17] Larsen, N., Mausser, H., and Uryasev, S., "Algorithms for Optimization of Value-at-Risk," *Financial Engineering, e-Commerce and Supply Chain*, edited by P. Pardalos and V. K. Tsitsiringos, Kluwer Academic Publishers, 2002, pp. 129–157.

[18] Rockafellar, R. T. and Uryasev, S., "Optimization of Conditional Value-at-Risk," *Journal of Risk*, Vol. 2, 2000, pp. 21–41.

[19] Rockafellar, R. T. and Uryasev, S., "Conditional Value-at-Risk for General Loss Distributions," *Journal of Banking and Finance*, Vol. 26, No. 7, 2002, pp. 1443–1471.

[20] Krokhmal, P., Palmquist, J., and Uryasev, S., "Portfolio Optimization with Conditional Value-At-Risk Objective and Constraints," *Journal of Risk*, Vol. 4, No. 2, 2002, pp. 43–68.

[21] Chekhlov, A., Uryasev, S., and Zabarankin, M., "Portfolio Optimization with Drawdown Constraints," Tech. Rep. 2000-5, University of Florida, ISE Dept., 2000.

[22] Krokhmal, P., Uryasev, S., and Zrazhevsky, G., "Risk Management for Hedge Fund Portfolios: A Comparative Analysis of Linear Rebalancing Strategies," *Journal of Alternative Investments*, Vol. 5, No. 1, 2002, pp. 10–29.

[23] Acerbi, C., Nordio, C., and Sirtori, C., "Expected shortfall as a tool for financial risk management," 2001, Working paper.

[24] Acerbi, C. and Tasche, D., "On the coherence of expected shortfall," *Journal of Banking and Finance*, Vol. 26, No. 7, 2002, pp. 1487–1503.

[25] Young, M. R., "A Minimax Portfolio Selection Rule with Linear Programming Solution," *Management Science*, Vol. 44, No. 5, 1998, pp. 673–683.

[26] Kroll, Y., Levy, H., and Markowitz, H. M., "Mean-Variance Versus Direct Utility Maximization," *Journal of Finance*, Vol. 39, No. 1, 1984, pp. 47–61.

[27] Fishburn, P. C., "Stochastic Dominance and Moments of Distributions," *Mathematics of Operations Research*, Vol. 5, 1964, pp. 94–100.

[28] Rothschild, M. and Stiglitz, J., "Increasing risk I: a definition," *Journal of Economic Theory*, Vol. 2, No. 3, 1970, pp. 225–243.

[29] Levy, H., *Stochastic Dominance*, Kluwer Academic Publishers, Boston-Dodrecht-London, 1998.

[30] Ruszczyński, A. and Dentcheva, D., "Optimization with Stochastic Dominance Constraints," *SIAM Journal on Optimization*, Vol. 14, No. 2, 2003, pp. 548–566.

[31] Artzner, P., Delbaen, F., Eber, J.-M., and Heath, D., "Thinking Coherently," *Risk*, Vol. 10, 1997, pp. 68–71.

[32] Artzner, P., Delbaen, F., Eber, J.-M., and Heath, D., "Coherent Measures of Risk," *Mathematical Finance*, Vol. 9, No. 3, 1999, pp. 203–228.

[33] Artzner, P., Delbaen, F., Eber, J.-M., Heath, D., and Ku, H., "Multiperiod Risk and Coherent Multiperiod Risk Measurement," 2001, Working paper.

[34] Delbaen, F., "Coherent risk measures on general probability spaces," 2000, Preprint.

[35] Pflug, G., "Some Remarks on the Value-at-Risk and the Conditional Value-at-Risk," *Probabilistic Constrained Optimization: Methodology and Applications*, edited by S. Uryasev, Kluwer Academic Publishers, 2000.

[36] Fischer, T., "Risk capital allocation by coherent risk measures based on one-sided moments," *Insurance: Mathematics and Economics*, Vol. 32, No. 1, 2003, pp. 135–146.

[37] Rockafellar, R. T., Uryasev, S., and Zabarankin, M., "Deviation Measures in Risk Analysis and Optimization," Tech. Rep. 2002–7, ISE Dept., University of Florida, 2002.

[38] Rockafellar, R. T., Uryasev, S., and Zabarankin, M., "Generalized Deviations in Risk Analysis," Tech. Rep. 2004-4, ISE Dept., University of Florida, 2004.

[39] Acerbi, C., "Spectral measures of risk: A coherent representation of subjective risk aversion," *Journal of Banking and Finance*, Vol. 26, No. 7, 2002, pp. 1487–1503.

[40] Ruszczyński, A. and Shapiro, A., "Optimization of Convex Risk Functions," *Working paper*, 2004.

[41] Rockafellar, R. T., *Convex Analysis*, Vol. 28 of *Princeton Mathematics*, Princeton University Press, 1970.

[42] Zălinescu, C., *Convex Analysis in General Vector Spaces*, World Scientific, Singapore, 2002.

[43] Ruszczyński, A. and Shapiro, A., *Stochastic Programming*, Vol. 10 of *Handbooks in Operations Research and Management Science*, Elsevier, 2003.

[44] Szegö, G., "Measures of Risk," *Journal of Banking and Finance*, Vol. 26, 2002, pp. 1253–1272.

[45] De Giorgi, E., "Reward-Risk Portfolio Selection and Stochastic Dominance," *Working paper*, 2002.

[46] Birge, J. R. and Louveaux, F., *Introduction to Stochastic Programming*, Springer, New York, 1997.

[47] Bretthauer, K. M. and Shetty, B., "The nonlinear knapsack problem – algorithms and applications," *European Journal of Operational Research*, Vol. 138, 2002, pp. 459–472.

[48] Hochbaum, D. S., "A nonlinear Knapsack problem," *Operations Research Letters*, Vol. 17, 1995, pp. 103–110.

[49] Manne, A. S., "A Target Assignment Problem," *Operations Research*, Vol. 6, 1958, pp. 346–351.

[50] denBroeger, G. G., Ellison, R. E., and Emerling, L., "On optimum target assignments," *Operations Research*, Vol. 7, 1959, pp. 322–326.

[51] Murphey, R., *An Approximate Algorithm For A Stochastic Weapon Target Assignment Problem*, 1999.

6

Shape Optimization of Electrodes for Piezoelectric Actuators

Andrew J. Kurdila, Weijian Wang, Yunfei Feng, and Richard J. Prazenica

Department of Mechanical and Aerospace Engineering
University of Florida
Gainesville, FL 32611-6250
kurdila@ufl.edu
prazenic@ufl.edu

Summary. This paper describes a method of shape optimization for electrodes used in piezo-electric actuators. This work is motivated by recent attempts to design piezoceramic bimorph actuators for the optimal control of two-dimensional flow fields. It has been observed experimentally that uniformly-electroded bimorphs induce undesirable three-dimensional effects into the flow. These effects complicate the optimal flow control problem considerably. Therefore, the objective is to design an electrode that is shaped in such a manner as to minimize three-dimensional effects while providing maximum actuator deflection. This paper considers electrode shape optimization for a typical two-layer composite transducer (unimorph) structure, which takes the form of a small, rigidly-clamped flap. The strong and weak forms of the governing equations of motion are derived for this system, which can be approximated as a thin plate. The finite element method is used to determine the dynamic flap response for a given electrode shape and voltage input. The optimization problem entails determining the electrode shape that, for a prescribed voltage input, minimizes the three-dimensional distortion at the trailing edge of the flap. Constraints are enforced that guarantee that the electrode shape is relatively smooth and that the electrode coverage is large enough to provide sufficient flap deflection. Theoretical results are presented that prove the existence of a solution when the shape optimization problem is formulated in this manner. Numerical results are given that validate the theory and illustrate the qualitative behavior of the shape optimization method. For the static case, the results are compared with those obtained in previous work using an ad hoc optimization algorithm that is not rigorously consistent with the shape optimization theory. Results are also given for the dynamic case in which the time-dependent response to a prescribed voltage input is considered.

Key words: piezoceramics, shape optimization, piezoelectrics, active composites

6.1 Introduction

Piezoceramic actuators have been recently employed in flow control applications. The design and synthesis of optimal flow control methodologies is greatly simplified if the flow responds in a two-dimensional regime. As noted in the previous companion paper [5], it has been observed experimentally that uniformly-electroded piezoceramic actuators can induce undesirable three-dimensional effects into the flow at various frequencies of optimization. This phenomenon has motivated recent research in the development of electrode shape optimization algorithms. The objective of this research has been to design electrode shapes that minimize three-dimensional distortions in piezoceramic laminates while achieving deflections that are large enough for effective flow control. In addition, continuous electrode shapes with smooth boundaries are desired for ease of fabrication.

Previous companion papers provide a rigorous formulation of the shape optimization problem [5; 4]. In particular, theoretical results are reviewed that guarantee the existence of a solution when the problem is posed in such a manner. In that paper, an ad hoc shape optimization algorithm is applied which, strictly speaking, is not consistent with the theoretical development. This approach entails optimizing a cost function that maximizes tip deflection while enforcing a constraint that limits the three-dimensional distortion. Numerical results were presented for the static case, which considers the static plate deflection under a fixed voltage. In low-dimensional parameterization cases, corresponding to coarse electrode boundaries, smooth boundaries were obtained using spline curve-fitting. As the number of design variables was increased, however, highly-oscillatory boundary solutions were obtained. This phenomenon is well-documented in the shape optimization literature and leads to shapes that are clearly undesirable for electrode fabrication. The oscillations in the optimized boundary shape may be a result of an ill-posed optimization problem, perhaps one that requires regularization, for example. Alternatively, the oscillations might result from a model associated with a mesh that is too coarse, so that the oscillation is a numerical artifact. One means for attempting to address the oscillation problem is to employ a denser finite element grid. Unfortunately, this approach is computationally intensive since, in order to achieve a smooth solution, the mesh must be at least five times as dense as the resolution of the boundary design variables.

This paper considers an alternative approach to the shape optimization problem. Specifically, the optimization problem is formulated using a cost function that is rigorously consistent with the theoretical results in [4]. This cost function is designed to minimize the three-dimensional distortion in the plate. In addition, constraints are enforced that ensure a relatively smooth electrode boundary and guarantee that sufficient plate deflection is achieved. Numerical results demonstrate that, in contrast to the ad hoc method, the theoretically-rigorous approach yields smooth boundary solutions in the high-dimensional parameterization case. In addition, the shape optimization theory generalizes to the dynamic case, which studies the dynamic response of the plate to a prescribed voltage input. It is then straightforward to modify the cost function in the shape optimization algorithm to consider the dynamic case.

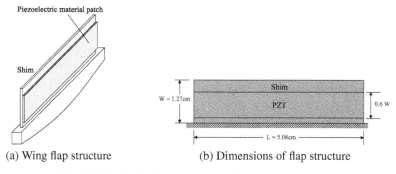

(a) Wing flap structure   (b) Dimensions of flap structure

Fig. 6.1. Model of piezoelectric wing flap structure.

This paper begins with a brief review of the physical plate model and the corresponding equations of motion. The relevant shape optimization theory, which is developed in [4], is likewise reviewed. The shape optimization algorithm is described for both the static and dynamic cases. Numerical results are presented to demonstrate the qualitative performance of the algorithms. In particular, static results generated using the theoretically-rigorous approach are compared with those obtained using the ad hoc approach. Numerical results are also presented for the dynamic case, which considers the time history of the transducer response.

6.2 Review of Physical Plate Model

The physical model of the unimorph piezoelectric transducer is depicted in Figure 6.1. The structure consists of a patch of piezoelectric material mounted to a rigidly-clamped metal shim. A voltage is applied across both layers of the structure. The outer layers of the piezoelectric material are covered with electrodes, forming a capacitor for electrostatic transduction. The electrode area is defined in terms of the characteristic function

$$\chi(x, y) = \begin{cases} 1, & \text{if } (x, y) \text{ is in the electrode region,} \\ 0, & \text{otherwise.} \end{cases} \tag{6.1}$$

6.2.1 Equations of Motion

The strong and weak forms of the equations of motion for the unimorph piezoelectric transducer have been derived in a previous paper [5]. Therefore, only the key results are reviewed here. The model is based on classical laminated plate theory [2], Kirchoff-Love thin plate theory, and linear piezoelectricity theory. In addition, we have the following assumptions for the composite transducer structures:

(1) Plane layers that are initially normal to the polarization of the piezoelectric material will remain normal during flexure.

(2) Different plies are perfectly bonded together, so that no slip occurs at ply interfaces [2].

(3) Interlaminar stresses are neglected.

(4) The thickness of the electrode is neglected.

It is further assumed that the material is homogeneous in each layer and that there is no stretch in the mid-plane. Under these conditions, and taking symmetry into account, the piezoelectric stress resultants can be written in the form

$$\mathbf{M} = \mathbf{D}\boldsymbol{\kappa} + \tilde{\mathbf{M}}, \tag{6.2}$$

where $\mathbf{M}$ is a vector of moment stress resultants, $\mathbf{D}$ is the laminate bending stiffness matrix, $\boldsymbol{\kappa}$ is a vector of curvature components, and $\tilde{\mathbf{M}}$ is induced by the piezoelectric effect. The structure of these matrices, which closely resemble the stress resultant equations for classical composites, can be found in [1], for example.

The strong form of the equations of motion can be derived by considering the equilibrium of force and moment components. Alternatively, the equations can be obtained via variational methods such as Hamilton's extended principle. Both derivations have been carried out in [5] and [6] and will not be repeated here. In deriving the strong form equations of motion, it is assumed that each layer of the composite material retains its inherent isotropic or orthotropic properties. The deformation is assumed to be continuous between adjacent layers and the stress distribution is assumed to be linear but not necessarily continuous. Then, the strong form equations of motion take the form

$$\rho h \frac{\partial^2 w}{\partial t^2} = \frac{\partial^2 M_x}{\partial x^2} + 2\frac{\partial^2 M_{xy}}{\partial x \partial y} + \frac{\partial^2 M_y}{\partial y^2}, \tag{6.3}$$

where $w(x, y, t)$ is the plate deflection, ρ is the material density, and h denotes the thickness of the layer. Equation (6.3) can alternatively be expressed as

$$\rho h \frac{\partial^2 w}{\partial t^2} + D_{11}\left(\frac{\partial^4 w}{\partial x^4} + 2\frac{\partial^4 w}{\partial x^2 \partial y^2} + \frac{\partial^4 w}{\partial y^4}\right) = \frac{\partial^2 \tilde{M}_1}{\partial x^2} + \frac{\partial^2 \tilde{M}_1}{\partial y^2} + 2\frac{\partial^2 \tilde{M}_6}{\partial x \partial y}, \tag{6.4}$$

where

$$D_{11} = h_s\left(c^2 + \frac{h_s^2}{12}\right)\frac{E_s}{1 - \nu_s^2} + h_p\left[\left(\frac{h_s}{2} - c\right)^2 + \left(\frac{h_s}{2} - c\right)h_p + \frac{h_p^2}{3}\right]\frac{E_p}{1 - \nu_p^2},$$

$$D_{12} = h_s\left(c^2 + \frac{h_s^2}{12}\right)\frac{E_s \nu_s}{1 - \nu_s^2} + h_p\left[\left(\frac{h_s}{2} - c\right)^2 + \left(\frac{h_s}{2} - c\right)h_p + \frac{h_p^2}{3}\right]\frac{E_p \nu_p}{1 - \nu_p^2},$$

$$D_{66} = h_s\left(c^2 + \frac{h_s^2}{12}\right)\frac{E_s}{1 + \nu_s} + h_p\left[\left(\frac{h_s}{2} - c\right)^2 + \left(\frac{h_s}{2} - c\right)h_p + \frac{h_p^2}{3}\right]\frac{E_p}{1 + \nu_p},$$

and

$$\tilde{M}_1 = -\frac{1}{2}e_{31}v(t)\chi(x,y)\left(h_s + h_p - 2c\right),$$

$$\tilde{M}_6 = -\frac{1}{2}e_{36}v(t)\chi(x,y)\left(h_s + h_p - 2c\right).$$

In these expressions, c represents the position of the neutral plane relative to the centroid of the metal shim, $\{e_{31}, e_{36}\}$ are components of the elastic matrix, and $v(t)$ is the time-dependent voltage input.

The theoretical developments in this paper employ the weak form of the governing equations. The weak form can be derived by multiplying (6.4) by a suitable test function $\phi(x,y)$ and integrating by parts. Using the fact that

$$D_{11} = D_{12} + D_{66},$$

the weak form of the equations of motion can be written as

$$\int_\Omega \rho h \frac{\partial^2 w}{\partial t^2}\,\phi\,dxdy + \int_\Omega \left\{ D_{66}\left(\frac{\partial^2 w}{\partial x^2}\frac{\partial^2 \phi}{\partial x^2} + 2\frac{\partial^2 w}{\partial x\partial y}\frac{\partial^2 \phi}{\partial x\partial y} + \frac{\partial^2 w}{\partial y^2}\frac{\partial^2 \phi}{\partial y^2}\right) \right.$$
$$\left. + D_{12}\left(\frac{\partial^2 w}{\partial x^2} + \frac{\partial^2 w}{\partial y^2}\right)\left(\frac{\partial^2 \phi}{\partial x^2} + \frac{\partial^2 \phi}{\partial y^2}\right) \right\} dxdy \quad (6.5)$$
$$= \int_\Omega \left\{ \tilde{M}_1(t)\left(\frac{\partial^2 \phi}{\partial x^2} + \frac{\partial^2 \phi}{\partial y^2}\right)dxdy + \tilde{M}_6(t)\frac{\partial^2 \phi}{\partial x\partial y} \right\} dxdy + BT.$$

More concisely, the governing equations can be written in a functional form where we seek $(w, \dot{w})$ such that

$$\int_\Omega \rho h \frac{\partial^2 w}{\partial t^2}\,\phi\,dxdy + a(q)(w,\phi) = f(\phi) + BT, \quad (6.6)$$

for all appropriately selected test functions ϕ. In this equation, BT refers to the boundary terms and we have introduced the parametrically-dependent bilinear form $a(q)(\cdot,\cdot)$

$$a(q)(w,\phi) = \int_\Omega \left\{ D_{66}\left(\frac{\partial^2 w}{\partial x^2}\frac{\partial^2 \phi}{\partial x^2} + 2\frac{\partial^2 w}{\partial x\partial y}\frac{\partial^2 \phi}{\partial x\partial y} + \frac{\partial^2 w}{\partial y^2}\frac{\partial^2 \phi}{\partial y^2}\right) \right.$$
$$\left. + D_{12}\left(\frac{\partial^2 w}{\partial x^2} + \frac{\partial^2 w}{\partial y^2}\right)\left(\frac{\partial^2 \phi}{\partial x^2} + \frac{\partial^2 \phi}{\partial y^2}\right) \right\} dxdy, \quad (6.7)$$

in which q is the electrode shape and

$$f(\phi) = \int_\Omega \left\{ \tilde{M}_1(t)\left(\frac{\partial^2 \phi}{\partial x^2} + \frac{\partial^2 \phi}{\partial y^2}\right)dxdy + \tilde{M}_6(t)\frac{\partial^2 \phi}{\partial x\partial y} \right\} dxdy. \quad (6.8)$$

In [4], it is proven that there exists an approximate and convergent solution of the optimization functional based on the governing equation with the form in (6.6).

6.2.2 Finite Element Procedure

In the shape optimization algorithm described later in this paper, the finite element method is used to solve the governing equations for the plate deflection $w(x, y, t)$. With this approach, we choose $\phi(x, y) = N_j(x, y)$, where N_j is any finite element satisfying the usual compatibility, smoothness, and approximation properties [3]. The approximation of the lateral deformation becomes

$$w(x, y, t) = \sum_{i=1}^{N} w_j(t) N_j(x, y). \tag{6.9}$$

Substituting this expression into the weak form of the actuator equation in (6.5), we obtain

$$M_{ij} \ddot{w}_i + K_{ij} w_i = B_j v(t), \tag{6.10}$$

where

$$M_{ij} = \int_{\Omega} \rho h N_i N_j dx dy,$$

$$K_{ij} = \int_{\Omega} \left\{ D_{11} \left(\frac{\partial^2 N_i}{\partial x^2} \frac{\partial^2 N_j}{\partial x^2} + \frac{\partial^2 N_i}{\partial y^2} \frac{\partial^2 N_j}{\partial y^2} \right) \right.$$
$$\left. + D_{12} \left(\frac{\partial^2 N_i}{\partial y^2} \frac{\partial^2 N_j}{\partial x^2} + \frac{\partial^2 N_i}{\partial x^2} \frac{\partial^2 N_j}{\partial y^2} \right) + 2 D_{66} \frac{\partial^2 N_i}{\partial x \partial y} \frac{\partial^2 N_j}{\partial x \partial y} \right\} dx dy,$$

$$B_j v(t) = - \int_{\Omega} \tilde{M}_1 \left(\frac{\partial^2 N_j}{\partial x^2} + \frac{\partial^2 N_j}{\partial y^2} \right) dx dy - 2 \int_{\Omega} \tilde{M}_6 \frac{\partial^2 N_j}{\partial x \partial y} dx dy.$$

The control influence vector on the right-hand side of (6.10) results from the piezo-electric effect. Using Green's formula, this vector can be alternatively written as

$$\mathbf{B} v(t) = \int_{\Gamma} \left(\tilde{M}_n \frac{\partial \mathbf{N}}{\partial n} + \tilde{M}_{ns} \frac{\partial \mathbf{N}}{\partial s} \right) d\Gamma. \tag{6.11}$$

The elemental consistent load vector is then

$$\mathbf{B}^e v(t) = \int_{\Gamma_e} \left(\tilde{M}_n \frac{\partial \mathbf{N}^e}{\partial n} + \tilde{M}_{ns} \frac{\partial \mathbf{N}^e}{\partial s} \right) d\Gamma. \tag{6.12}$$

That is, the piezoelectric laminate yields load terms that are equivalent to the line moment along the boundary of the electrode. This fact is used in the optimization algorithm described later in this paper.

6.3 Shape Optimization Theory

In this section, we review the primary theoretical results from [4] and indicate their relevance to this paper. To make the statement of our optimal shape control problem precise, it remains to a) define the set of electrode shapes over which the optimization will proceed and b) define the functional to be minimized.

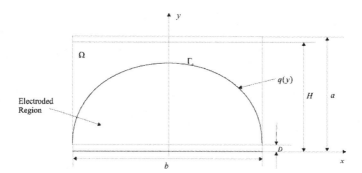

Fig. 6.2. Top view of the composite laminate.

Figure 6.2 depicts the top view of the composite laminate, and the electrode pattern on the surface. From the figure, it is clear that the electroded region specified by the characteristic function $\chi(x,y)$ depends on the boundary of the electrode Γ_e. We assume that the electrode boundary is in fact given by the graph of a curve q. The set $\mathcal{Q}$ of all admissible curves q is given by

$$\mathcal{Q} = \left\{ q \in C^{0,1}([-b/2, b/2]) : D \leq q(\xi) \leq H < a, |q(\xi) - q(\eta)| \leq \lambda|\xi - \eta| \right.$$
$$\left. \forall \xi, \eta \in [-b/2, b/2] \quad \text{and} \quad q\left(-\frac{b}{2}\right) = q\left(\frac{b}{2}\right) = D \right\}.$$

(6.13)

The boundary of the electroded region can be written in terms of any element $q \in \mathcal{Q}$:

$$\Gamma_e(q) \triangleq \left\{ (x,y) \in \left[-\frac{b}{2}, \frac{b}{2}\right] \times [D, H] \ \ s.t. \ y = q(x), q \in \mathcal{Q} \right\}.$$

It should be carefully noted that not only are the graphs Lipschitz continuous

$$q \in C^{0,1}\left[-b/2, b/2\right],$$

but the maximum modulus of continuity is some fixed constant $\lambda \in \mathbb{R}$. The set $\mathcal{Q}$ is a metric space, with the usual metric

$$d(p,q) \triangleq \sup_{\xi \in [-\frac{b}{2}, \frac{b}{2}]} |p(\xi) - q(\xi)|.$$

The optimal electrode shape is determined with the introduction of a performance functional

$$J(q) = \mathcal{J}(w(q), \dot{w}(q))$$
$$\triangleq \frac{1}{2} \|C_0 w - \tilde{y}_0\|_{Z_0}^2 + \frac{1}{2} \|C_1 \dot{w} - \tilde{y}_1\|_{Z_1}^2.$$

(6.14)

This equation introduces the functional notation $(w(q), \dot{w}(q))$ for the solution of the governing equation, (6.6), generated by a choice of $q \in Q$. In this equation, Z_0 and Z_1 are Banach spaces of observations corresponding to the displacements w and velocities $\dot{w}$ respectively. The bounded linear operators C_0 and C_1 map the state $(w, \dot{w})$ to the observation spaces. They depend on the specific nature of the measurements $\tilde{y}_0$ and $\tilde{y}_1$, which are assumed given. For examples C_0 might represent point evaluations of the displacement field w, and $\tilde{y}_0$ would correspond to time histories of measured displacements. Numerous other possibilities exist, some of which are discussed in [1].

In this paper, we study functionals that yield electrode designs that approximate specific shape distributions along the outboard free edge of the cantilever active composite. One reasonable functional is a specific case of the general form in (6.14):

$$
\begin{aligned}
J(q) &\triangleq \frac{1}{2}\|\gamma_0 w - \tilde{y}_0\|^2_{L^2(0,T;\Gamma_{out})} \\
&= \frac{1}{2}\int_0^T \int_{-b/2}^{b/2} |(\gamma_0 w)(x,t) - \tilde{y}_0(x,t)|^2 dx dt.
\end{aligned}
\tag{6.15}
$$

In this equation, Γ_{out} is the outboard edge of the composite cantilever and γ_0 is the trace operator that evaluates the state along the boundary. In any event, the optimal electrode shape will be given as the solution of the infinite dimensional optimization problem: Find $q_0 \in Q$ such that

$$
J(q_0) = \inf_{q \in Q} J(q) = \inf_{q \in Q} \mathcal{J}(w(q), \dot{w}(q)),
\tag{6.16}
$$

where $(w(q), \dot{w}(q))$ is the solution of (6.5) generated by $q \in Q$. The following theorem shows that this problem is indeed well posed.

Theorem 1. *There is a solution $q_0 \in Q$ that solves the abstract optimization problem (6.16).*

Proof. The proof is a straightforward consequence of generalized Weierstrass theorem. The mapping

$$
q \to J(q)
$$

is continuous by the construction of $J(q)$ in (6.14) and the continuity of the solution map in (6.6). Moreover, the set Q is a family of uniformly bounded equicontinuous functions. By the Arzela-Ascoli theorem the set Q is sequentially compact and the result follows. Details can be found in [5]. □

6.4 Optimization Algorithm

In this section we present the details of the electrode shape optimization algorithms. For a given FEM grid, it is assumed that an element is either fully electroded or unelectroded. Thus, the optimization can be viewed as a discrete optimization or combinatorial optimization. Unfortunately, these problems are exceptionally costly numerically. To reduce the number of design variables, we cast the problem as a continuous optimization method based on conventional gradient search methods. Since the electrode shape is symmetric in this practical application, we only consider the half-structure for FEM computation.

In the previous companion paper [5], an ad hoc formulation of the optimization problem is described for the static case. In this section, an alternative approach is presented that is rigorously consistent with the conditions of Theorem 1. Furthermore, it is shown that it is straightforward to modify this formulation to consider the dynamic case. First, however, it is necessary to review the essential features of the electrode shape design strategy, which was originally presented in [5].

6.4.1 Electrode Shape Design

Consider, as an example, the case where we have only two design variables, corresponding to a point on the center line and another point on the top edge of the active laminate. A simple approach for defining the electroded region is to connect these two points by a line. As illustrated in Figure 6.3, the design variables define a line that determines the right boundary of the electrode while the left boundary is assumed to be a fixed vertical line. Of course, the right boundary shape is more accurately described as the number of control points is increased. Smoother boundary shapes can be obtained using higher-order polynomials, such as B-splines, to connect the points. Recall that, in calculating the optimal electrode shape, each element is assumed to be either fully electroded or fully unelectroded. A more accurate method would compute the integration over the exact area in each element located on the electrode boundary, as shown in Figure 6.4. The difference between these two approaches diminishes, however, as the finite element mesh resolution increases.

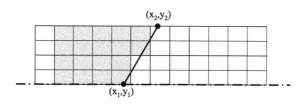

Fig. 6.3. Approximate determination of electrode boundary shape.

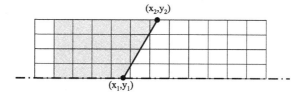

Fig. 6.4. Precise determination of electrode boundary shape.

6.4.2 Ad Hoc Approach: Objective Function and Constraints

As discussed in the previous paper [5], an economic design is sought that maximizes the deflection of the structure under a given applied voltage. Therefore, one objective function can be defined as

$$J(q) = \tilde{w}/\tilde{w}_{max}, \qquad (6.17)$$

where

$$\tilde{w} = \left(\sum_i^n w_i^2/n \right)^{\frac{1}{2}}. \qquad (6.18)$$

In the preceding equation, $\tilde{w}_{max}$ is the maximum displacement obtained from (6.18) when all elements on the piezoelectric material are actuated. The integer n represents the number of nodes, and the objective function is normalized so that its range is $[0, 1]$. In addition, the following constraint is imposed to minimize the lateral distortion in the deflection shape:

$$g(\Lambda) = \left[\frac{\sum_{i \in \Lambda} (w_i - w_{ave,i})^2 / n_\Lambda}{\tilde{w}} \right]^{\frac{1}{2}} /C_s - 1. \qquad (6.19)$$

This constraint effectively places an upper limit on the standard deviation of the displacement of the selected nodes. In this equation, Λ is a set of nodes that determines the portion of the composite over which the constraint is imposed. For example, Λ may be taken to be the last row of outboard nodes in the finite element mesh that are farthest from the root of the cantilever plate. Minimizing the standard deviation of the displacement of this line of nodes results in an overall deflection that is more nearly two dimensional. The integer n_Λ is simply the number of nodes in the set Λ and C_s is a constraint ratio weighting factor selected to guarantee that the range of the constraint function is $[0, 1]$. The value $w_{ave,i}$ is the average displacement along the lateral line of nodes in which node i is a member.

6.4.3 Theoretically-Consistent Approach: Cost Function and Constraints

Strictly speaking, the previously-described optimization approach is ad hoc in the sense that it is not rigorously consistent with the conditions of Theorem 1. For the static case, a theoretically-consistent objective function can be defined as

$$J(q) = \frac{1}{2} \int_{\bar{\Omega}} (w(x,y) - \tilde{y}_0)^2 \, dx \, dy, \qquad (6.20)$$

where $\bar{\Omega}$ represents the domain of interest. Once again, the set of nodes over which this function is optimized is denoted as Λ. For the piezoelectric transducer considered in this study, Λ is simply the trailing edge of the flap. In this case, (6.20) takes the form

$$J(q) = \frac{1}{2}\int_{-L/2}^{L/2} (w(x,\bar{y}) - \tilde{y}_0)^2\, dx$$

$$= \int_0^{L/2} (w(x,\bar{y}) - \tilde{y}_0)^2\, dx, \tag{6.21}$$

where $\bar{y}$ denotes the width of the structure. Note that symmetry has been taken into account in (6.21).

In practice, the objective function in (6.21) is evaluated using the finite element approximation of the deflection:

$$w(x,y) = \sum_{j=1}^{N} w_j N_j(x,y), \tag{6.22}$$

where N is the total number of degree-of-freedom in the model. Then, (6.21) can be calculated as

$$J(q) = \int_0^{L/2} \left(\sum_{j=1}^{N} w_j N_j(x,\bar{y}) - \tilde{y}_0 \right)^2 dx \tag{6.23}$$

$$= \sum_{i=1}^{\tilde{N}} \int_{-1}^{1} \left(\sum_{j=1}^{N} w_j^e N_j(\xi,\bar{y}) - \tilde{y}_0 \right)^2 \bar{x}\, d\xi.$$

In (6.23), $\tilde{N}$ denotes the number of elements on the trailing edge. The finite element mesh is chosen to be uniformly dense, and $\bar{x}$ represents the elemental length. Finally, w_j^e represents the displacement of the jth node in the ith element. The integral in (6.23) can then be computed numerically using Gauss-Legendre quadrature. The highest order of x in the integrand is 6 since, when y is fixed, the shape functions $\{N_j\}$ are 3rd-order polynomials in x. Therefore, a 4-point Gauss-Legendre quadrature is sufficient to accurately evaluate the integral as

$$J(q) = \sum_{i=1}^{\tilde{N}} \bar{x} \sum_{k=1}^{4} H_k \left(\sum_{j=1}^{N} w_j^e N_j(\xi_k,\bar{y}) - \tilde{y}_0 \right)^2, \tag{6.24}$$

where $\{\xi_k\}$ and $\{H_k\}$ are the nodes and weights in the 4-point integration formula.

In addition to the objective function that is optimized, a number of constraints are enforced. First, the electrode shape must satisfy the boundary condition

$$q\left(-\frac{L}{2}\right) = q\left(\frac{L}{2}\right) = D, \tag{6.25}$$

where D is the amount of offset between the piezoelectric patch and the clamped edge. In addition, the following smoothness constraint is imposed:

$$\|q(\xi) - q(\eta)\| \le \lambda\|\xi - \eta\|. \tag{6.26}$$

Equation (6.26) limits the slope between two adjacent points on the electrode boundary, ensuring that a reasonably smooth solution is obtained. A final constraint is needed
in order to ensure a tip deflection that is sufficiently large for flow control. Since the
maximum deflection is dependent on the electrode area, an area constraint is defined
as

$$\int_{-L/2}^{L/2} q(x)dx \geq cA_{\Omega_p},\tag{6.27}$$

where A_{Ω_p} is the area of the piezoelectric domain Ω_p. The parameter $c \in [0,1]$ is
the ratio of the area covered by the electrode to the total area of the piezoelectric material. The shape optimization procedure then entails finding the electrode boundary
shape that minimizes the cost function in (6.21) subject to the constraints in (6.25)
through (6.27).

The preceding development is for the static case. In order to consider the dynamic
case, the cost function must be modified to account for the time-dependent response
of the structure. Then, the following cost function is obtained:

$$J(q) = \frac{2}{T} \int_0^T \int_0^{L/2} (w(x,\bar{y},t) - w_{ave}(t))^2 \, dxdt,\tag{6.28}$$

where w_{ave} denotes the time-dependent average deflection of the tip across the length
of the flap:

$$w_{ave}(t) = \frac{2}{L} \int_0^{L/2} |w(\xi,\bar{y},t)|d\xi.\tag{6.29}$$

In (6.28) and (6.29), T represents the total time period of interest. The objective
function defined in (6.28) encourages minimal lateral deflection in the plate over the
entire time history. The constraints for the dynamic case are the same as in the static
case since they represent restrictions on the electrode boundary shape, which is not a
function of time.

Practically speaking, the calculation of the dynamic cost function in (6.28) is performed in a similar manner as the static case. The following finite element approximation of the deflection is used:

$$w(x,y,t) = \sum_{j=1}^N w_j(t)N_j(x,y),\tag{6.30}$$

which is the same as in (6.22) except that the coefficients $\{w_j\}$ are now functions
of time. The calculation of the integrals in (6.28) and (6.29) then proceeds using
four-point Gauss-Legendre quadrature.

6.5 Numerical Results and Discussion

6.5.1 Static Case

In this section, the static response of the structure is considered for the shape optimization problem. All results presented here have been obtained using a 50×100 finite

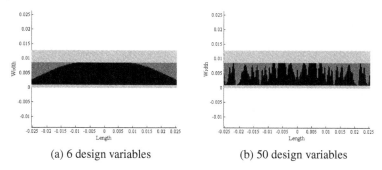

(a) 6 design variables   (b) 50 design variables

Fig. 6.5. Optimal electrode shapes obtained using the ad hoc approach with B-spline smoothing.

element mesh. In all cases, the Bogner-Fox-Schmidt element is employed, which is composed of 4 nodes and 4 degrees-of-freedom per node. A detailed description of this element can be found in [6], for example. It is well known that the Bogner-Fox-Schmidt element satisfies all the minimal requirements of smoothness, compatibility, and approximation order required for our model. While more refined elements are available, their consideration is not salient to the study outlined in this paper. In this section, the performance of the shape optimization approach derived in this paper is compared to that of the ad hoc approach that was previously employed.

Figure 6.5 depicts two examples of optimal shapes obtained using the ad hoc optimization procedure. In these examples, the set of constraint nodes Λ consists of 5 rows of nodes along the outboard free edge of the composite. The first electrode shape was obtained using a low-dimensional parameterization of the boundary. A B-spline curve-fit has been employed in order to obtain a smoother boundary shape, which is desirable from a fabrication standpoint. It has been shown in [5] that B-spline smoothing has a negligible effect on the value of the objective function. Therefore, smoothing does not adversely affect the performance of the actuator. The second electrode shape depicted in Figure 6.5, on the other hand, was obtained using a high-dimensional parameterization of the boundary consisting of 51 design variables. The resulting shape is highly oscillatory and is clearly undesirable for electrode fabrication. This oscillation phenomenon, which cannot be alleviated using B-spline smoothing, has been commonly observed in the shape optimization literature.

In comparison, Figure 6.6 depicts a series of optimal electrode shapes obtained using the theoretically-consistent optimization algorithm. In these examples, the number of design variables has been varied over the set $\{5, 10, 25, 50\}$ and the minimum area constraint c has been varied over $\{0.85, 0.75, 0.65, 0.6\}$. A number of observations can be made from these representative results. First, in contrast to the ad hoc approach, relatively smooth electrode shapes are obtained even as the number of design variables becomes large. This is readily apparent from a comparison of the 50 design variable cases in Figures 6.5 and 6.6. It is also worth noting that, with the theoretically-consistent method, no spline fitting is necessary to smooth the boundary shapes.

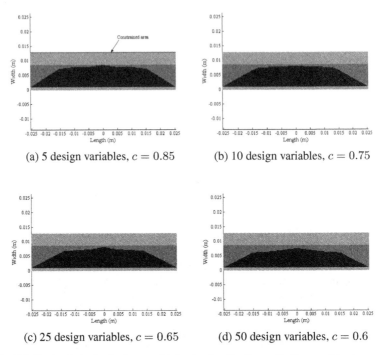

(a) 5 design variables, $c = 0.85$

(b) 10 design variables, $c = 0.75$

(c) 25 design variables, $c = 0.65$

(d) 50 design variables, $c = 0.6$

Fig. 6.6. Optimal electrode shapes obtained using the theoretically-consistent approach.

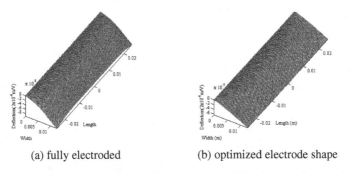

(a) fully electroded

(b) optimized electrode shape

Fig. 6.7. Three-dimensional static deflection shapes: fully electroded and optimized electrode shape with 5 design variables, $c = 0.85$.

Figure 6.7 depicts the three-dimensional static flap deflection generated by the optimal electrode shape from the theoretically-consistent method with 5 design variables, $c = 0.85$. Also shown, for comparison, is the static deflection obtained without electrode shape optimization. In this case, the piezoelectric material is fully covered by the electrode. Clearly, a more uniform tip deflection is obtained from the optimized

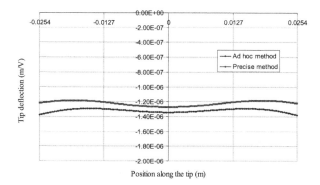

Fig. 6.8. Comparison of tip deflections from the ad hoc method with 6 design variables and the theoretically-consistent method with 5 design variables, $c = 0.85$.

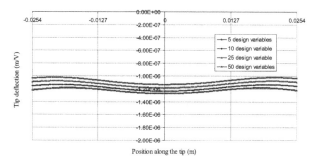

Fig. 6.9. Comparison of tip deflections generated by the optimal electrode shapes with varying design variables.

electrode shape. The performance of the two optimization methods is compared in Figure 6.8, which depicts the tip deflections from the optimal electrode shapes obtained using the ad hoc method with 6 design variables and the theoretically-consistent method with 5 design variables, $c = 0.85$. While the ad hoc approach yields a slightly higher tip deflection, the deflection shape from the theoretically-consistent method exhibits less lateral distortion. It is apparent, then, that comparable results are obtained from the two optimization methods for coarse representations of the electrode boundary. As discussed earlier, however, the ad hoc approach is not suitable for high parameterizations of the boundary due to the oscillation problem.

As a final observation regarding the theoretically-consistent algorithm, recall that c denotes the fraction of the piezoelectric material that is covered by the electrode. Therefore, smaller values of c result in less electrode coverage and, consequently, less tip deflection. Figure 6.6 illustrates that as the number of design variables is increased, the electrode coverage must be decreased in order to obtain smooth boundary shapes. Figure 6.9 compares the static tip deflections for the electrode shapes depicted in Figure 6.6. These results show that, as expected, the tip deflection decreases as the number of design variables is increased.

6.5.2 Dynamic Case

This section focuses attention on the dynamic case, in which the time history of the actuator response to a prescribed voltage input is considered. All results in this section have been obtained using a 25×50 finite element mesh. As in the static case, the Bogner-Fox-Schmidt element has been employed. The dynamic optimization was performed with respect to a 1 kHz sinusoidal voltage input over a time duration of $0.05\ s$. The time discretization step was taken as $5 \times 10^{-5}\ s$ so that there were a total of 1000 time steps in the simulation. The initial deflection and slope of the plate were set to zero. For ease of computation, the optimization was performed using modal coordinates and a reduced-order model that included the first 30 modes. The frequencies corresponding to these modes ranged from 676 Hz to 32.4 kHz.

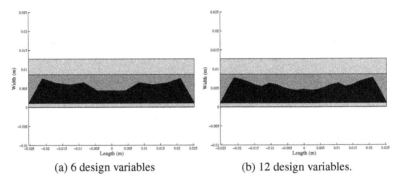

(a) 6 design variables   (b) 12 design variables.

Fig. 6.10. Optimal electrode shapes for the dynamic case.

Optimal electrode shapes were computed using 6 and 12 design variables. Figure 6.10 depicts the resulting electrode shapes, which clearly resemble each other. The tip deflection shapes at several different time steps have been plotted in Figure 6.11. In each case, the fully-electroded deflection shape has been plotted along with the deflection shapes from the two optimized shapes. The results show that the fully-electroded actuator exhibits significantly more lateral distortion than either of the optimized electrodes. In addition, the optimal electrode shape with 12 design variables generates less lateral distortion than the 6 design variable shape.

6.6 Conclusion

This paper has investigated electrode shape optimization for piezoelectric actuators with the goal of obtaining two-dimensional flap deflections suitable for flow control. Theoretical results have been reviewed that guarantee a convergent solution to the shape optimization problem under certain conditions. An optimization approach has been introduced that is consistent with the optimization theory. This approach entails

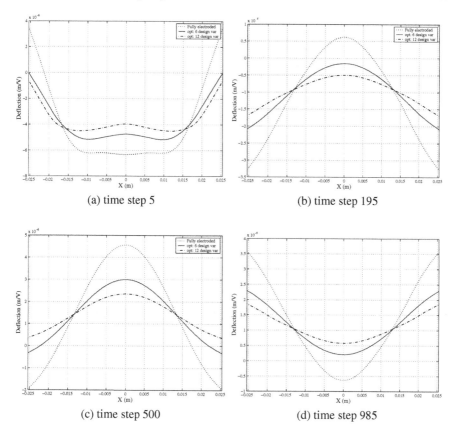

Fig. 6.11. Tip deflection shapes obtained at several time steps from the fully-electroded and optimized electrode shapes.

optimizing an objective function designed to minimize three-dimensional flap distortion. Constraints are also imposed in order to obtain relatively smooth boundary shapes and sufficiently large flap deflection.

Static results were presented and compared to those obtained in previous work using an ad hoc approach that is not, strictly-speaking, consistent with the theory. The static results from the two methods were shown to be comparable if the number of design variables was small; however, the ad hoc approach suffers from an oscillation problem as the number of design variables is increased. In contrast, the theoretically-consistent approach does not exhibit this phenomenon. In addition, it has been demonstrated that the theoretically-consistent method can be applied to treat the dynamic case. The results presented in the paper show that, in the dynamic case, the computed optimal electrode shapes yield less lateral distortion than a fully-electroded actuator.

References

[1] Banks H.T., Smith R.C., Yang Y. (1996) Smart material structures: modeling, estimation, and control. John Wiley & Sons, Paris
[2] Gibson R.F. (1994) Principles of composite material mechanics. McGraw-Hill, Inc., New York
[3] Hughes Thomas J.R. (1987) The finite element method: linear static and dynamic finite element analysis. Prentice-Hall, Englewood Cliffs New Jersey
[4] Kurdila A.J., Hager W., Prazenica R.J. (2005) Optimization of shape and approximation of piezoelectric composites. preprint
[5] Wang W., Kurdila A.J., Venkataraman S. (2003) Shape optimization of electrodes for piezoelectric actuators – static analysis. 44th AIAA/ASME/ASCE/AHS Structures, Structural Dynamics, and Materials Conference, Norfolk Virginia
[6] Wang W. (2003) Shape optimization of piezoelectric transducers. PhD Dissertation, University of Florida, Gainesville

7

Robust Static Super-Replication of Barrier Options in the Black-Scholes model

Jan H. Maruhn[1] and Ekkehard W. Sachs[2]

[1] Universität Trier
Department of Mathematics
D–54286 Trier, Germany
maruhn@uni-trier.de
[2] Virginia Tech
ICAM
Department of Mathematics
Blacksburg, VA 24061, USA
and
Universität Trier
Department of Mathematics
D–54286 Trier, Germany
sachs@uni-trier.de

Summary. Static hedge portfolios for barrier options usually depend on the parameters of the financial market model under consideration. In this paper we propose a linear semi-infinite programming formulation of a hedging strategy recently developed in the literature and reduce the sensitivity of the portfolio with respect to model parameters by solving an appropriate robust optimization problem. The resulting robust hedge portfolio offers protection for a broad range of model parameters and is only marginally more expensive than its non-robust counterpart.

Key words: robust optimization, static hedging, super–replication, barrier options, Black-Scholes Model

7.1 Introduction

The financial products which are available on the capital market have grown into fairly complex tools. Well known simple examples of these products include European options like puts and calls. In these cases, the client pays the bank a fixed amount at time $t = 0$ and obtains the right to realize some insecure payments in the future. These future payments are usually based on the performance of a so-called underlying, for example the stock price, from time $t = 0$ to $t = T > 0$.

However, once the bank has sold an option, it is exposed to the risk of the insecure future payment guaranteed to the customer. Thus traders try to set up a portfolio of tradable financial products which closely matches the insecure future payment in all possible states of the economy. The problem is to compose such an efficient *hedge portfolio* for a given claim sold by the bank.

In this paper we consider the problem of hedging a barrier option, more precisely an up-and-out call, by a buy-and-hold portfolio of European options which is liquidated once the barrier is hit. Recent research has shown, that these so called static strategies are particularly efficient in the context of hedging barrier options.

For example, Bowie and Carr show in [2], that the payoff of a barrier option can be replicated perfectly in the Black-Scholes model by a specific buy-and-hold portfolio if the difference between the interest rate and the dividend yield equals zero. However, as for instance Giese and Maruhn show in [10], the Bowie and Carr hedge portfolio fails to hedge the target option if the interest rate and dividend yield differ. Furthermore, the hedge portfolio includes rarely traded options.

Derman, Ergener and Kani [6] take an alternative approach by exactly replicating the payoff of the barrier option by an infinite number of European options. However, a reasonable finite approximation of this strategy will contain a large number of options which makes the strategy less practical. Moreover, as for example Toft, Xuan [16] and Fink [8] show, this particular trading strategy is very sensitive to the assumption of constant volatility.

To overcome these problems, Giese and Maruhn [10] extend the idea of cost-optimal super-replication to static hedging and solve the resulting stochastic optimization problem by a sample average approach. The resulting hedge portfolio has attractive properties, consisting only of a handful of European options and being only marginally more expensive than the barrier option itself.

However, the derived strategy is based on the crucial assumption, that the hedge portfolio can be liquidated for model prices based on parameters chosen at time $t = 0$. In this paper we show, that the assumption of constant model parameters can lead to large hedging losses if these parameters are subject to change. We propose to include these parameters in a robust optimization formulation in the form of a worst case design. This yields a linear semi-infinite programming formulation of the stochastic optimization problem derived in [10] for the Black Scholes model and extends to a robust optimization problem.

In Section 7.2 we introduce the problem in detail and derive the robust linear semi-infinite programming problem. Furthermore, we prove the existence of a robust solution under mild conditions. In Section 7.3 we analyze the structure of the semi-infinite optimization problem and describe the algorithm we use to solve the problem. In Section 7.4 we present a detailed numerical example, computing and comparing the non-robust and the robust solution for given real world data. As it turns out, the robustness of the solution can be gained by relatively little additional cost such that the portfolio is attractive to option traders.

7.2 Description of the Problem

7.2.1 Classical Approach

In this section we pose the problem of setting up a robust static hedge portfolio for barrier options and show that it can be described as a semi-infinite optimization problem. We give a brief introduction to the concept of barrier options, more precisely an up-and-out call, and the associated static hedge portfolios.

An up-and-out call is characterized as follows: The client obtains the right to buy a specific stock S at time $t = T$ for a predefined price $K > 0$. However, if the stock price touches or crosses the barrier $D > 0$ at some time from $t = 0$ to $t = T$, the call expires without payoff (*knock-out*). The time instance when this happens is defined as

$$t_D = \begin{cases} \min\{t \in [0, T] \mid S_t = D\}, & \text{if } \max_{0 \le t \le T} S_t \ge D, \\ +\infty, & \text{if } \max_{0 \le t \le T} S_t < D, \end{cases} \quad (7.1)$$

where we denote the stock price at time t by S_t. The payoff of the up-and-out call C_{uo} for the customer at time $t = T$ is hence given by

$$C_{uo} = \begin{cases} 0, & \text{if } t_D \in (0, T], \\ \max\{S_T - K, 0\}, & \text{if } t_D = +\infty. \end{cases}$$

Figure 7.1 illustrates the payoff of an up-and-out call for two possible future realizations of the stock price $(S_t)_{t \in [0, T]}$.

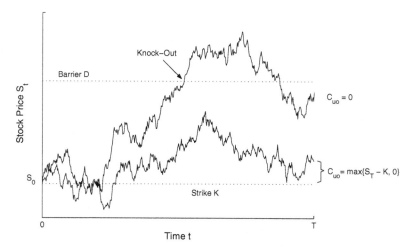

Fig. 7.1. Payoff of the up-and-out call for two possible stock-price paths.

It is obvious, that the value of an up-and out call is extremely sensitive in case the stock-price is close to the barrier. That is also why the usual dynamic hedging strategies (a continuous realigning of the stock position in the hedge portfolio) could lead to substantial losses for the bank. In fact, traders are mostly concerned about the case of hitting the barrier close to maturity T of the target option C_{uo}. To overcome these problems, research in the last decade has focussed on static trading strategies. Instead of a continuous adjustment of the stock position, these trading strategies consist of a buy-and-hold position of a given set of other financial instruments. Research has shown (see for example [2], [6]), that European calls are a particularly efficient choice as hedge instruments for an up-and-out call.

Definition 1. *Let the payoff of an* up-and-out call *with maturity $T > 0$ be given by*

$$C_{uo} = \begin{cases} 0, & \text{if } \max_{0 \leq t \leq T} S_t \geq D, \\ \max\{S_T - K, 0\}, & \text{if } \max_{0 \leq t \leq T} S_t < D, \end{cases}$$

where $K > 0$ denotes the strike price, $D > S_0 > 0$ the barrier satisfying $D > K$ and $(S_t)_{t \in [0,T]}$ a stock price history over $[0, T]$.
Let the price of European calls with maturities $0 < T_i \leq T$ and strikes $K_i > 0$, $K_i \neq K_j$ for $T_i = T_j$, $i \neq j$, be denoted by $C_i(t, S_t)$, $i = 1, ..., n$ and $0 \leq t \leq T_i$. The payoff at maturity T_i of these options is given by

$$C_i(T_i, S_{T_i}) = \max\{S_{T_i} - K_i, 0\}.$$

A *static trading strategy* is defined as follows: At time $t = 0$ the bank buys α_i units of the European call C_i. This portfolio is held constant until either the barrier D is hit or the maturity T of the barrier option is reached. In case the barrier is hit, we immediately liquidate the hedge portfolio. Otherwise the value of our portfolio is given by the value of the European calls at time $t = T$. Note that such a strategy can be uniquely identified by the portfolio weights $\alpha \in \mathbb{R}^n$. In the following the value of the hedge portfolio α at time t will be denoted by

$$\Pi(t, \alpha) = \sum_{i=1}^{n} \alpha_i C_i(t, S_t). \tag{7.2}$$

A bank prefers to set up a hedge portfolio whose payoff is greater or equal to the payoff of the up-and-out call in all possible states of the stock market. Such a portfolio is called a *super-replication* portfolio. As an up-and-out call either knocks out until time T or provides a payoff at terminal time, we can characterize a super-replication portfolio by the following two conditions:

$$\Pi(t_D, \alpha) \text{ is non-negative for all times } t_D \text{ the barrier might be hit.} \tag{7.3}$$
$$\Pi(T, \alpha) \geq \max\{S_T - K, 0\} \quad \text{if the barrier is not hit at all.} \tag{7.4}$$

Figure 7.2 illustrates, that the value of the hedge portfolio $\Pi(t_D, \alpha)$ at time t_D the barrier is hit in general depends on the payoff of the calls with maturities $T_i < t$.

This payoff in turn depends on the stochastic development of the stock price. Hence, if the evolution of the stock price is given by a specific stochastic model, conditions (7.3) and (7.4) are in fact stochastic constraints. In [10] Giese and Maruhn analyze these stochastic constraints and solve the corresponding optimization problem by appropriate discretization methods.

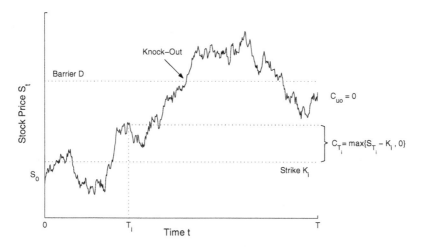

Fig. 7.2. Payoff of a European call with strike $K_i < D$ at time $T_i < T$.

We show that this hedging problem can be rephrased and hence simplified by choosing particular call options for our hedging portfolio.

Definition 2. *Consider n European calls as given in definition 1 whose strike prices satisfy in addition $K_i \geq D$ for $T_i < T$. The set of these options shall be denoted by $\mathcal{C} = \{C_1, ..., C_n\}$.*

Let us note that calls with strike $K_i < D$ and maturity $T_i < T$ are inefficient hedge instruments as they can quite frequently result in positive payoffs although the up-and-out call expires worthless (see the example in figure 7.2). This is why we only include the calls listed in definition 2 in our hedge portfolio. In the next Lemma we will show, that the value of such a portfolio at a given time is independent of the payoff of previously expired calls.

Lemma 1. *Consider a static hedge portfolio consisting of calls from $\mathcal{C}$. If the barrier has not been hit before time t, i.e. $t \leq t_D$, the value of the portfolio at time t is independent of the calls C_i with maturities $T_i < t$:*

$$\Pi(t, \alpha) = \sum_{C_i \in \mathcal{C}, T_i \geq t} \alpha_i C_i(t, S_t).$$

Proof. Consider European calls C_i in $\mathcal{C}$, i.e. with strikes $K_i \geq D$ for $T_i < T$. If $T_i < t \leq t_D$ and hence $S_{T_i} < D$, the call has already expired without payoff due to

$$\max\{S_{T_i} - K_i, 0\} \leq \max\{D - K_i, 0\} = 0.$$

For the rest of the calls in the portfolio with $T_i \geq t$, the value is $C_i(t, S_t)$ which yields the statement. □

Based on Lemma 1 we can restate conditions (7.3) and (7.4) in the following way:

$$\sum_{C_i \in \mathcal{C}, T_i \geq t_D} \alpha_i C_i(t_D, D) \geq 0, \tag{7.5}$$

$$\sum_{C_i \in \mathcal{C}, T_i = T} \alpha_i C_i(T, S_T) \geq \max\{S_T - K, 0\}. \tag{7.6}$$

To determine whether and when the barrier is hit and to calculate the option prices $C_i(t, S_t)$ for $T_i \geq t$, the future stock prices have to be described by an appropriate stochastic model. In this paper our analysis will be based on the Black-Scholes model.

Definition 3. *The dynamics of the stock-price S and a risk-free bond B are described by the Black-Scholes model:*

$$dS_t = (r - \delta)S_t dt + \sigma S_t dW_t, \quad t \in [0, T], \tag{7.7}$$
$$dB_t = rB_t dt,$$

with given initial values $S_0, B_0 > 0$. Here $r > 0$ is the risk-free rate, $\delta > 0$ is an instantaneous dividend rate and $\sigma > 0$ is the volatility of $(S_t)_{t \in [0,T]}$. $(W_t)_{t \in [0,T]}$ is a Brownian motion with corresponding filtration $(\mathcal{F}_t)_{t \in [0,T]}$, where $\mathcal{F}_0$ is assumed to be trivial.

In this model, the value of the European calls $C_i(t, S)$ at time $t < T_i$, given a stock price S at time t and volatility σ, can be computed via the Black Scholes formula (see for example [17])

$$C_i(t, S; \sigma) = Se^{-\delta(T_i - t)} \Phi\left(\frac{\log(\frac{S}{K_i}) + (r - \delta + \frac{\sigma^2}{2})(T_i - t)}{\sigma\sqrt{T_i - t}}\right)$$

$$-K_i e^{-r(T_i - t)} \Phi\left(\frac{\log(\frac{S}{K_i}) + (r - \delta - \frac{\sigma^2}{2})(T_i - t)}{\sigma\sqrt{T_i - t}}\right) \tag{7.8}$$

as the solution of the Cauchy problem for the parabolic differential equation

$$\begin{aligned} C_t + \tfrac{1}{2}\sigma^2 S^2 C_{SS} + (r - \delta)SC_S - rC = 0, & \quad (t, S) \in (0, T_i) \times (0, \infty), \\ C(T_i, S) = \max\{S - K_i, 0\}, & \quad S \in (0, \infty), \\ C(t, 0) = 0, & \quad t \in (0, T_i). \end{aligned} \tag{7.9}$$

In (7.8) $\Phi(\cdot)$ denotes the cumulative standard normal distribution.

Including the volatility as a fixed parameter conditions (7.5) and (7.6) hence yield:

$$\sum_{C_i \in \mathcal{C}, T_i \geq t} \alpha_i C_i(t, D; \sigma) \geq 0, \quad \forall t \in (0, T], \tag{7.10}$$

$$\sum_{C_i \in \mathcal{C}, T_i = T} \alpha_i \max\{S - K_i, 0\} \geq \max\{S - K, 0\}, \quad \forall S \in (0, D). \tag{7.11}$$

Due to the continuity of the functions involved, these inequalities also hold for the compact intervals $[0, T]$ and $[0, D]$, respectively[3].

By construction, static strategies $\alpha \in \mathbb{R}^n$ satisfying inequalities (7.10), (7.11) super-replicate the payoff of an up-and-out call in all states of the economy. The goal is to find the strategy of lowest cost, where the cost of $\alpha \in \mathbb{R}^n$ is

$$\Pi(0, \alpha) = \sum_{i=1}^{n} \alpha_i C_i(0, S_0; \sigma),$$

with $C_i(0, S_0; \sigma)$ given by the Black-Scholes formula (7.8).

7.2.2 Robust Approach

The key assumption in the derivation above is, that one is able to sell the hedge portfolio in the future at the model prices (7.8). Since the model prices in the Black-Scholes model correspond in a bijective way to the volatility σ specified in the set-up of the model, the assumption of a constant volatility over time may lead to significant deviations from prices traded in reality. In order to make the hedge portfolio more robust with respect to changes in the volatility we solve the following robust counterpart of the optimization problem sketched above.

Definition 4 (Robust hedging problem). *Let $\sigma_0 > 0$ be the implied volatility corresponding to current option prices in $t = 0$ and let $0 < \sigma_{min} \leq \sigma_0 \leq \sigma_{max}$ be a given volatility interval. A cost-optimal robust super-replicating strategy is defined as a solution of the following robust optimization problem:*

$$\min_{\alpha \in \mathbb{R}^n} \sum_{C_i \in \mathcal{C}} \alpha_i C_i(0, S_0; \sigma_0)$$
$$\text{s.t.} \sum_{C_i \in \mathcal{C}, T_i \geq t} \alpha_i C_i(t, D; \sigma) \geq 0, \quad \forall (t, \sigma) \in [0, T] \times [\sigma_{min}, \sigma_{max}],$$
$$\sum_{C_i \in \mathcal{C}, T_i = T} \alpha_i \max\{S - K_i, 0\} \geq \max\{S - K, 0\}, \quad \forall S \in [0, D], \tag{7.12}$$
$$C_i(t, S; \sigma) = Se^{-\delta(T_i - t)} \Phi\left(\frac{\log(S/K_i) + (r - \delta + \sigma^2/2)(T_i - t)}{\sigma\sqrt{T_i - t}}\right)$$
$$- K_i e^{-r(T_i - t)} \Phi\left(\frac{\log(S/K_i) + (r - \delta - \sigma^2/2)(T_i - t)}{\sigma\sqrt{T_i - t}}\right).$$

The feasible set of this optimization problem shall be denoted by $\mathcal{SR} = \mathcal{SR}(\sigma_{min}, \sigma_{max})$. A strategy $\alpha \in \mathcal{SR}(\sigma_{min}, \sigma_{max})$ is called a robust super-replicating strategy.

---
[3] Note that the continuity is proved in detail in Lemma 3.

Note that (7.12) is a linear semi-infinite optimization problem of the form

$$\min_{\alpha \in \mathbb{R}^n} c^T \alpha$$
$$\text{s.t. } a_1(t, \sigma)^T \alpha \geq 0, \quad \forall (t, \sigma) \in [0, T] \times [\sigma_{min}, \sigma_{max}], \qquad (7.13)$$
$$a_2(s)^T \alpha \geq b_2(s), \quad \forall s \in [0, D],$$

where $a_1 : [0, T] \times [\sigma_{min}, \sigma_{max}] \to \mathbb{R}^n$ is a nonlinear map depending on the Black-Scholes formula and $a_2 : [0, D] \to \mathbb{R}^n$, $b_2 : [0, D] \to \mathbb{R}$ are continuous and piecewise linear.

Clearly, the feasible set $\mathcal{SR}(\sigma_{min}, \sigma_{max})$ of optimization problem (7.13) is closed and convex, but generally unbounded. However, the specific structure of the problem allows us to prove the existence of solutions.

Theorem 1. *If $\mathcal{SR}(\sigma_{min}, \sigma_{max})$ is nonempty, a solution of optimization problem (7.12) exists. Furthermore, the set of optimal solutions is convex and compact.*

Proof. Due to well-known theorems in linear semi-infinite optimization (see for example [12]), it is sufficient to show, that the objective function and $\mathcal{SR}(\sigma_{min}, \sigma_{max})$ have no direction of recession in common. Assume, that $d \in \mathbb{R}^n$ is such a direction of recession. In the following, we will show, that d can only be equal to the zero vector.

Clearly, d is also a common direction of recession of the objective and the non-robust feasible set $\mathcal{SR}(\sigma_0, \sigma_0)$. This means, that d satisfies

$$c^T d \leq 0,$$
$$a_1(t, \sigma_0)^T d \geq 0, \quad \forall t \in [0, T],$$
$$a_2(s)^T d \geq 0, \quad \forall s \in [0, D].$$

Economically, d can hence be interpreted as a portfolio with cost $c^T d \leq 0$ that has a payoff greater or equal to zero in all possible states of the economy. As the Black Scholes model is arbitrage-free in the set of dynamic trading strategies satisfying the usual regularity assumptions, it is also arbitrage-free in the considered set of static trading strategies[4]. Hence the portfolio d cannot be traded in the market with negative cost which implies $c^T d = 0$. On the other hand, a portfolio with zero cost and non-negative payoff in all states of the economy cannot produce a positive payoff with positive probability, because this would also be an arbitrage opportunity. Due to the continuity[5] of a_1, a_2 the vector d hence satisfies

$$c^T d = 0,$$
$$a_1(t, \sigma_0)^T d = 0, \quad \forall t \in [0, T], \qquad (7.14)$$
$$a_2(s)^T d = 0, \quad \forall s \in [0, D].$$

---

[4] A more detailed discussion can be found in [13] and [10].
[5] The continuity is proved in detail in Lemma 3.

As $a_1(t, \sigma)^T d$ and $a_2(s)^T d$ represent the value of the portfolio d, these equalities would imply, that the payoffs of the calls in the portfolio along the barrier and in case the barrier is not hit at all, are linearly dependent. The next lemma shows, that this is not the case, hence $d = 0$.   □

Lemma 2. *If for some vector d equations (7.14) hold, then $d = 0$.*

Proof. Without loss of generality, we group the European calls into sets $I_1,...,I_r$ with equal maturities $\bar{T}_1 < \bar{T}_2 < ... < \bar{T}_r$. Then condition (7.14) implies

$$\sum_{i \in I_r} d_i C_i(t, D; \sigma_0) = 0, \quad \forall t \in (\bar{T}_{r-1}, \bar{T}_r]. \tag{7.15}$$

By the superposition principle, the function

$$c(t, S) = \sum_{i \in I_r} d_i C_i(t, S; \sigma_0)$$

also satisfies the Black-Scholes partial differential equation (7.9) with end condition

$$c(\bar{T}_r, S) = \sum_{i \in I_r} d_i \max\{S - K_i, 0\}, \quad \forall S \in (0, \infty).$$

Based on (7.9) and (7.14) we can conclude that $c(t, S)$ vanishes on $(\bar{T}_{r-1}, \bar{T}_r] \times \{0\}$ and $(\bar{T}_{r-1}, \bar{T}_r] \times \{D\}$ and, in addition, we have $c(\bar{T}_r, S) = 0$ for $S \in [0, D]$. By the maximum principle, see e.g. Friedman [9], this implies $c(t, S) = 0$ on the strip $[\bar{T}_{r-1}, \bar{T}_r] \times [0, D]$. Since $c(t, D) = \frac{\partial}{\partial S} c(t, D) = 0$ on $(\bar{T}_{r-1}, \bar{T}_r]$, by Mizohata's uniqueness theorem [14] this implies that c vanishes everywhere on $[\bar{T}_{r-1}, \bar{T}_r] \times [0, \infty)$. In particular this implies $\sum_{i \in I_r} d_i \max\{S - K_i, 0\} = 0 \; \forall S \in (0, \infty)$ and hence $d_i = 0$ for $i \in I_r$.

This argument is repeated on the time strip $(\bar{T}_j, \bar{T}_{j+1}]$ with c defined analogously. By definition 2 we have $K_i \geq D$ for $i \in I_{j+1}, j+1 < r$ which guarantees $c(\bar{T}_{j+1}, S) = 0$ for $S \in [0, D]$. Hence, proceeding recursively, all coefficients d_i have to vanish. This proves the statement of the lemma.   □

Note that the existence of a robust super-replicating strategy is assured, if a call with the same strike and maturity as the up-and-out call is included in the hedge portfolio, because $\max\{S_T - K, 0\} \geq C_{uo}$. Thus the assumption $\mathcal{SR}(\sigma_{min}, \sigma_{max}) \neq \emptyset$ can easily be fulfilled which implies the existence of a solution of the robust optimization problem. In the next section we will discuss the numerical solution of problem (7.12).

7.3 Numerical Solution

Optimization problem (7.12) is a linear semi-infinite programming problem where the parameters (t, σ) and s vary within compact sets. The numerical solution of such a problem will in general depend on appropriate discretizations of these parameter sets. Let $M_1 \subset [0, T] \times [\sigma_{min}, \sigma_{max}]$, $|M_1| < \infty$ and $M_2 \subset [0, D]$, $|M_2| < \infty$ denote such discretizations, for example an equidistant grid in $[0, T] \times [\sigma_{min}, \sigma_{max}]$ and $[0, D]$, respectively. Then the discretized robust optimization problem is defined as follows:

$$\min_{\alpha \in \mathbb{R}^n} \sum_{C_i \in \mathcal{C}} \alpha_i C_i(0, S_0; \sigma_0)$$

$$\text{s.t.} \sum_{C_i \in \mathcal{C}, T_i \geq t} \alpha_i C_i(t, D; \sigma) \geq 0, \quad \forall\, (t, \sigma) \in M_1, \tag{7.16}$$

$$\sum_{C_i \in \mathcal{C}, T_i = T} \alpha_i \max\{s - K_i, 0\} \geq \max\{s - K, 0\}, \quad \forall\, s \in M_2.$$

An algorithm will proceed by successively solving the subproblems (7.16), refining M_1, M_2 or exchanging points within these sets from one iteration to the next one. To guarantee convergence of such a procedure, it is crucial to verify the continuity of the vector valued functions a_1, a_2 with respect to their parameters $(t, \sigma) \in [0, T] \times [\sigma_{min}, \sigma_{max}]$ and $s \in [0, D]$. The following Lemma states the corresponding result.

Lemma 3. *The semi-infinite constraint coefficients* $a_1 = (a_1^1, ..., a_1^n)^T$, $a_2 = (a_2^1, ..., a_2^n)^T$ *of optimization problem (7.12), (7.13) given by*

$$a_1^i(t, \sigma) = \begin{cases} C_i(t, D; \sigma), & t \leq T_i, \\ 0, & t > T_i, \end{cases}$$

$$a_2^i(s) = \begin{cases} \max\{s - K_i, 0\}, & T_i = T, \\ 0, & T_i < T, \end{cases}$$

as well as $b_2(s) = \max\{s - K, 0\}$ *are continuous for all* $(t, \sigma) \in [0, T] \times [\sigma_{min}, \sigma_{max}]$ *and* $s \in [0, D]$, *respectively.*

Proof. For a_2^i and b_2 the continuity is obvious. Further, for $T_i = T$ the Black-Scholes formula (7.8) immediately implies the continuity if a_1^i.

Now assume that $T_i < T$. In this case we will make use of definition 2 ($K_i \geq D$) to prove the continuity. It is easy to see, that $a_1^i(t, \sigma)$ is continuous for $(t, \sigma) \in [0, T_i) \times [\sigma_{min}, \sigma_{max}]$ and $(t, \sigma) \in (T_i, T] \times [\sigma_{min}, \sigma_{max}]$.

To verify, that a_1^i is also continuous in (T_i, σ), $\sigma \in [\sigma_{min}, \sigma_{max}]$, it is sufficient to show that the left hand limit $\lim_{t \to T_i-} C_i(t, D; \sigma)$ equals zero for fixed $\sigma \in [\sigma_{min}, \sigma_{max}]$. Note that

$$\lim_{t \to T_i-} \Phi\left(\frac{\log(D/K_i) + (r - \delta \pm \sigma^2/2)(T_i - t)}{\sigma \sqrt{T_i - t}}\right) =$$

$$= \lim_{t \to T_i-} \Phi\left(\frac{\log(D/K_i)}{\sigma \sqrt{T_i - t}}\right) = \begin{cases} \Phi(0) = \frac{1}{2}, & K_i = D, \\ \Phi(-\infty) = 0, & K_i > D. \end{cases}$$

This in turn implies

$$\lim_{t \to T_i-} C_i(t, D; \sigma) = \begin{cases} D \cdot \frac{1}{2} - K_i \cdot \frac{1}{2} = 0, & K_i = D, \\ D \cdot 0 - K_i \cdot 0 = 0, & K_i > D, \end{cases}$$

which concludes the proof.   □

As an implication of Lemma 3, problem (7.12) is a continuous linear semi-infinite programming problem with compact parameter sets. Applying the general theory of linear semi-infinite optimization, we can derive the following convergence theorem for the discretized problems.

Theorem 2. *Assume that* $\mathcal{SR}(\sigma_{min}, \sigma_{max}) \neq \varnothing$. *Then for each* $\epsilon > 0$ *there exist* $\delta_1, \delta_2 > 0$ *such that for each discrete subset* $M_1 \subset [0, T] \times [\sigma_{min}, \sigma_{max}]$, $|M_1| < \infty$ *and* $M_2 \subset [0, D]$, $|M_2| < \infty$ *satisfying*

$$\Delta(M_1) = \max_{(t,\sigma) \in [0,T] \times [\sigma_{min}, \sigma_{max}]} \min_{(t_1, \sigma_1) \in M_1} \|(t, \sigma) - (t_1, \sigma_1)\|_2 \leq \delta_1,$$

$$\Delta(M_2) = \max_{s \in [0,D]} \min_{s_2 \in M_2} |s - s_2| \leq \delta_2,$$

the discretized problem (7.16) is solvable and for each solution α_d^* *of (7.16) there exists a solution* α^* *of (7.12) such that* $\|\alpha_d^* - \alpha^*\|_2 \leq \epsilon$.

Proof. Note that, as we have shown in the proof of Theorem 1, the feasible set $\mathcal{SR}(\sigma_{min}, \sigma_{max})$ and the objective have no direction of recession in common. Hence the theorem is an immediate consequence of linear semi-infinite optimization theory, see for example Hettich [12].   □

Theorem 2 guarantees the convergence of solutions of the discretized problems (7.16) to solutions of the original problem (7.12) as we refine the grids M_1, M_2. Similar to Goberna and Lopez [11] we avoid unnecessary efforts at each iteration by only refining the meshes locally in the neighborhood of nearly active constraints. This leads to the following algorithm:

Algorithm 1 *Let* $M_1 \subset [0, T] \times [\sigma_{min}, \sigma_{max}]$, $|M_1| < \infty$ *and* $M_2 \subset [0, D]$, $|M_2| < \infty$ *be given initial grids. Further let* ϵ_1, ϵ_2 *and* $TOL > 0$ *be given error tolerances.* $d_1, d_2 > 1$ *shall denote refinement parameters, let* $i = 0$ *be the initial iteration index.*

(S1) Calculate an optimal solution $\alpha^i \in \mathbb{R}^n$ *of the discretized problem (7.16) with given sets* M_1, M_2.

(S2) Calculate the slack at α^i:

$$\delta_1 = \min_{(t,\sigma) \in [0,T] \times [\sigma_{min}, \sigma_{max}]} a_1(t, \sigma)^T \alpha^i$$

$$\delta_2 = \min_{s \in [0,D]} a_2(s)^T \alpha^i - b_2(s)$$

If $\min\{\delta_1, \delta_2\} \geq -TOL$ *then stop.*

(S3) For each $(t, \sigma) \in M_1$, $s \in M_2$ Do

   If $a_1(t, \sigma)^T \alpha^i < \epsilon_1$ locally refine M_1 around (t, σ).

   If $a_2(s)^T \alpha^i - b_2(s) < \epsilon_2$ locally refine M_2 around s.

End Do
Set $i \leftarrow i + 1$, $\epsilon_1 \leftarrow \epsilon_1/d_1$, $\epsilon_2 \leftarrow \epsilon_2/d_2$ and go to step (S1).

Note that due to Theorem 1 the existence of a solution of the discretized problem in step (S1) of the algorithm is guaranteed, if the initial grids M_1, M_2 are sufficiently dense in the corresponding parameter sets.

Furthermore, we can prove in analogy to Lemma 3, that a_1^i is twice continuously differentiable in case $K_i > D$. If $K_i = D$ the smoothness is preserved on $([0, T_i) \cup (T_i, T]) \times [\sigma_{min}, \sigma_{max}]$ - however, a corner occurs at $\{T_i\} \times [\sigma_{min}, \sigma_{max}]$. If we smooth out the non-differentiabilities, the minimization in step (S2) can be carried out by an appropriate Newton-based method. Of course, the points resulting from this optimization can also be included in the sets M_1, M_2 for the next iteration.

Regarding the mesh refinement in step (S3), we decided to add the following points to the sets M_1, M_2:

$$M_1 \leftarrow M_1 \cup \left\{ \begin{pmatrix} t - \Delta t \\ \sigma - \Delta \sigma \end{pmatrix}, \begin{pmatrix} t - \Delta t \\ \sigma + \Delta \sigma \end{pmatrix}, \begin{pmatrix} t + \Delta t \\ \sigma - \Delta \sigma \end{pmatrix}, \begin{pmatrix} t + \Delta t \\ \sigma + \Delta \sigma \end{pmatrix} \right\},$$
$$M_2 \leftarrow M_2 \cup \{s - \Delta s, s + \Delta s\},$$

where Δt, $\Delta \sigma$ and Δs are successively reduced during the iteration process.

In the next section we present several numerical examples that show the efficiency of the proposed method and of the obtained hedge portfolios.

7.4 An Example of a Hedging Portfolio

In this section we solve optimization problem (7.12) with the proposed algorithm to compute static hedging strategies with varying degree of robustness. Following Giese and Maruhn [10], our goal is to hedge an up-and-out call with strike $K = 2750$, barrier $D = 3300$ and maturity $T = 1$. As the underlying of the barrier option we choose the EuroStoxx50 index with price $S_0 = 2750$ in September 2004. Furthermore, the risk-free rate and the dividend yield are assumed to satisfy $r = 5.5\%$ and $\delta = 2.5\%$. The implied volatility at time $t = 0$ shall be given by $\sigma_0 = 20\%$.

For these parameters, the value of the up-and-out call C_{uo} can be calculated via the following closed form formula (see for instance [17]) to be approximately $\Pi(C_{uo}) = 31.00$.

$$\Pi(C_{uo}) = \text{Call}(S_0, K) - \text{Call}(S_0, D) - (D - K)e^{-rT}\Phi(d_1) -$$

$$- \left(\frac{D}{S_0}\right)^{\frac{2\nu}{\sigma^2}} \left\{\text{Call}\left(\frac{D^2}{S_0}, K\right) - \text{Call}\left(\frac{D^2}{S_0}, D\right) - (D - K)e^{-rT}\Phi(d_2)\right\},$$

$$\nu = r - \delta - \frac{\sigma^2}{2}, \qquad d_1 = \frac{\log\left(\frac{S_0}{D}\right) + \nu T}{\sigma\sqrt{T}}, \qquad d_2 = \frac{\log\left(\frac{D}{S_0}\right) + \nu T}{\sigma\sqrt{T}}.$$

Here Call(A, B) denotes the Black Scholes value of a European call at time $t = 0$ given by (7.8) with strike B, maturity $T = 1$, value of the underlying A at $t = 0$ and $\sigma_0 = 20\%$.

To compute a static hedging strategy for the up-and-out call, we have to decide which calls C_i should be included in the hedge portfolio. Instead of selecting artificial instruments, we choose C_i to be the EuroStoxx50 European calls listed in table 7.1. These calls are liquidly traded on the EUREX[6] such that the hedge portfolio can in fact be bought in reality. Note that the strikes and maturities satisfy Definition 2. In addition the call with maturity 1.00 and strike 2750 guarantees, that a super-replicating static strategy exists - that means the feasible set $\mathcal{SR}(\sigma_{min}, \sigma_{max})$ is non-empty for any choice of $0 < \sigma_{min} \leq \sigma_0 \leq \sigma_{max}$. Hence, by Theorem 1 the linear semi-infinite optimization problem (7.12) has a solution.

Table 7.1. European calls C_i included in the hedge portfolio.

	C_1	C_2	C_3	C_4	C_5	C_6	C_7	C_8	C_9	C_{10}
T_i	0.50	0.50	0.75	0.75	0.75	1.00	1.00	1.00	1.00	1.00
K_i	3300	3500	3300	3400	3600	2750	3300	3350	3450	3600

In the following we will first compute the non-robust static hedge portfolio ($\sigma_{min} = \sigma_0 = \sigma_{max}$) and numerically test its sensitivity with respect to the volatility parameter σ. Afterwards, we increase the degree of robustness by gradually extending the robustness interval $[\sigma_{min}, \sigma_{max}]$.

7.4.1 Solution of the Non-Robust Problem

To solve the non-robust optimization problem, we choose the following parameters in algorithm 1: $\epsilon_1 = 10^{-1}$, $d_1 = 10$ and an error tolerance of $TOL = 10^{-4}$ which is a sufficient accuracy for the hedge error in Euros. For the non-robust problem the compact set $[0, T] \times [\sigma_{min}, \sigma_{max}]$ is in fact just the line $[0, T] \times \{\sigma_0\}$, hence it suffices to discretize the time interval. We start with an initial grid M_1 consisting of 41 time steps.

---

[6] For further information see http://www.eurexchange.com.

A quick analysis[7] shows, that we can replace the semi-infinite constraint $a_2(s)^T \alpha \geq b_2(s) \ \forall \ s \in [0, D]$ by the single constraint $a_2(D)^T \alpha \geq b_2(D)$ due to the particular choice of the calls in table 7.1. Hence we can omit the grid M_2 as well as the parameters ϵ_2 and d_2 during the optimization.

Table 7.2 presents the results from the iteration process. Algorithm 1 successfully terminates after five outer iterations with the optimal solution given in table 7.3, reaching a slack of less than 10^{-4}. The cost of the hedge portfolio at iteration i, denoted by $f(\alpha^i)$ clearly converges to an optimal value of 31.39, which is only marginally more expensive than the barrier option itself $(\Pi(C_{uo}) = 31.00)$.

Table 7.2. Iteration Process for the Non-Robust Problem.

Iteration	Slack	$f(\alpha^i)$	$|M_1|$	ϵ_1
0	-5.690659e-001	3.1373255e+001	41	—
1	-1.340881e-001	3.1389511e+001	68	1.00000e-001
2	-9.928853e-003	3.1391149e+001	95	1.00000e-002
3	-1.095008e-003	3.1391429e+001	122	1.00000e-003
4	-1.216220e-004	3.1391459e+001	149	1.00000e-004
5	-3.335607e-005	3.1391461e+001	208	1.00000e-004

Table 7.3. Optimal Portfolio Weights α_i for the European Calls C_i.

	C_1	C_2	C_3	C_4	C_5	C_6	C_7	C_8	C_9	C_{10}
T_i	0.50	0.50	0.75	0.75	0.75	1.00	1.00	1.00	1.00	1.00
K_i	3300	3500	3300	3400	3600	2750	3300	3350	3450	3600
α_i	0.020	-0.09	-0.04	0.04	-0.25	1.00	-33.02	40.06	-8.20	0.79

The number of grid points $|M_1|$ increases by a factor of five until a sufficient accuracy is reached - however, the mesh is only locally refined in the vicinity of active constraints. This property is clearly visible in figure 7.3, which plots the hedge error of the optimal portfolio for all times $t \in [0, T]$ in case the barrier is hit.

Figure 7.3 also reveals the qualitative behavior of the static hedge portfolio. The hedge error is almost always zero along the barrier, however, it jumps to $550 = D - K$ close to maturity of the barrier option[8]. In fact a *perfect* super-replication portfolio solely

---

[7] The only call satisfying $T_i = T$ and $K_i < D$ is the European call with maturity 1.00 and strike $K_i = 2750 = K$. Hence the linear system $a_2(s)^T \alpha = \alpha_i \max\{s - 2750, 0\} \geq b_2(s) = \max\{s - 2750, 0\}$ is linearly dependent and reduces to the single constraint $\alpha_i \max\{D - 2750, 0\} \geq \max\{D - 2750, 0\}$ which is equivalent to $\alpha_i \geq 1$.

[8] Note that the hedge portfolio has to guarantee the payout of 550 at $T = 1$ to satisfy the constraint $a_2(D)^T \alpha \geq b_2(D)$, that means to super-replicate the barrier option in case the barrier is not hit until time $T = 1$.

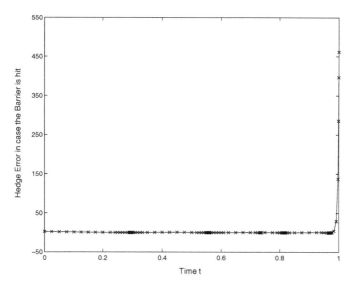

Fig. 7.3. Hedge Error $a_1(t, \sigma_0)^T \alpha$ on the Barrier for the Optimal Portfolio.

consisting of calls would deliver a hedge error of zero for $t \in [0, 1)$ and of 550 at time $T = 1$. However, as the value of the hedge portfolio and hence the hedge error evolves continuously in time (see Lemma 3), the graph exhibits a steep slope towards 550 close to maturity. Obviously, the optimal hedge portfolio is *nearly perfect*.

As an additional benefit, the adaptive grid refinement illustrated in Figure 7.3 identifies the points in time, in which a portfolio trader has to pay particular attention to prevent losses in case the barrier is hit. In our case these seem to be five time points, where four of them occur after half of the maturity of the barrier option.

In [10] Giese and Maruhn also derived the optimal hedge portfolio listed in table 7.3. However, they obtain the portfolio by a sample average approach to solve the underlying stochastic optimization problem which requires the simulation of M sample paths of the stock price, where M usually has to be chosen quite large. In contrast to this, the solution of the proposed linear semi-infinite optimization problem only takes a fraction of the time it takes to solve the stochastic optimization problem. In fact, the optimization above was carried out in less than five seconds on a usual PC with 3 GHz CPU.

As mentioned before, the key assumption of the hedge portfolio derived so far is, that the portfolio can be sold anytime the barrier is hit for model prices based on the volatility σ_0 chosen at $t = 0$. If the volatility at the hitting time differs from the initial volatility, the hedge portfolio might not offer the protection expected from a super-replication portfolio. To analyze this, we simulated $M = 100,000$ sample paths[9] of the Black-Scholes model equation (7.7) to determine the simulated hedge error for different volatility states σ. In particular we focus on the loss the bank might encounter by selling the barrier option and buying the hedge portfolio consisting of European calls. The results are shown in table 7.4.

Table 7.4. Simulation results for the non-robust optimal portfolio and various volatility states σ of the Black Scholes model. The results are based on $100,000$ sample paths of the underlying.

	5%	10%	15%	20%	25%	30%
Average Loss	−0.07	−2.89	−2.80	0.00	0.00	0.00
Largest Loss	−128.06	−42.35	−16.72	0.00	0.00	0.00
Probability of Loss	0.10%	9.35%	25%	0.00	0.00	0.00

Obviously, the non-robust super-replication portfolio offers protection for volatility states $\sigma \geq \sigma_0 = 20\%$. However, as the volatility gradually decreases below 20%, large hedge errors may occur for a significant number of future states of the economy. Although the probability of hitting the barrier and encountering portfolio losses decreases as we reduce the volatility, a probability of approximately 9% for quite large losses signalizes a problem of the hedge portfolio. Moreover, compared to the price of the barrier option $\Pi(C_{uo}) = 31.00$, a largest loss of for example -42.35 in case $\sigma = 10\%$ is unacceptable and contradicts the notion of a super-replication strategy.

Figure 7.4 gives some further insight into the structure of the error, showing a continuous plot of the hedge error on the barrier for volatilities ranging from 0% to 40%. Although it looks on first sight as if the characteristics of the hedge error plotted in figure 7.3 carry over to the other volatility states, a closer look at the minimal hedge error over time reveals, that the portfolio is exactly fitted for the volatility state $\sigma = 20\%$. An increase in volatility also increases the hedge error, but does not lead to a loss. In contrast to this, a decreasing volatility may lead to significant losses that can be multiples of the initial value of the barrier option.

In summary we can conclude that the optimal non-robust hedge portfolio, although one is tempted to assume a general super-replication property, only offers protection for volatility states greater or equal to the initial volatility $\sigma_0 = 20\%$. Volatility shocks below this initial volatility may result in large hedging losses which are undesired in practice. To avoid these losses, we have to compute an optimal solution which is robust with respect to changes in the volatility.

---

[9] For the time discretization of the stochastic differential equation, we used the Euler-Maruyama Scheme with step size $\Delta t = 0.001$.

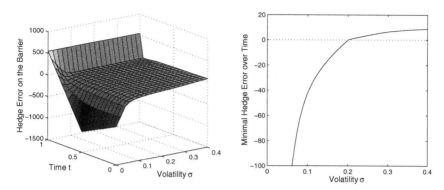

Fig. 7.4. Hedge Error on the barrier for the non-robust optimal portfolio for a wide variety of volatility states

7.4.2 Adding Robustness to the Solution

Having experienced the sensitivity of the hedge portfolio computed in the previous subsection with respect to changes in the model parameters, we now aim at adding robustness to the super replication strategy. Mathematically, we will solve optimization problem (7.12) with $\sigma_{min} < \sigma_0 = 20\% < \sigma_{max}$. Economically, this guarantees the super-replication property for volatility states $\sigma \in [\sigma_{min}, \sigma_{max}]$. We begin by demanding robustness for the interval $[15\%, 20\%]$, hence $\sigma_{min} = 15\%$, $\sigma_{max} = 25\%$.

To solve the robust optimization problem (7.12) we choose the following parameters in algorithm 1: $\epsilon_1 = 1.0$, $d_1 = 5$ and $TOL = 10^{-4}$. In contrast to the non-robust problem, the set $[0, T] \times [\sigma_{min}, \sigma_{max}]$ is now two-dimensional, such that we have to discretize in both the time- and the volatility-dimension. We start with an equidistant grid M_1, consisting of a total of $21 \cdot 21 = 441$ points. By the same argument as in the previous subsection, we can omit M_2, ϵ_2 and d_2 from the optimization.

The output of algorithm 1 and the corresponding optimal solution is listed in tables 7.5 and 7.6, respectively.

The algorithm terminates after eight iterations with zero slack. The optimal portfolio has a cost of 32.69 which is only slightly more expensive than the non-robust portfolio (31.39) and the barrier option (31.00). However, in contrast to the non-robust portfolio, the robust solution offers protection for the desired volatility range $[15\%, 25\%]$ as can be seen in figure 7.5. The graph of the hedge error also shows the local mesh refinement around the nearly active constraints. During the iteration process, the number of grid points increases approximately by a factor of three.

Figure 7.5 also reveals the points of non-differentiability of the hedge-error function on the barrier. Recall from the end of Section 7.3, that these points are given by $\{T_i\} \times [\sigma_{min}, \sigma_{max}]$ for calls C_i with $K_i = D$ and $T_i < T$. In our case, these are the two calls C_1 and C_3 leading to non-differentiabilities along the lines $\{0.5\} \times [15\%, 25\%]$

Table 7.5. Iteration Process for the Robust Problem.

Iteration	Slack	$f(\alpha^i)$	$|M_1|$	ϵ_1
0	-4.411300e-001	3.260217e+001	441	—
1	-3.082539e-001	3.265463e+001	650	1.00000e+000
2	-8.193626e-002	3.268421e+001	745	2.00000e-001
3	-1.482588e-001	3.268934e+001	823	4.00000e-002
4	-1.976512e-001	3.269014e+001	891	8.00000e-003
5	-2.593012e-002	3.269058e+001	967	1.60000e-003
6	-1.910988e-001	3.269072e+001	1035	3.20000e-004
7	-1.901378e-003	3.269076e+001	1103	6.40000e-005
8	0.000000e+000	3.269078e+001	1171	1.28000e-005

Table 7.6. Optimal Portfolio Weights α_i for the European Calls C_i.

	C_1	C_2	C_3	C_4	C_5	C_6	C_7	C_8	C_9	C_{10}
T_i	0.50	0.50	0.75	0.75	0.75	1.00	1.00	1.00	1.00	1.00
K_i	3300	3500	3300	3400	3600	2750	3300	3350	3450	3600
α_i	0.00	0.06	0.01	0.15	-0.10	1.00	-30.87	35.66	-4.31	-1.46

and $\{0.75\} \times [15\%, 25\%]$. However, as we smoothed out the objective at these points, the non-differentiabilities do not pose a severe problem.

Compared to the non-robust solution, the qualitative behavior of the hedge observed in figure 7.3 now carries over to the whole volatility interval. The robust portfolio has a relatively small hedge error for small t and offers a comfortable safety cushion close to maturity. The price for this additional protection, preventing possibly large hedging losses, seems to be quite low. In comparison to the non-robust portfolio, the price only increases by about 4%.

Of course the obtained robust hedge portfolio might still deteriorate if the volatility drops below $\sigma_{min} = 15\%$. This is clearly visible in figure 7.6 which shows the minimal hedge error for the robust portfolio. However, the probability that the portfolio fails to hedge the barrier option is much lower than in the non-robust case.

To offer protection for the additional volatility states not covered by the robust hedge portfolio derived so far, we have to increase the volatility interval underlying the robust optimization problem (7.12). Intuitively, the cost of a robust portfolio will increase as we enhance the protection to a larger volatility interval. Hence the question of how large to choose the volatility interval is merely a question of how risk-averse the portfolio trader is. As is common in utility theory, there will be an optimal risk/cost-combination for every trader.

As examples for the possible degrees of robustness and the associated cost-effects we list several risk/cost combinations in table 7.7. A classical risk-averter would hence choose the robust portfolio offering protection for the set of volatility states

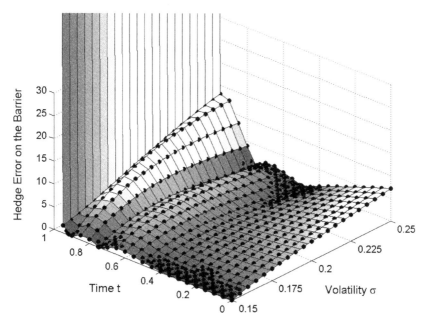

Fig. 7.5. Hedge Error on the barrier for the robust optimal portfolio for a wide variety of volatility states.

$[0\%, 100\%]$ with an additional cost of about 9 Euros in comparison to the non-robust portfolio. This risk-averter portfolio is about 36 basispoints[10] more expensive than the barrier option itself, which is still not too expensive for a robust portfolio also covering extreme cases.

Table 7.7. Cost of robust hedge portfolios with varying degree of robustness.

σ_{min}	20%	15%	10%	5%	0%
σ_{max}	20%	25%	30%	50%	100%
Cost	31.39	32.69	35.08	37.08	40.77

7.5 Conclusions

In this paper we extended the static super-replication approach developed by Giese and Maruhn in [10] to a robust semi-infinite linear programming formulation. Due

---

[10] One basispoint is 0.01% of the underlying stock price at $t = 0$, in our case $S_0 = 2750$.

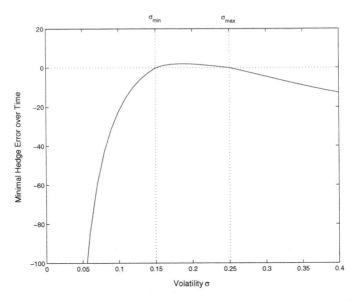

Fig. 7.6. Minimal Hedge Error on the barrier for the robust optimal portfolio.

to the structure of the problem, we are able to prove existence of an optimal solution and the convergence of solutions of the discretized problem. The proposed algorithm quickly solves the semi-infinite problem and hence is superior to the sample average approach used by Giese and Maruhn to solve the equivalent stochastic optimization problem.

As the numerical results show, the non-robust portfolio may result in large hedging losses for portfolio traders if model parameters change over time. In contrast to this, the robust static hedge portfolio preserves the super replication property for a broad range of model parameters although it is not much more expensive than the non-robust portfolio. This makes the portfolio attractive for option traders in practice.

Although we derived our results in the Black Scholes model, the robust hedging approach is likely to succeed in more advanced models that calibrate better to a given volatility surface. This will be the focus of further investigation. Under suitable assumptions, jumps in the stock price can also be included in the semi-infinite programming formulation to add robustness with respect to possible liquidation delays at the hitting time.

Acknowledgments

The authors wish to thank A. M. Giese (HypoVereinsbank AG, HVB Group, Corporates & Markets, Equity Linked Products, Munich, Germany) for his encouragement and support.

References

[1] Andersen, L. B. G., Andreasen, J. and Eliezer, D. *Static Replication of Barrier Options: Some General Results.* Journal of Computational Finance, Volume 5, number 4, 2002.

[2] Bowie, J. and Carr, P. *Static Simplicity.* Risk 7, pp.45-49, 1994.

[3] Boyle, A., Coleman, T. and Li, Y. *Hedging a Portfolio of Derivatives by Modeling Cost.* Technical Report, Cornell University, 2003.

[4] Carr, P., Ellis, K. and Gupta, V. *Static Hedging of Exotic Options.* Journal of Finance, Volume 53, pp. 1165-1190, 1998.

[5] Coleman, T., Li, Y. and Patron, M. *Discrete Hedging under Piecewise Linear Risk Minimization.* Cornell University, Technical Report, 2002.

[6] Derman, E., Ergener, D. and Kani, I. *Static options replication.* The Journal of Derivatives, Volume 2, pp. 78-95, 1995.

[7] Dupont, D. Y. *Hedging Barrier Options: Current Methods and Alternatives.* EURANDOM - TUE. Technical Report. 2001.

[8] Fink, J. *An Examination of the Effectiveness of Static Hedging in the Presence of Stochastic Volatility.* The Journal of Futures Markets, Volume 23, number 9, pp. 859-890, 2003.

[9] Friedman, A. *Partial Differential Equations of Parabolic Systems.* Prentice-Hall, 1964.

[10] Giese, A. M. and Maruhn, J. H. *Cost-Optimal Static Super-Replication of Barrier Options - An Optimization Approach.* Preprint, University of Trier, 2005.

[11] Goberna, M. A. and López, M. A. *Linear Semi-Infinite Optimization.* Wiley, 1998.

[12] Hettich, R. *Numerische Methoden der Approximation und semi-infiniten Optimierung.* Teubner, Stuttgart, 1982.

[13] Karatzas, I. and Shreve, S. E. *Methods of Mathematical Finance.* Springer, 1998.

[14] Mizohata, S. *Unicité du prolongement des solutions pour quelques operatéurs differentiels paraboliques.* Mem. Coll. Sci. Univ. Kyoto Ser. A Math., Volume 31, pp. 219-239, 1958.

[15] Rockafellar, R. T. *Convex Analysis.* Princeton University Press, 1970.

[16] Toft, K. B. and Xuan, C. *How Well Can Barrier Options Be Hedged by a Static Portfolio of Standard Options?* The Journal of Financial Engineering, Volume 7, number 2, pp. 147-175, 1998

[17] Zhang, P. G. *Exotic Options - A Guide to Second Generation Options.* World Scientific, 1998.

8

Numerical Techniques in Relaxed Optimization Problems [*]

Tomáš Roubíček[1,2]

[1] Mathematical Institute
Charles University
Sokolovská 83
CZ-186 75 Praha 8, Czech Republic
[2] Institute of Information Theory and Automation
Academy of Sciences
Pod vodárenskou věží 4
CZ-182 08 Praha 8, Czech Republic
`tomas.roubicek@mff.cuni.cz`

Summary. Young measures and their various generalizations are a basic analytical tool for extension (=relaxation) of optimization problems that may lack of solutions because of various oscillation/concentration effects. Various numerical techniques designed directly for the relaxed problems are surveyed together with some specific applications.

Key words: Young measures, variational problems, numerical approximation

8.1 Introduction

Let us begin with a rather elementary motivating example, assigned to Bolza, which consists in the following optimal-control problem governed by an initial-value problem for an ordinary differential equation on a fixed time horizon $T > 0$:

$$
\begin{aligned}
\min \quad & J(y, u) := \int_0^T \left((1 - u^2)^2 + y^2 \right) \mathrm{d}t \\
\text{s.t.} \quad & \frac{\mathrm{d}y}{\mathrm{d}t} = u, \ \ y(0) = 0, \\
& y \in W^{1,4}(0, T), \quad u \in L^4(0, T),
\end{aligned}
\tag{8.1}
$$

[*] This research was partly covered by the grants IAA 1075402 (GA AV ČR), and MSM 0021620839 (MŠMT ČR). Special thanks are to dr. Martin Kružík for his comments and for the calculations that have been exploited for Figures 8.3 and 8.4.

where $L^p(0,T)$ and $W^{1,p}(0,T)$ denote the standard Lebesgue space of measurable functions $[0,T] \to \mathbb{R}$ with p-power integrable and the Sobolev space of functions $[0,T] \to \mathbb{R}$ whose distributional derivative belongs to $L^p(0,T)$, respectively; here $p = 4$. An example of a minimizing sequence $\{(u_k, y_k)\}_{k \in \mathbb{N}}$ for (8.1) is on the following Figure 8.1:

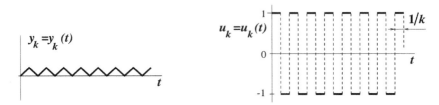

Fig. 8.1. An example of a minimizing sequence $\{(y_k, u_k)\}_{k \in \mathbb{N}}$ for (8.1).

The original problem (8.1) has no solution, which can easily be seen by a contradiction argument: as $J \geq 0$, the infimum of (8.1) is obviously non-negative, and Figure 8.1 shows that it is even zero, but if $J(y,u) = 0$, then $y = 0$, hence also $\frac{d}{dt}y = 0$, hence $u = 0$, but then $J(y,u) = T$, a contradiction. One can intuitively imagine that the minimizing sequence $\{u_k\}_{k \in \mathbb{N}}$ wants to stand simultaneously around 1 and -1 with an equal probability, and we say that, in a sense, $\{u_k\}_{k \in \mathbb{N}}$ converges to a parameterized probability measure (a so-called *Young measure*) $\{\nu_t\}_{t \in [0,T]}$, which is here constant (i.e. independent of t):

$$\nu_t = \frac{1}{2}\delta_{-1} + \frac{1}{2}\delta_1, \tag{8.2}$$

where δ denotes the Dirac measure. The corresponding extended (so-called *relaxed*) problem

$$\min \quad J(y,\nu) := \int_0^T \left(\int_{\mathbb{R}} (1-s^2)^2 \nu_t(\mathrm{d}s) + y^2 \right) \mathrm{d}t$$
$$\text{s.t.} \quad \frac{\mathrm{d}y}{\mathrm{d}t} = \int_{\mathbb{R}} s\, \nu_t(\mathrm{d}s), \quad y(0) = 0, \tag{8.3}$$
$$y \in W^{1,4}(0,T), \quad \nu \in \mathcal{Y}^4(0,T),$$

where $\mathcal{Y}^p(0,T) \equiv \mathcal{Y}^p(0,T;\mathbb{R})$ is the set of the so-called L^p-Young measures on $[0,T]$ valued in $\mathbb{R}$, defined for a general Ω and $\mathbb{R}^m$ as

$$\mathcal{Y}^p(\Omega;\mathbb{R}^m) := \Big\{ \nu : \Omega \to \mathcal{M}_1^+(\mathbb{R}^m); \ t \mapsto \nu(t) \equiv \nu_t \text{ weakly*}$$
$$\text{measurable}, \ \int_\Omega |s|^p \nu_t(\mathrm{d}s)\mathrm{d}t < +\infty \Big\}, \tag{8.4}$$

where $\mathcal{M}_1^+(\mathbb{R}^m)$ denotes the set of probability measures on $\mathbb{R}^m$. The above problem (8.1) obviously lacked solution because of fast-oscillation effects in (all!) minimizing sequences. Let us remark that, quite mysteriously, if the term $+y^2$ in (8.1) is replaced by $-y^2$, (8.1) has got a solution; for a nonconstructive proof using relaxation, Bauer's extremal principle, and the "Dirac-measure structure" of extreme Young measures see [47].

Other reason for nonexistence of a solution might be concentration effects, as shown on another model optimal control problem:

$$
\begin{aligned}
\min \quad & J(y,u) := \int_0^T \left(2-2t+t^2\right)\left|u(t)\right|\,\mathrm{d}t + \left(y(T){-}1\right)^2 \\
\text{s.t.} \quad & \frac{\mathrm{d}y}{\mathrm{d}t} = u, \ \ y(0) = 0, \\
& y \in W^{1,1}(0,T), \ \ u \in L^1(0,T).
\end{aligned}
\tag{8.5}
$$

For $T > 1$, a minimizing sequence for (8.5) is, e.g.,

$$
y_k(t) = \begin{cases} 0, & \text{if } t \in (0,1], \\ \tfrac{1}{2}k(t{-}1), & \text{if } t \in (1,1+1/k], \\ 1/2, & \text{if } t \in (1+1/k,T], \end{cases} \qquad u_k(t) = \begin{cases} 0, \\ \tfrac{1}{2}k, \\ 0, \end{cases}
\tag{8.6}
$$

as illustrated on Figure 8.2.

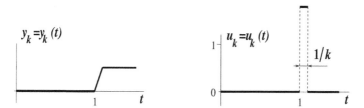

Fig. 8.2. An example of a minimizing sequence $\{(y_k, u_k)\}_{k\in\mathbb{N}}$ for (8.5) from (8.6).

Here the construction of a relaxed problem is incidentally quite simple because there are no oscillation effects, neither any complicated interplay between the state y and the resulting impulse control: it can use a conventional measure $\sigma \in \mathcal{M}([0,T])$, the controlled system $\frac{\mathrm{d}}{\mathrm{d}t}y = \sigma$ in terms of distributions with the initial condition $y|_{t=0} = 0$, and the cost functional $J(\sigma,y) = \int_0^T (2-2t+t^2)|\sigma|(\mathrm{d}t) + (y|_{t=T}-1)^2$ where $|\sigma|$ denotes the total variation of σ and $y|_{t=0}$ and $y|_{t=T}$ have a meaning of traces of y at the points 0 and T, respectively.

Oscillation and concentration effects can be combined together and the corresponding relaxation theory then must use objects that can record more "limit" information than Young measures or than measures on $[0,T]$, cf. [25; 26; 46; 49].

Classical Young measures has been introduced in [59] in late 30ties and soon developed e.g. by McShane [37]. Intensive usage of the Young measures started in optimal control theory [17; 18; 56], cf. e.g. [15; 16; 46] for recent research. Then it started to be popular in variational calculus and partial differential equations [3; 32; 40; 53]. We do not want to give any thorough historical survey (rather referring the reader e.g. to [2; 9; 46; 55]) because we want to focus on the direct numerical implementation of Young measures (and their possible generalization). Early, but only conceptual attempt for an iterative solution directly of the relaxed problems is due to Warga [57]. Real computation attempt then followed much later due to Nicolaides and Walkington [41], and recently quite intensive development has taken place in numerical analysis, algorithms, and applications. We want to present its survey from a general viewpoint of a theory of convex local compactifications of Lebesgue L^p-spaces, as well as to point out some specific applications.

8.2 A General Framework: Convex Local Compactifications of L^p-spaces

The Lebesgue spaces are definitely the most occurring function spaces in applications. Following [46; 48], we will briefly present a fairly universal construction of their locally compact envelopes that are also convex in a natural linear space and allow a continuous and affine extension of Nemytskiĭ mappings.

Let $\Omega \subset \mathbb{R}^n$ bounded, measurable set, and let us consider the Lebesgue space $L^p(\Omega; \mathbb{R}^m) = \{u : \Omega \to \mathbb{R}^m \text{ measurable}; \int_\Omega |u(x)|^p \mathrm{d}x < +\infty\}$; in the previous Section 8.1, $n = 1$ and $\Omega = (0, T)$. We define a normed linear space

$$\mathrm{Car}^p(\Omega; \mathbb{R}^m) := \Big\{ h : \Omega \times \mathbb{R}^m \to \mathbb{R},$$

$$h(\cdot, s) \text{ measurable}, \ h(x, \cdot) \text{ continuous},$$

$$\exists a \in L^1(\Omega), \ b \in \mathbb{R} : \ |h(x, s)| \le a(x) + b|s|^p \Big\} \qquad (8.7)$$

of "test integrands", and equip it with the norm

$$\|h\|_{\mathrm{Car}^p(\Omega; \mathbb{R}^m)} := \inf_{|h(x,s)| \le a(x)+b|s|^p} \|a\|_{L^1(\Omega)} + b. \qquad (8.8)$$

The essential trick is to consider a sufficiently large (but preferably still separable) linear subspace $H \subset \mathrm{Car}^p(\Omega; \mathbb{R}^m)$, to define the embedding

$$i_H : L^p(\Omega; \mathbb{R}^m) \to H^* : u \mapsto \Big(h \mapsto \int_\Omega h(x, u(x)) \mathrm{d}x \Big), \qquad (8.9)$$

and eventually to put

$$Y_H^p(\Omega; \mathbb{R}^m) := \text{ the weak* closure of } i_H(L^p(\Omega; \mathbb{R}^m)). \qquad (8.10)$$

One can show that $Y_H^p(\Omega; \mathbb{R}^m)$ is always convex in H^*. Assuming, rather for simplicity, that H contains at least one coercive integrand, i.e. $H \ni h_0$ with $h_0(x, s) \ge |s|^p$, then $Y_H^p(\Omega; \mathbb{R}^m)$ is a *convex locally compact hull* of $L^p(\Omega; \mathbb{R}^m)$ and $L^p(\Omega; \mathbb{R}^m)$ itself is embedded into it (norm,weak*)-continuously. Moreover, if H is rich enough (cf. [46; 48] for details), then this embeddeding i_H is even homeomorphical. If H is separable, then $Y_H^p(\Omega; \mathbb{R}^m)$ is locally sequentially compact. Thus $Y_H^p(\Omega; \mathbb{R}^m)$ may be a very natural envelope of $L^p(\Omega; \mathbb{R}^m)$.

Due to a large variety of such envelopes, it is worth thinking about an ordering of them. We say that, for H_1 and H_2 two subspaces of $\mathrm{Car}^p(\Omega; \mathbb{R}^m)$, $Y_{H_1}^p(\Omega; \mathbb{R}^m)$ is finer than $Y_{H_2}^p(\Omega; \mathbb{R}^m)$ (or the latter one is coarser than the former one) if there exists an affine continuous surjection fixing $L^p(\Omega; \mathbb{R}^m)$. If $Y_{H_1}^p(\Omega; \mathbb{R}^m)$ is finer and coarser than $Y_{H_2}^p(\Omega; \mathbb{R}^m)$, then we write $Y_{H_1}^p(\Omega; \mathbb{R}^m) \cong Y_{H_2}^p(\Omega; \mathbb{R}^m)$. In this context, a natural topology of $\mathrm{Car}^p(\Omega; \mathbb{R}^m)$ is the locally convex topology on $\mathrm{Car}^p(\Omega; \mathbb{R}^m)$ induced via seminorms $\{|\cdot|_\varrho\}_{\varrho \in \mathbb{N}}$ with $|h|_\varrho := \sup_{\|u\| \leq \varrho} |\int_\Omega h(x, u(x)) dx|$, rather than the norm topology induced by (8.8). For a fixed coercive h_0, the class $\{Y_H^p(\Omega; \mathbb{R}^m); h_0 \in H \subset \mathrm{Car}^p(\Omega; \mathbb{R}^m)\}$ is a *lattice*, the supremum and the infimum being given respectively by

$$\sup \left\{ Y_{H_1}^p(\Omega; \mathbb{R}^m), Y_{H_2}^p(\Omega; \mathbb{R}^m) \right\} = Y_{H_1 + H_2}^p(\Omega; \mathbb{R}^m), \qquad (8.11a)$$

$$\inf \left\{ Y_{H_1}^p(\Omega; \mathbb{R}^m), Y_{H_2}^p(\Omega; \mathbb{R}^m) \right\} = Y_{\bar{H}_1 \cap \bar{H}_2}^p(\Omega; \mathbb{R}^m), \qquad (8.11b)$$

where $\bar{H}_j =$ the closure of H_j in $\mathrm{Car}^p(\Omega; \mathbb{R}^m)$ with respect to its natural topology.

8.3 General Approximation Theory

To present the theory in an elegant and general framework based on an adjoint-operator technique, we will follow the construction developed in [35; 46; 45]. The first attempt exploiting the adjoint-operator technique for a particular case was due to Tartar [54] and independently in general cases in [45]. Exploiting the abstract construction of $Y_H^p(\Omega; \mathbb{R}^m)$ developed in Section 8.2, we adopt a "finite-element like" trick by considering some (often finite-dimensional) subspace H_d of $H \subset \mathrm{Car}^p(\Omega; \mathbb{R}^m)$. Here $d = (d_1, d_2)$ denotes the discretization bi-parameter, $d_1 > 0$ referring to a discretization of the "physical" domain Ω while $d_2 > 0$ is for a discretization of $\mathbb{R}^m$. We will consider H_d in a form

$$H_d = P_{d_1} Q_{d_2} H, \quad P_{d_1}, Q_{d_2} : H \to H \text{ linear, bounded.} \qquad (8.12)$$

To perform a rate-of-error analysis, we introduce a space of smooth integrands

$$\mathrm{Car}^{k,l,p}(\Omega; \mathbb{R}^m) := \left\{ h \in \mathrm{Car}^p(\Omega; \mathbb{R}^m); \ \|h\|_{\mathrm{Car}^{k,l,p}(\Omega; \mathbb{R}^m)} < +\infty \right\} \qquad (8.13)$$

with k referring to the order of differentiability in $x \in \Omega$ and l to the order of differentiability in $s \in \mathbb{R}^m$. A natural choice of the norm in (8.13) is

$$\|h\|_{\mathrm{Car}^{k,l,p}(\Omega;\mathbb{R}^m)} := \|h\|_{\mathrm{Car}^p(\Omega;\mathbb{R}^m)}$$

$$+ \left\|\frac{\partial^k h}{\partial x^k}\right\|_{\mathrm{Car}^p(\Omega;\mathbb{R}^m)} + \left\|\frac{\partial^l h}{\partial s^l}\right\|_{\mathrm{Car}^{p-l}(\Omega;\mathbb{R}^m)}$$

provided $p \geq l$, with the convention that, if $n > 1$, $\partial^k/\partial x^k$ means all kth-order derivatives and analogously, if $m > 1$, $\partial^l/\partial s^l$ means all lth-order derivatives.

As already outlined, an *approximation "in the x-variable"* is via an abstract linear $(\mathrm{Car}^p(\Omega;\mathbb{R}^m), \mathrm{Car}^p(\Omega;\mathbb{R}^m))$-bounded operator $P_{d_1} : H \to H$, with $0 < d_1$ a discretization parameter. Often (but not necessarily), P_{d_1} will be a projector, i.e. $P_{d_1} P_{d_1} = P_{d_1}$. The following abstract approximation property is desirable:

$$\|h - P_{d_1} h\|_{\mathrm{Car}^{0,0,p}(\Omega;\mathbb{R}^m)} \leq C_1 d_1^k \|h\|_{\mathrm{Car}^{k,0,p}(\Omega;\mathbb{R}^m)}. \tag{8.14}$$

Likewise, an *approximation "in the s-variable"* is via an abstract linear $(\mathrm{Car}^p(\Omega;\mathbb{R}^m), \mathrm{Car}^p(\Omega;\mathbb{R}^m))$-bounded operator $Q_{d_2} : H \to H$ (often being a projector) with $0 < d_2$ another discretization parameter, and, like in (8.14), we assume the approximation property:

$$\|h - Q_{d_2} h\|_{\mathrm{Car}^{0,0,p}(\Omega;\mathbb{R}^m)} \leq C_2 d_2^l \|h\|_{\mathrm{Car}^{0,l,p}(\Omega;\mathbb{R}^m)}. \tag{8.15}$$

By combining (8.14) and (8.15), we get the estimate:

$$\begin{aligned}
\|h - P_{d_1} Q_{d_2} h\|_{\mathrm{Car}^p(\Omega;\mathbb{R}^m)} &= \|h - P_{d_1} Q_{d_2} h\|_{\mathrm{Car}^{0,0,p}(\Omega;\mathbb{R}^m)} \\
&\leq \|h - P_{d_1} h\|_{\mathrm{Car}^{0,0,p}(\Omega;\mathbb{R}^m)} + \|P_{d_1} h - P_{d_1} Q_{d_2} h\|_{\mathrm{Car}^{0,0,p}(\Omega;\mathbb{R}^m)} \\
&\leq C_1 d_1^k \|h\|_{\mathrm{Car}^{k,0,p}(\Omega;\mathbb{R}^m)} + C_2 d_2^l N_{d_1} \|h\|_{\mathrm{Car}^{0,l,p}(\Omega;\mathbb{R}^m)} \\
&\leq C(d_1^k + d_2^l) \|h\|_{\mathrm{Car}^{k,l,p}(\Omega;\mathbb{R}^m)}, \tag{8.16}
\end{aligned}$$

provided

$$\limsup_{d_1 \to 0} N_{d_1} < +\infty, \quad N_{d_1} := \|P_{d_1}\|_{\mathcal{L}(\mathrm{Car}^p(\Omega;\mathbb{R}^m), \mathrm{Car}^p(\Omega;\mathbb{R}^m))}. \tag{8.17}$$

Then the set $Q_{d_2}^* P_{d_1}^* Y_H^p(\Omega;\mathbb{R}^m)$ approximates $Y_H^p(\Omega;\mathbb{R}^m)$ in some sense, which gives rise to approximate problems simply by replacing $Y_H^p(\Omega;\mathbb{R}^m)$ in the original problem in question by $Q_{d_2}^* P_{d_1}^* Y_H^p(\Omega;\mathbb{R}^m)$. Worth mentioning, this set is always *convex* because $Y_H^p(\Omega;\mathbb{R}^m)$ itself is convex and because both $P_{d_1}^*$ are $Q_{d_2}^*$ linear. A quite desirable (but not automatic!) property is $Q_{d_2}^* P_{d_1}^* Y_H^p(\Omega;\mathbb{R}^m) \subset Y_H^p(\Omega;\mathbb{R}^m)$; in analog to the conventional finite-element theory, we call such an approximation as *conformal*, otherwise *nonconformal*. Approximate "solutions" obtained by non-conformal approximations need not be attainable by "original controls", and loose thus a reasonable interpretation. If such an approximation is conformal, than it can be identified by a coarser convex locally compact hull obtained by replacing H with H_d from (8.12), i.e.

$$Q_{d_2}^* P_{d_1}^* Y_H^p(\Omega; \mathbb{R}^m) \cong Y_{H_d}^p(\Omega; \mathbb{R}^m) := Y_{P_{d_1} Q_{d_2} H}^p(\Omega; \mathbb{R}^m), \qquad (8.18)$$

cf. [46, Lemma 3.5.1].

Anyhow, both conformal and nonconformal approximations can be exploited to the rate-of-error estimate at least as far as the optimal value of a minimized functional concerns according to the following scheme: Let us put

$$\|\eta\|_{-k,-l,p} := \sup_{\substack{h \in H \\ \|h\|_{\mathrm{Car}^{k,l,p}(\Omega; \mathbb{R}^m)} \leq 1}} \langle \eta, h \rangle. \qquad (8.19)$$

Then, by a *transposition*, one obtains the error estimate:

$$
\begin{aligned}
\|\eta - Q_{d_2}^* P_{d_1}^* \eta\|_{-k,-l,p} &= \sup_{\substack{h \in H \\ \|h\|_{\mathrm{Car}^{k,l,p}(\Omega; \mathbb{R}^m)} \leq 1}} \langle \eta - Q_{d_2}^* P_{d_1}^* \eta, h \rangle \\
&= \sup_{\substack{h \in H \\ \|h\|_{\mathrm{Car}^{k,l,p}(\Omega; \mathbb{R}^m)} \leq 1}} \langle \eta, h - P_{d_1} Q_{d_2} h \rangle \\
&\leq \|\eta\|_{0,0,p} \sup_{\substack{h \in H \\ \|h\|_{\mathrm{Car}^{k,l,p}(\Omega; \mathbb{R}^m)} \leq 1}} \|h - P_{d_1} Q_{d_2} h\|_{\mathrm{Car}^p(\Omega; \mathbb{R}^m)} \\
&\leq C\big(d_1^k + d_2^l\big) \|\eta\|_{0,0,p}. \qquad (8.20)
\end{aligned}
$$

The general "energy-error" estimation scheme can then be based on the assumption that a function $\Phi : Y_H^p(\Omega; \mathbb{R}^m) \to \mathbb{R}$ to minimize is Lipschitz continuous ($\ell :=$ the corresponding Lipschitz constant) with respect to the norm (8.19), which represents a requirement of some regularity of the data involved in Φ (but not any regularity of the minimizers themselves).

Proposition 1 (General "energy-error" estimate).

Let $|\Phi(\eta_1) - \Phi(\eta_2)| \leq \ell \|\eta_1 - \eta_2\|_{-k,-l,p}$, and (8.14), (8.15), and (8.17) hold. Moreover, let at least one from the following two conditions holds:

(i) $Q_{d_2}^* P_{d_1}^* Y_H^p(\Omega; \mathbb{R}^m) \subset Y_H^p(\Omega; \mathbb{R}^m)$, *i.e. the approximation is conformal, or*

(ii) $\exists C \; \forall d_1 > 0, \, d_2 > 0 : \|\eta_d\|_{0,0,p} \leq C$ *with η_d being some minimizer of Φ on $Q_{d_2}^* P_{d_1}^* Y_H^p(\Omega; \mathbb{R}^m)$.*

Then

$$\big| \min \Phi\big(Q_{d_2}^* P_{d_1}^* Y_H^p(\Omega; \mathbb{R}^m)\big) - \min \Phi\big(Y_H^p(\Omega; \mathbb{R}^m)\big) \big| = \mathcal{O}\big(d_1^k + d_2^l\big). \qquad (8.21)$$

Proof. Always it holds

$$
\begin{aligned}
\min \Phi\big(Q_{d_2}^* &P_{d_1}^* Y_H^p(\Omega; \mathbb{R}^m)\big) - \min \Phi\big(Y_H^p(\Omega; \mathbb{R}^m)\big) \\
&\leq \Phi\big(Q_{d_2}^* P_{d_1}^* \eta\big) - \Phi(\eta) \leq \ell \big\|\eta - Q_{d_2}^* P_{d_1}^* \eta\big\|_{-k,-l,p} \\
&\leq \ell C\big(d_1^k + d_2^l\big) \|\eta\|_{0,0,p} = \mathcal{O}\big(d_1^k + d_2^l\big) \qquad (8.22)
\end{aligned}
$$

with $\eta \in Y_H^p(\Omega; \mathbb{R}^m)$ being some minimizer of Φ on $Y_H^p(\Omega; \mathbb{R}^m)$. If the approximation is conformal, we have automatically the left-hand side lower bounded by 0. In the nonconformal case, we can make the lower estimate analogously by using $\eta = \eta_d$ a minimizer of Φ on $Q_{d_2}^* P_{d_1}^* Y_H^p(\Omega; \mathbb{R}^m)$ provided that, by a coercivity that has to be assumed, $\{\|\eta_d\|_{0,0,p}\}_{d_1>0,\, d_2>0}$ is bounded.   □

Of course, in nontrivial problems, also Φ itself usually involves complicated initial or/and boundary value problems and has to be approximated too.

8.4 Concrete Discretizations and Their Implementation

The theory from Section 8.3 works for general problems where oscillations can combine with concentration effects, and has been really implemented for the lowest-order case in [26; 49]. The presentation from Sections 8.2 and 8.3 can straightforwardly be generalized by replacing $\mathbb{R}^m$ with a subset $S \subset \mathbb{R}^m$; this may be convenient, e.g., when the optimal control problems like (8.1) or (8.5) would contain also a control constraint of the type $u(t) \in S$. Since higher-order discretizations has not been yet scrutinized in general context, hence we confine ourselves to the case when concentration effects are a-priori excluded by considering S a compact subset of $\mathbb{R}^m$ (so that p is irrelevant and we will omit it) and, in Sections 8.4.1 and 8.4.3 briefly present some general construction from [35]. Note that S compact ensures $\mathrm{Car}(\Omega; S) \cong L^1(\Omega; C(S))$.

8.4.1 The operator P_{d_1}

Considering a finite collection of ansatz functions $\{g_i\}_{i=1}^{I(d_1)} \subset L^1(\Omega)$ and a "dual" collection $\{g_i^*\}_{i=1}^{I(d_1)} \subset L^\infty(\Omega)$, we construct P_{d_1} by a *quasi-interpolation* with respect to these bases, i.e.

$$[P_{d_1} h](x,s) := \sum_{i=1}^{I(d_1)} \alpha_i(s) g_i(x), \qquad \alpha_i(s) := \frac{\int_\Omega g_i^*(x) h(x,s)\, dx}{\int_\Omega g_i^*(x)\, dx}. \quad (8.23)$$

The desired approximation property (8.14) (with S instead of $\mathbb{R}^m$) now reads as

$$\|h - P_{d_1} h\|_{L^1(\Omega; C(S))} \le C_1 d_1^k \|h\|_{W^{k,1}(\Omega; C(S))}, \quad (8.24)$$

which, in fact, needs the collections $\{g_i\}_{i=1}^{I(d_1)}$ and $\{g_i^*\}_{i=1}^{I(d_1)}$ to be linked with each other to some extent. In some cases, the following L^2-type orthogonality holds:

$$\int_\Omega g_i(x) g_j^*(x)\, dx \begin{cases} = 0 & \text{for } i \neq j \\ > 0 & \text{for } i = j. \end{cases} \quad (8.25)$$

Proposition 2 (The adjoint operator $P_{d_1}^*$, see [35]). *The following formula holds for $P_{d_1}^* \nu$:*

$$[P_{d_1}^* \nu]_x = \sum_{i=1}^{I(d_1)} g_i^*(x) \frac{\int_\Omega g_i(\xi)\nu_\xi d\xi}{\int_\Omega g_i^*(\xi)d\xi}. \tag{8.26}$$

Proof. By applying Fubini's theorem, one obtains:

$$\langle P_{d_1}^* \nu, h \rangle = \langle \nu, P_{d_1} h \rangle = \int_\Omega \int_S \sum_{i=1}^{I(d_1)} \frac{\int_\Omega g_i^*(\xi)h(\xi,s)d\xi}{\int_\Omega g_i^*(\xi)d\xi} g_i(x)\nu_x(ds)dx$$

$$= \sum_{i=1}^{I(d_1)} \frac{1}{\int_\Omega g_i^*(\xi)d\xi} \int_{\Omega\times\Omega} \int_S g_i^*(\xi)h(\xi,s)g_i(x)\nu_x(ds)d(x,\xi)$$

$$= \int_\Omega \int_S h(x,s) \left(\sum_{i=1}^{I(d_1)} \frac{g_i^*(x)}{\int_\Omega g_i^*(\xi)d\xi} \int_\Omega g_i(\xi)\nu_\xi d\xi \right)(ds)dx. \tag{8.27}$$

$\square$

Corollary 1 (Conformal approximation, see [35]). *If*

$$g_i \geq 0, \quad g_i^* \geq 0, \quad i = 1, ..., I(d_1), \tag{8.28a}$$

$$\sum_{i=1}^{I(d_1)} g_i^*(x) = 1 \quad \text{for a.a. } x \in \Omega, \tag{8.28b}$$

$$\int_\Omega g_i(\xi)d\xi = \int_\Omega g_i^*(\xi)d\xi, \quad i = 1, ..., I(d_1), \tag{8.28c}$$

then $P_{d_1}^* \mathcal{Y}(\Omega; S) \subset \mathcal{Y}(\Omega; S)$.

Proof. Obviously, $[P_{d_1}^* \nu]_x$ is a positive measure because of (8.28a), and

$$\int_S [P_{d_1}^* \nu]_x(ds) = \sum_{i=1}^{I(d_1)} g_i^*(x) \frac{\int_\Omega g_i(\xi) \int_S \nu_\xi(ds)d\xi}{\int_\Omega g_i^*(\xi)d\xi} = \sum_{i=1}^{I(d_1)} g_i^*(x) \frac{\int_\Omega g_i(\xi)d\xi}{\int_\Omega g_i^*(\xi)d\xi} = 1$$

because of (8.28b,c). Hence we proved that this approximation is conformal. $\square$

8.4.2 Spatial Discretization by Using P_{d_1}

The condition (8.28) is quite restrictive and, unlike conventional finite elements, it seems to prevent us to consider approximations of an order k higher than 2. Intuitively, this is related with the fact that any interpolation other then either constant of affine between probability measures is not a probability measure in general.

8.4.2.1 Concrete options concerning P_{d_1}

When discretizing the, say, polyhedral domain Ω by simplexes with the diameter less than d_1, one can consider both the collections $g_i = g_i^*$ as the piecewise affine continuous hat functions. Then (8.24) with $k = 2$ holds, although P_{d_1} is not a projector and (8.25) does not hold either. The same properties can be obtained by allowing the piecewise affine functions $g_i = g_i^*$ to be discontinuous, but as more degrees of freedom is admitted, the error (although having still the same rate $k = 2$) might be less.

Even a simpler option, relying on an arbitrary "finite-volume-like" partition of an arbitrarily shaped Ω, consists in putting $g_i = g_i^*$ equal to the characteristic function of a particular "finite volume". Then (8.25) and (8.24) with $k = 1$ hold, and P_{d_1} is a projector.

Beside these three options, if Ω has a form of a Cartesian product of lower-dimensional domains, a tensorial product of the functions considered above is possible, too. Then these Q1-elements still enrich the variety of all possibilities leading to conformal approximation. Yet, anyhow, $k \leq 2$.

8.4.2.2 Discretization of classical Young measures by P_{d_1}

The operator P_{d_1} itself can already be used for efficient numerical schemes. The conventional choice $H = L^1(\Omega; C(S))$ which leads to the classical Young measures $Y_H^p(\Omega; S) \cong \mathcal{Y}(\Omega; S)$ and the piece-wise constant approximation (leading to $k = 1$) give piece-wise homogenous Young measures. Such measures still form a finite-dimensional subset in $H^* = L_w^\infty(\Omega; \mathcal{M}(S))$, the set of weakly* measurable mappings $\Omega \to \mathcal{M}(S) \cong C(S)^*$. However, typical optimization problems involve only a finite number, say j, of integrands, e.g. the problem (8.1) involves only two integrands: $h_1(x, s) = (1 - s^2)^2$ and $h_2(x, s) = s$. By the celebrated Carathéodory theorem, at a given $x \in \Omega$, each probability measure ν_x can be replaced by a measure that is a convex combination of only $j + 1$ Diracs and that has the same momenta with respect to a collection of nonlinearities $\{h_i(x, \cdot)\}_{i=1}^j$. In view of this, one can work with piece-wise constant (depending on the discretization parameter d_1) Young measures composed by a limited number of Dirac measures only. Such a set can already be implemented on computers. For the first time, such an algorithm (based on the steepest-descent strategy) has conceptually been proposed by Warga [57], and later used also in [10; 11; 12]. Further modification of this idea, suggested in [46], consists in analysis of the optimality condition (i.e. a maximum principle) that allows to say that at least one optimal control is composed from only j Dirac measures. Thus the used optimization routine can work still with smaller number of variables. This idea was implemented by Mátlová in [46].

In principle, if we would use still less number of Dirac measures, the convergence can still be preserve because even a single Dirac measure on each element, i.e. the original control, guarantees the mere convergence just by a density argument. Yet, of course, the order of convergence can be destroyed in general, although in particular case it can still be observed.

8.4.2.3 Discretization of coarse convex compactifications by P_{d_1}

It is often a case that only a finite number, say j, of nonlinearities $v_i : S \to \mathbb{R}$ occurs in the problem; e.g. (8.1) involves $v_1(s) = (1 - s^2)^2$ and $v_2(s) = s$, hence $j = 2$ would suffices. Then, instead of Young measures used in (8.1), one can take $Y_H^4([0,T];\mathbb{R})$ with, e.g., $H = C([0,T]) \otimes \{v_i\}_{i=1}^j := \{(t,s) \mapsto \sum_{i=1}^j g_i(t)v_i(s); g_i \in C([0,T])\}$. In a general case, let us consider $Y_H^p(\Omega;\mathbb{R}^m)$ with $H = C(\bar{\Omega}) \otimes \{v_i\}_{i=1}^j$ where $\bar{\Omega}$ denotes the closure of Ω. The elements from $Y_H^p(\Omega;\mathbb{R}^m)$ that can be attained by sequences $\{u_k\}_{k\in\mathbb{N}}$ whose "energy" $\{|u_k|^p\}_{k\in\mathbb{N}}$ is relatively weakly compact in $L^1(\Omega)$ can be identified as the set of momenta of Young measures from $\mathcal{Y}^p(\Omega;\mathbb{R}^m)$. After discretizing $Y_H^p(\Omega;\mathbb{R}^m)$ by using P_{d_1}, the resulting convex set $P_{d_1}^* Y_H^p(\Omega;\mathbb{R}^m)$ is already finite dimensional and can thus be implemented on computers. As such, Warga's conceptual approach [57] mentioned in Section 8.4.2.2 can alternatively be understood as a steepest-descent iterative method working just on such a coarse convex local compactification involving finite number of momenta.

Explicit characterization of such $P_{d_1}^* Y_H^p(\Omega;\mathbb{R}^m)$ is not easy, however. Therefore avoiding Young measures, on which [57] is ultimately based, is not easy in general, if possible at all. However, in special cases, it is possible. The simplest case is $m = 1$, $S = \mathbb{R}$ and $v_i(s) = s^{i-1}$, that allows to handle polynomial nonlinearities up to the order j. E.g., (8.1) would need $j = 5$ now. In the series of works [14; 38; 39], it has been implemented the noteworthy result from the probability theory, namely that, with j even, a measure $\mu \in \mathcal{M}(\mathbb{R})$ is a probability measure if and only if the so-called Henkel's matrix $[\int_{\mathbb{R}} s^{i+k-2}\mu(ds)]_{i,k=1}^{(j+1)/2}$ is positive semidefinite. Therefore, in this special case, $P_{d_1}^* Y_H^p(\Omega;\mathbb{R})$ can be characterized by a finite number (depending on d_1) of constraints on the Henkel's matrices to be positive semidefinite. This leads to the so-called semi-definite programming for which efficient optimization routines are available, cf. [20; 58]. The generalization of $S \neq \mathbb{R}$ is, however, difficult, although in special cases possible.

8.4.2.4 Variational problems and gradient Young measures

Considering $\Omega \subset \mathbb{R}^n$ a bounded Lipschitz domain, $\Gamma \subset \partial\Omega$, the following vectorial variational problem received huge attention during past many decades:

$$\min \quad \Phi(y) := \int_\Omega \varphi(x, y(x), \nabla y(x))\, dx$$
$$\text{s.t.} \quad y \in W^{1,p}(\Omega;\mathbb{R}^m), \quad y|_\Gamma = 0, \tag{8.29}$$

cf. references in [43; 46]. If $p > 1$, the corresponding relaxed problem takes the form:

$$\min \quad \bar{\Phi}(y,\nu) := \int_\Omega \int_{\mathbb{R}^{m\times n}} \varphi(x, y(x), A)\, \nu_x(dA)dx$$
$$\text{s.t.} \quad \int_{\mathbb{R}^{m\times n}} A\, \nu_x(dA) = \nabla y(x) \text{ for a.a. } x \in \Omega,$$
$$y \in W^{1,p}(\Omega;\mathbb{R}^m), \quad \nu \in \mathcal{G}^p(\Omega;\mathbb{R}^{m\times n}), \quad y|_\Gamma = 0, \tag{8.30}$$

where $\mathcal{G}^p(\Omega; \mathbb{R}^{m \times n}) := \{\nu \in \mathcal{Y}^p(\Omega; \mathbb{R}^{m \times n}); \exists \{y_k\}_{k \in \mathbb{N}} \subset W^{1,p}(\Omega; \mathbb{R}^m) : \delta_{\nabla y_k} \to \nu$ weakly*$\}$ denotes the set of the so-called gradient L^p-Young measures. Let us remark that, in fact, choosing $m = n = 1, p = 4, \varphi(x, r, s) = (1 - s^2)^2 + r^2$, $\Omega := [0, T], \Gamma = \{0\}$, we obtain just (8.1) and Figure 8.1 then shows a minimizing sequence with a highly oscillating gradient.

An example for a two-atomic gradient Young measure is

$$\nu_x = \lambda(x) \delta_{A_1(x)} + (1 - \lambda(x)) \delta_{A_2(x)}, \tag{8.31a}$$
$$\lambda \in L^\infty(\Omega), \quad \lambda(x) \in [0, 1], \tag{8.31b}$$
$$A_1, A_2 \in L^p(\Omega; \mathbb{R}^{m \times n}), \quad \text{Rank}(A_1(x) - A_2(x)) \leq 1, \tag{8.31c}$$

more precisely, this measure is two-atomic only if $\lambda \neq 0$ or $\lambda \neq 1$ and $\text{Rank}(A_1(x) - A_2(x)) \neq 0$ identically. Such a measure is called a (1st-order) laminate in connection with an interpretation as a microstructure occurring in nonlinear elasticity where y is a deformation and ν_x describes a limit behavior of fast spatial oscillations of the deformation gradient; cf. Figure 8.3 below. In fact, the conditions (8.31b,c) are also necessary for a measure of the form (8.31a) to belong to $\mathcal{G}^p(\Omega; \mathbb{R}^{m \times n})$. This procedure can be iterated to obtain a 4-atomic gradient Young measure as a 2-nd order laminate, etc., cf. [42; 43; 46].

Taking P_{d_1} in the form (8.23) satisfying (8.28), the conceptual discretization of (8.30) consists in replacing $\mathcal{G}^p(\Omega; \mathbb{R}^{m \times n})$ by a smaller set $\mathcal{G}^p(\Omega; \mathbb{R}^{m \times n}) \cap P_{d_1}^* \mathcal{Y}^p(\Omega; \mathbb{R}^{m \times n})$.

Unfortunately, if $m > 1$ and $n > 1$, the set $G^p(\Omega; \mathbb{R}^{m \times n})$ cannot be characterized explicitly and not every $\nu \in G^p(\Omega; \mathbb{R}^{m \times n})$ is of the form of a laminate, and in homogeneous situation even need not be attainable by homogenous laminates, which is related with the celebrated Šverák's counterexample [51]. This causes that, although the mere convergence is always guaranteed by spatially fast oscillating laminates, the order of convergence corresponding to the particular operator P_{d_1} is guaranteed only if $m = 1$ or $n = 1$ or the minimizing Young measure is a laminate of the same order as used to implement the discretized version of (8.30).

This laminate-form Young measures have been implemented together with P_{d_1} constructed by element-wise constant anzatz functions $g_i = g_i^*$ (leading to piecewise-homogenous laminate-type Young measure and therefore element-wise affine deformation y) for applications in nonlinear elasticity for so-called shape-memory alloys where the deformation gradient exhibits fast spatial oscillations very typically due to a multiwell character of the stored energy, and the Young measure then quite naturally describes a *microstructure*, cf. [1; 4; 5; 21; 23; 29; 51].

8.4.3 The operator Q_{d_2}

Like in Section 8.4.1 (but not precisely as this) we consider a finite collection of ansatz functions $\{v_j\}_{j=1}^{J(d_2)} \subset C(S)$ and $\{v_j^*\}_{j=1}^{J(d_1)} \subset \text{rca}(S) \cong C(S)^*$ and construct the projector Q_{d_2} as a quasi-interpolation with respect to these bases, i.e.

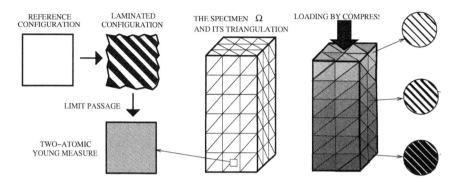

Fig. 8.3. A schematic situation of a material preferred to stay, instead of a reference cubic configuration, at two configurations orthorhombic configurations (displayed as black and white) forming a first-order laminate that can be described, in a limit, by the Young measure (8.31), and a macroscopical specimen of the undeformed shape Ω triangulated by a simplectic mesh which, under a compression, exhibits a nonhomogeneous microstructure described by an element-wise constant two-atomic Young measure from which laminates can schematically be reconstructed a-posteriori; based on [51] modelling re-orientation of martensite in a NiMnGa single crystal.

$$[Q_{d_2}h](x,s) := \sum_{j=1}^{J(d_2)} \alpha_j(x)v_j(s), \quad \alpha_j(x) := \frac{\int_S h(x,s)v_j^*(\mathrm{d}s)}{\int_S v_j^*(\mathrm{d}s)}. \qquad (8.32)$$

The desired approximation property (8.14) (with S instead of $\mathbb{R}^m$) now reads as

$$\|h - Q_{d_2}h\|_{L^1(\Omega;C(S))} \leq C_2 d_2^l \|h\|_{L^1(\Omega;C^l(S))}, \qquad (8.33)$$

and again it represents a certain linkage between $\{v_j\}_{j=1}^{J(d_2)}$ and $\{v_j^*\}_{j=1}^{J(d_1)}$. In some cases, the following L^2-type orthonormality, related with the property that Q_{d_2} is a projector, holds:

$$\frac{\int_S v_i(s)v_j^*(\mathrm{d}s)}{\int_S v_j^*(\mathrm{d}s)} = \begin{cases} 0 & \text{for } i \neq j, \\ 1 & \text{for } i = j. \end{cases} \qquad (8.34)$$

Proposition 3 (The adjoint operator $Q_{d_2}^*$, see [35]). *The following formula holds for $Q_{d_2}^*\nu$:*

$$[Q_{d_2}^*\nu]_x = \sum_{j=1}^{J(d_2)} \frac{\int_S v_j(\sigma)\nu_x(\mathrm{d}\sigma)}{\int_S v_j^*(\mathrm{d}\sigma)} v_j^*. \qquad (8.35)$$

Proof. By (a continuously extended) Fubini's theorem, one obtains:

$$\langle Q_{d_2}^* \nu, h \rangle = \langle \nu, Q_{d_2} h \rangle = \int_\Omega \int_S \sum_{j=1}^{J(d_2)} \frac{\int_S h(x,\sigma) v_j^*(d\sigma)}{\int_S v_j^*(d\sigma)} v_j(s) \nu_x(ds) dx$$

$$= \int_\Omega \sum_{j=1}^{J(d_2)} \frac{1}{\int_S v_j^*(d\sigma)} \int_{S \times S} h(x,\sigma) v_j(s) [v_j^* \times \nu_x] d(\sigma, s) dx$$

$$= \int_\Omega \int_S h(x,s) \left(\sum_{j=1}^{J(d_2)} \frac{\int_S v_j(\sigma) \nu_x(d\sigma)}{\int_S v_j^*(d\sigma)} v_j^*(ds) \right) dx.$$

$\square$

Corollary 2 (Conformal approximation, see [35]). *Let*

$$v_j \geq 0, \quad v_j^* \geq 0, \quad j = 1, ..., J(d_2), \tag{8.36a}$$

$$\sum_{j=1}^{J(d_2)} v_j(s) = 1 \quad \text{for all } s \in S, \tag{8.36b}$$

then $Q_{d_2}^* \mathcal{Y}(\Omega; S) \subset \mathcal{Y}(\Omega; S)$.

Proof. Obviously, $[Q_{d_2}^* \nu]_x$ is non-negative and

$$\int_S [Q_{d_2}^* \nu]_x(s) ds = \sum_{j=1}^{J(d_2)} \frac{\int_S v_j(\sigma) \nu_x(d\sigma)}{\int_S v_j^*(d\sigma)} \int_S v_j^*(ds)$$

$$= \int_S \left(\sum_{j=1}^{J(d_2)} v_j(\sigma) \right) \nu_x(d\sigma) = \int_S \nu_x(d\sigma) = 1. \tag{8.37}$$

$\square$

8.4.4 Full discretization by using $P_{d_1} Q_{d_2}$

Like (8.28), also the condition (8.36) is quite restrictive and it seems to prevent us to consider approximations of an order l higher than 2. Anyhow, it gives several options for Q_{d_2}, whose usage opens some other numerical strategies.

8.4.4.1 Concrete options concerning Q_{d_2}

When discretizing the, say, polyhedral domain S by simplexes with the diameter less than d_2, one can consider both the collections v_i as the piecewise affine continuous hat functions corresponding to the particular node $s_j \in S$, and the measures v_i^* as absolutely continuous with the density $v_i(s) ds$, or even $c_i v_i(s) ds$ with $c_i > 0$). Then (8.33) with $l = 2$ and (8.36) hold, although Q_{d_2} is not a projector, i.e. (8.34) does not hold.

Another option is to take v_i as above but $v_i^* = \delta_{s_i}$. Then, in addition to (8.33) with $l = 2$ and (8.36), also (8.34) holds, and Q_{d_2} is therefore a projector.

Beside these two options, if S has a form of a Cartesian product of lower-dimensional domains, a tensorial product of the functions considered above is possible, too. Then these Q1-elements still enrich the variety of all possibilities leading to conformal approximation. Yet, anyhow, $l > 3$ is not reached by this way.

All these operators Q_{d_2} commute with P_{d_1}; in fact, this follows from the general forms (8.23) and (8.32).

This strategy, with the piecewise-constant elements for P_{d_1} and the Dirac measures for the dual base used for Q_{d_2}, has been used for a nontrivial application in modelling of a microstructure of magnetization in ferromagnetic materials. In an isothermal case, by the so-called Heisenberg-Weiss constraint, the magnetization has a fixed magnitude M_s so that the Young measure ν_x describing, at each point x of the magnet $\Omega \subset \mathbb{R}^3$, a probability distribution of the magnetization vector is supported on the sphere $S := \{|m| = M_s\}$ with M_s a given saturation magnetization. In the simplest static case, the "mesoscopically described" magnetization $\nu \in \mathcal{Y}(\Omega; S)$ solves, together with the magnetic potential $u_m \in W^{1,2}(\mathbb{R}^3)$, the problem:

$$
\begin{aligned}
\min \quad & \int_\Omega \int_S \Big(\varphi(s) - h(x) \cdot s\Big) \nu_x(\mathrm{d}s)\mathrm{d}x + \int_{\mathbb{R}^3} \frac{|\nabla u_m(x)|^2}{2}\, \mathrm{d}x \\
\text{s.t.} \quad & \operatorname{div}(-\mu_0 \nabla u_m + m\chi_\Omega) = 0 \quad \text{on } \mathbb{R}^3, \\
& m(x) = \int_S s\, \nu_x(\mathrm{d}s), \quad u_m \in W^{1,2}(\mathbb{R}^3), \quad \nu \in \mathcal{Y}(\Omega; S),
\end{aligned}
\tag{8.38}
$$

where χ_Ω is the characteristic function of Ω, m the macroscopical magnetization, $h : \Omega \to \mathbb{R}^3$ here denotes the outer (given) magnetic field, $\varphi : S \to \mathbb{R}^+$ the anisotropy energy having minima at the easy direction of magnetization, and μ_0 the vacuum permeability, cf. [22; 24; 27; 28; 50] for details about this model and its discretization and efficient implementation (by using also the method from Section 8.4.4.2).

It should be, however, pointed out that the discretization by means of both P_{d_1} and Q_{d_2} leads usually to problems of very high dimensions, especially if $\Omega \subset \mathbb{R}^n$ with $n \geq 2$ and also $m \geq 2$. Therefore, some additional tricks are desirable, as outlined in Sections 8.4.4.2–3.

8.4.4.2 Active-set strategy

The discrete probability measures, if solving optimization problems, are usually composed only with a relatively small number of Dirac measures, hence running optimization routines with full number of variables resulted from the discretization $P_{d_1}^* Q_{d_2}^* \mathcal{Y}(\Omega; S)$ would only waste computer memory by huge amount of zeros and enormously slow down the calculations because the complexity of optimization routines is usually proportional to a power of number of variables. Thus a reliable estimation of "active atoms" (i.e. Dirac measures with nonvanishing coefficients) is desirable. The strategy, first time proposed in [8], is based on the maximum principle for the approximate problem in the cases when this principle is not only necessary

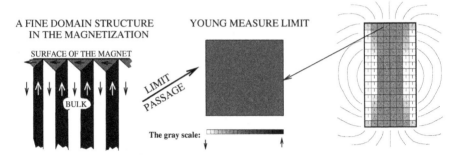

Fig. 8.4. The schematic situation of a fine structure of the magnetization in a uniaxial ferromagnet with vertical easy-magnetization axis ("north" black, "south" white), its "Young-measure" limit, and calculation of the distribution of the volume fraction of the magnetization (the gray levels in Ω) and the magnetic line outside the ferromagnet; based on [27] modelling a CoZrDy cylindrical ferromagnet.

but also sufficient. Just for illustration, let us remind the maximum principle for the continuous problem (8.3):

$$\int_{\mathbb{R}} h_\lambda(t,s)\nu_t(ds) = \max_{s\in\mathbb{R}} h_\lambda(t,s) \quad \text{for a.a. } t \in [0,T], \tag{8.39a}$$

$$h_\lambda(t,s) = \lambda(t)s - (1-s^2)^2, \tag{8.39b}$$

$$\frac{d\lambda}{dt} = 2y, \quad \lambda(T) = 0, \tag{8.39c}$$

$$\frac{dy}{dt} = \int_{\mathbb{R}} s\,\nu_t(ds), \quad y(0) = 0. \tag{8.39d}$$

As (8.3) is convex, (8.39) is also sufficient. This property is preserved also for the approximate problem where, instead of the "Hamiltonian" h_λ from (8.39b), one is to deal with $P_{d_1}Q_{d_2}h_\lambda$ with λ solving a discrete version of the adjoint problem (8.39c). Furthermore, even the "discrete Hamiltonian" $P_{d_1}Q_{d_2}h_\lambda$ is not known a-priori because λ corresponding through (8.39d) to an optimal y is not known either. Then the philosophy is to take some of its estimate, to use (8.39a) with a certain tolerance to "activate" such atoms where this Hamiltonian is nearly maximal, execute an optimization routine with relatively very small number of variables, and the a-posteriori check the maximum principle for atoms (which is relatively easy) and, if not satisfy, then repeat the procedure with a greater tolerance, otherwise we got already the approximate solution.

The mentioned estimate (i.e. an approximation) of the discrete Hamiltonian $P_{d_1}Q_{d_2}h_\lambda$ can be obtained when considering a hierarchy of nested discretizations and using λ calculated on a coarser discretization. This multigrid technique has been proposed in [8] and later used also in [6]. Another possibility is naturally suggested if the relaxed problem is parameterized (e.g. by time) and is evolving, and then one can obtain a good estimate from a previous time level; this option has also been employed in

[22; 24] for calculations in Figure 8.4. Other option is offered if an iterative procedure as in Section 8.4.4.4 is involved; then the previous iteration can naturally offer a good estimate for the Hamiltonian. The last option was employed in [7; 30; 31]. Of course, in relevant cases, those strategies can be combined mutually.

8.4.4.3 Sparse-interpolation strategy

Recently, a special sparse multi-scale finite-element technique for homogenization problems has been developed in series of works [19; 33; 34]. This relies on a two-scale regularity of solutions and then a resolution of the fine scale is possible with a substantially fewer number of variables when "thinning out" the multi-scale full tensor product space by using a sparse interpolation. In this way the order of error can be preserved. It seems possible [36] to apply such approach to approximation of Young measures by exploiting the natural tensorial-product structure created by the operator $P_{d_1}^* Q_{d_2}^*$ and a regularity assumed for test functions like we implicitly did for Proposition 1.

8.4.4.4 Various iterative techniques for lower-order terms

During relaxation, the original convex problems are partly or fully convexified. E.g., the nonconvex problem (8.1) led to the problem (8.3) which involve a convex quadratic functional and a linear constraint. After the full discretization as in Section 8.4.4 it gives a quadratic programming problem (QP) for which efficient finite solvers do exists.

If the term y^2 would be replaced by y, (8.3) would lead, after discretization, even to a linear programming problem (LP) for which even much more efficient solvers do exists, which allows for much finer discretizations, cf. [6].

However, even the quadratic problem (8.3) itself can, after discretization, by solved by LP-algorithms if organized iteratively, and at each iteration (8.3) is linearized. This technique is called a sequential linear programming (SLP) and was analyzed in a special case of the scalar (i.e. $m = 1$) variation problem (8.29) in [7].

In fact, the previous SLP-method does not need the discrete relaxed problem to be quadratic. E.g., in case of (8.1), it would be the case if the dynamics in (8.1) would be nonlinear. In this case, the iterative procedure need not work with a linearized problem but with only a quadratic approximation. This leads to sequential quadratic programming problems (SQP), scrutinized in this context in [30].

An additional iterative strategy is a so-called proximal-point method (PPM), see [44], which can iteratively treat problems which are nonconvex by convex techniques. In this context, it applies to problems that remains nonconvex even after the relaxation (e.g. (8.1) with y^2 replaced by $-y^2$). Thus PMM can be combined with the previous method, say QP or SQP. The comparison of all mentioned approaches has been done in [31]. The following table summarizes only their mere applicability:

Table 8.1. Usability of numerical methods (see [31]).

Properties of the relaxed problem	Numerical approach					
	LP	QP	SQP	SLP	PPM+QP	PPM+SQP
Linear	x	x	x	x	x	x
Quadratic	–	x	x	x	x	x
Convex, non-quadratic	–	–	x	x	x	x
Quadratic, non-convex	–	–	–	x	x	x
Non-convex, non-quadratic	–	–	–	x	–	x

8.5 Conclusion, Outlook for Further Research

Only rather recently the relaxed optimization problems had started to be realized as not only an analytical tool to describe limits of minimizing sequences but also, after a suitable discretization, as a certain tool for a numerical approach to *multiscale* problems. Variety of approximation techniques leading to implementable numerical strategies and allowing various mutual combinations has been presented in this contribution with an attempt to summarize rather recent research undertaken during past decade.

Some gaps in the above presented achievements can easily be seen. Let us point out at least some of them.

First, the relaxed problems with concentration/oscillation has been discretized, to the author's knowledge, only by the lowest-order approximation in [49]. Application of the above higher-order approximations has not been done yet, as well as combination with, e.g., the positive semidefinite programming technique as mentioned in Sect. 8.4.2.3. Besides, it should be mentioned that the relaxation theory of such problems, related with the impulse-control theory, is very difficult for a general coupling of the concentrating control with the state response.

The sparse technique itself as outlined in Section 8.4.4.3 was not developed yet in the context of Young measures. Its possible combination with the other methods would definitely be worth investigating, too.

In case of "laminate" Young measures of the 1st-order type (8.31) or of a higher order, only the element-wise constant approximation has ever been implemented and analyzed, which corresponds through the constraints in (8.30) to element-wise affine deformation y. This is however known to cause some extra numerical rigidity of the elastic structure unless the discretization is very fine, hence higher-order elements would be very desirable at this occasion. E.g. simplectic P2-elements might lead to a choice of λ in (8.31) as piecewise linear on each simplectic element while A_1 and A_2 would be constant. In case of a rectangular grid and $Q1$ elements, the 3rd-order polynomial occurs if $n = 3 = m$ and then one can either take higher-order polynomial for λ or play also with A_1 and A_2. In higher-order laminates the options are even wider.

Usage of nonconformal approximations, i.e. (8.28) and (8.36) can be avoided to open possibilities for higher-order approximation, has not been scrutinized at all, except Proposition 1(ii). The approximate optimal solutions in $P_{d_1}^* Q_{d_2}^* Y_H^p(\Omega; \mathbb{R}^m)$ then need not belong to $Y_H^p(\Omega; \mathbb{R}^m)$ and, to obtain the attainability by ordinary controls, one would need to make some a-posteriori modification which would live in $Y_H^p(\Omega; \mathbb{R}^m)$. It is not clear, however, whether it can be done in particular cases by a computationally "inexpensive" way and by such a way that does not deteriorate the rate of error (8.21).

References

[1] Aubri, S., Fago, M., Ortiz, M.: A constrained sequential-lamination algorithm for the simulation of sub-grid microstructure in martensitic materials. *Comp. Meth. in Appl. Mech. Engr.* **192** (2003), 2823–2843.

[2] Balder, E.J.: Lectures on Young measure theory and its applications in economics. *Rend. Ist. Mat. Univ. Trieste* **31**, Suppl.1 (2000), 1–69.

[3] Ball, J.M.: A version of the fundamental theorem for Young measures. In: *PDEs and Continuum Models of Phase Transition.* (Eds. M.Rascle, D.Serre, M.Slemrod.) Lecture Notes in Physics **344**, Springer, Berlin, 1989, pp.207–215.

[4] Ball, J.M., James, R.D.: Fine phase mixtures as minimizers of energy. *Archive Rat. Mech. Anal.* **100** (1988), 13–52.

[5] Bhattacharya, K.: *Microstructure of martensite. Why it forms and how it gives rise to the shape-memory effect.* Oxford Univ. Press, 2003.

[6] Bartels, S.: Adaptive approximation of Young measure solution in scalar nonconvex variational problems. *SIAM J. Numer. Anal.* **42** (2004), 505-529.

[7] Bartels, S., Roubíček, T.: Linear-programming approach to nonconvex variational problems. (Preprint no.74, DFG SPP 1095, Stuttgart, 2002) *Numerische Math.* **99** (2004), 251-287.

[8] Carstensen, C., Roubíček, T.: Numerical approximation of Young measure in nonconvex variational problems. *Numerische Mathematik* **84** (2000), 395-415.

[9] Castaing, C., Raynaud de Fitte, P., Valadier, M.: *Young Measures on Topological Spaces. With Applications in Control Theory and Probability Theory.* To appear.

[10] Chryssoverghi, I., Numerical approximation of nonconvex optimal control problems defined by parabolic equations. J. Optim. Theory Appl. **45** (1985), 73–88.

[11] Chryssoverghi, I., Bacopoulos, A., Kokkinis, B., Coletsos, J.: Mixed Frank-Wolfe penalty method with applications to nonconvex optimal control problems. *J. Optimization Theory Appl.* **94** (1997), 311-334.

[12] Chryssoverghi, I.; Coletsos, J.; Kokkinis, B.: Approximate relaxed descent method for optimal control problems. *Control Cybern.* **30** (2001), 385-404.

[13] DiPerna, R.J., Majda, A.J.: Oscillations and concentrations in weak solutions of the incompressible fluid equations. *Comm. Math. Physics* **108** (1987), 667–689.

[14] Egozcue, J.J.; Meziat, R.; Pedregal, P.: From a nonlinear, nonconvex variational problem to a linear, convex formulation. *Appl. Math. Optimization* **47** (2002), 27-44.

[15] FATTORINI, H.O.: *Infinite Dimensional Optimization Theory and Optimal Control.* Cambridge Univ. Press, Cambridge, 1999.

[16] Frankowska, H., Rampazzo, F.: Relaxation of control systems under state constraints. *SIAM J. Control Optimization* **37** (1999), 1291-1309.

[17] Gamkrelidze, R.V.: On sliding optimal regimes. *Dokl. Akad. Nauk SSSR* **143** (1962), 1243–1245; Engl. transl.: *Soviet Math. Dokl.* **3** (1962), 390–395.

[18] Ghouila-Houri, A.: Sur la géneralisation de la notion de commande d'un systéme guidable. *Rev. Francaise Informat. Recherche Operationnelle* **1** (1967), No.4, 7–32.

[19] Hoang, V.H., Schwab, C.: High-dimensional finite elements for elliptic problems with multiples scales. *SIAM J. Multiscale Analysis* 2004, to appear.

[20] Klerk, E. de : *Aspects of Semidefinite Programming.* Kluwer Acad Publ., Dordrecht, 2002.

[21] Kružík, M.: Numerical approach to double-well problem. *SIAM J. Numer. Anal.* **35** (1998), 1833-1849.

[22] Kružík, M.: Maximum principle based algorithm for hysteresis in micromagnetics. *Adv. Math. Sci. Appl.* **13** (2003), 461-485.

[23] Kružík, M., Luskin, M.: The computation of martensitic microstructure with piecewise laminates. *J. Sci. Comput.* **19** (2003), 293-308.

[24] Kružík, M., Prohl, A.: Young measures approximation in micromagnetics, *Numer. Math.* **90** (2001), 291–307

[25] Kružík, M., Roubíček, T.: On the measures of DiPerna and Majda. *Mathematica Bohemica* **122** (1997), 383-399.

[26] Kružík, M., Roubíček, T.:Optimization problems with concentration and oscillation effects: relaxation theory and numerical approximation, *Numer. Funct. Anal. Optim.* **20** (1999), 511-530.

[27] Kružík, M., Roubíček, T.: Specimen shape influence on hysteretic response of bulk ferromagnets. *J. Magnetism and Magn. Mater.* **256** (2003), 158–167.

[28] Kružík, M., Roubíček, T.: Interactions between demagnetizing field and minorloop development in bulk ferromagnets. *J. Magnetism and Magn. Mater.* **277** (2004), 192-200.

[29] Luskin, M.: On the computation of crystalline microstructure. *Acta Numerica* **5** (1996), 191-257.

[30] Mach, J.: Numerical solution of a class of nonconvex variational problems by SQP. *Numer. Funct. Anal. Optim.* **23** (2002), 573-587.

[31] Mach, J.: Methods of numerical solution of a class of non-convex variational problems. PhD thesis, Math.-Phys. Faculty, Charles University, Prague, 2004.

[32] Málek, J., Nečas, J., Rokyta, M., Růžička, M.: *Weak and measure-valued solutions to evolution partial differential equations.* Chapman & Hall, 1996.

[33] Mataché, A.-M.: Sparse two-scale FEM for homogenization problems. *J. Sci. Comput.* **17** (2002), 659-669.

[34] Mataché, A.-M., Schwab, C.: Two-scale FEM for homogenization problems. *RAIRO Anal. Numerique* **36** (2002), 537-572.

[35] Mataché, A.-M., Roubíček, T., Schwab, C.: Higher-order convex approximations of Young measures in optimal control. *Adv. in Comput. Math.* **19** (2003), 73–97.

[36] Mataché, A.-M., Schwab, C., at al.: in preparation.

[37] McShane, E.J.: Generalized curves. *Duke Math. J.* **6** (1940), 513–536.

[38] Meziat, R.J.: Analysis of non convex polynomial programs by the method of moments. In: *Frontiers in global optimization.* (C.A.Floudas et al., eds.) Kluwer, Boston, 2004, pp.353-371.

[39] Meziat, R., Egozcue, J.J., Pedregal, P.: The method of moments for non-convex variational problems. In: *Advances in Convex Analysis and Global Optimization* (N.Hadjisavvas et al., eds.) Kluwer, Dordrecht, 2001, pp.371-382.

[40] Müller, S.: Variational models for microstructure and phase transitions. (Lect.Notes No.2, Max-Planck-Institut für Math., Leipzig, 1998). In: *Calculus of variations and geometric evolution problems.* (Eds.: S.Hildebrandt et al.) Lect. Notes in Math. **1713** (1999), Springer, Berlin, pp.85–210.

[41] Nicolaides, R.A., Walkington, N.J.: Computation of microstructure utilizing Young measure representations. *J. Intel. Materials System Struct.* **4** (1993), 457–462.

[42] Pedregal, P.: Numerical approximation of parametrized measures. *Numer. Funct. Anal. Opt.* **16** (1995), 1049–1066.

[43] Pedregal, P.: *Parametrized Measures and Variational Principles.* Birkhäuser, Basel, 1997.

[44] Rockafellar, R.T.: Monotone operators and the proximal point algorithm. *SIAM J. Control Optim.* **14** (1976), 877-898.

[45] Roubíček, T.: Approximation theory for generalized Young measures. (Preprint 1992 submited to SIAM J. Numer. Anal.) *Numer. Funct. Anal. Opt.* **16** (1995), 1233-1253.

[46] Roubíček, T.: *Relaxation in Optimization Theory and Variational Calculus,* W. de Gruyter, Berlin, 1997.

[47] Roubíček, T.: *Existence results for some nonconvex optimization problems governed by nonlinear processes,* In: Proc. 12th Conf. on *Variational Calculus, Optimal Control and Applications* (W.H.Schmidt, K.Heier, L.Bittner, R.Bulirsch, eds.) Birkäuser, Basel, 1998, pp. 87-96.

[48] Roubíček, T.: Convex locally compact extensions of Lebesgue spaces and their applications. In: *Calculus of Variations and Optimal Control.* (A.Ioffe, S.Reich, I.Shafrir, eds.) Chapman & Hall / CRC Res. Notes in Math. **411**, CRC Press, Boca Raton, FL, 1999, pp.237-250.

[49] Roubíček, T., Kružík, M.: Adaptive approximation algorithm for relaxed optimization problems. In: *Fast solution of discretized optimization problems* (K.-H.Hoffmann, R.H.W.Hoppe, V.Schultz, eds.), ISNM **138**, Birkhäuser, Basel, 2001, pp.242-254.

[50] Roubíček, T., Kružík, M.: Microstructure evolution model in micromagnetics. Zeit. für angew. Math. und Physik, **55** (2004), 159-182.

[51] Roubíček, T., Kružík, M.: Mesoscopic model of microstructure evolution in shape memory alloys, its numerical analysis and computer implementation. 3rd *GAMM Seminar on microstructures* 2004. (Ed. C. Miehe), GAMM Mitteilungen, J. Wiley, in print.

[51] Šverák, V.: Rank-one convexity does not imply quasiconvexity. *Proc. R. Soc. Edinb.* **120 A** (1992), 185-189.

[53] Tartar, L.: On mathematical tools for studying partial differential equations of continuum physics: H-measures and Young measures. In: *Developments in Partial Differential Equations and Applications to Mathematical Physics.* (Eds. G.Butazzo, G.P.Galdi, L.Zanghirati.) Plenum Press, New York, 1992, pp.201–217.

[54] Tartar, L.: Some remarks on separately convex functions. In: *Microstructure and Phase Transition.* IMA Vol. 54 (Eds. D.Kinderlehrer et al.), Springer, New York, 1993, pp.192-204.

[55] Valadier, M.: Young measures. In: *Methods of Nonconvex Analysis.* (A.Cellina, ed.) Lecture Notes in Math. **1446**, Springer, Berlin, 1990, pp. 152–188.

[56] Warga, J.: *Optimal Control of Differential and Functional Equations.* Academic Press, New York, 1972.

[57] Warga, J.: Steepest descent with relaxed controls. *SIAM J. Control Optim.* **15** (1977), 674–682.

[58] Wolkowicz, H., Saigal, R., Vandenberghe, L., eds: *Hindbook of Semidefinite Programming.* Kluwcr Acad Publishers, Norwell, MA, 2000.

[59] Young, L.C.: Generalized curves and the existence of an attained absolute minimum in the calculus of variations. *Comptes Rendus de la Société des Sciences et des Lettres de Varsovie*, Classe III **30** (1937), 212–234.

Combining Model and Test Data for Optimal Determination of Percentiles and Allowables: CVaR Regression Approach, Part I

Stan Uryasev[1] and A. Alexandre Trindade[2]

[1] American Optimal Decisions, Inc.
and Department of Industrial and Systems Engineering
University of Florida
uryasev@ufl.edu
[2] Department of Statistics
University of Florida

Summary. We propose a coherent methodology for integrating various sources of variability on properties of materials in order to accurately predict percentiles of their failure load distribution. The approach involves the linear combination of factors that are associated with failure load, into a statistical factor model. This model directly estimates percentiles of the failure load distribution (rather than mean values as in ordinary least squares regression). A regression framework with CVaR deviation as the measure of optimality, is used in constructing the estimates. We consider estimates of confidence intervals for the estimates of percentiles, and adopt the most promising of these to compute A-Basis and B-Basis values. Numerical experiments with the available dataset show that the approach is quite robust, and can lead to a significant savings in number of actual testings. The approach pools together information from earlier experiments and model runs, with new experiments and model predictions, resulting in accurate inferences even in the presence of relatively small datasets.

Key words: information integration, quantile regression, conditional value-at-risk, B-basis, tolerance limit

Executive Summary

We propose a coherent methodology for integrating various sources of variability on properties of materials in order to accurately predict percentiles of their failure load distribution. The approach involves the linear combination of factors that are associated with failure load, into a statistical regression or *factor model*. The methodology can take as inputs various factors which can predict percentiles. For instance, for a given model, the difference between the 90th and 50th percentiles of the distribution generated by this model via Monte Carlo, can be viewed as a factor predicting the 90th percentile of the actual failure load distribution. Also, several model estimates, expert opinion estimates, and actual measurement estimates, can be pooled to estimate percentiles. However, in this particular study we were limited to the available failure dataset on one test type for a composite material, including the Model S predicted failure load limits, and actual experimental failure load data, for various stacking sequences. Specifically, and in the context of a supplied dataset we fit a model of the form,

$$Y = c_0 + c_1\mu + c_2\sigma + c_3m + c_4s + \varepsilon,$$

where Y is the failure load, $c_0, \ldots, c_4$ are unknown coefficients to be estimated from the data, and the pairs (μ, σ) are the midpoints and spreads of Model S limits (upper and lower bounds of failure loads) and (m, s) are estimates of the mean and standard deviation of the test data. ε is the residual error that accounts for the variability in failure loads that cannot be explained through the factors. We develop the statistical tool of *CVaR regression*, a special case of which is *quantile regression*, first introduced by Koenker and Basset (1978). These allow us to estimate various percentile characteristics of the failure load distribution as a function of the factors. Recent related work on quantile regression provides methods for calculating confidence intervals for estimates of such percentiles. We adopt the most promising of these methods for the calculation of one-sided upper confidence intervals for the first and tenth percentiles. The lower limits of such intervals are by definition A-Basis and B-Basis values, respectively.

The so called *CVaR deviation measure* commonly used in financial risk management, is becoming increasingly popular in other areas such as supply chain and military risks. For a random value Y representing losses, Conditional Value-at-Risk (CVaR) measures the average of some fraction (typically 5%) of the highest losses. We show how CVaR emerges naturally in the objective function whose minimizer determines the quantile regression estimators. This means that it is equivalent to think about minimizing residual error tail risk when procuring percentile estimates. For A-Basis and B-basis calculation, it is more natural to think in terms of controlling the risk on the lower tail of the failure load distribution, and the CVaR regression objective function can be readily adapted for that purpose. As a consequence, we obtain a minimum that is more appropriate for assessing the goodness-of-fit of the data to the factor model.

The methodology is applied to the provided dataset in order to incorporate information from analytical model (Model S) predictions and actual open-coupon failure test data. For this dataset, the results, based on a combination of in-sample analyses, out-of-sample analyses, and Monte Carlo simulations, suggest the following.

(i) Model S information analyzed via CVaR regression, provides plausible percentile estimates, even in the absence of any experimental test information.

(ii) For each stacking sequence, the available dataset contained measurements only from one lab, which resulted in relatively small variability of failure loads among experiments. For such a dataset with (a small variability in actual measurements) the contribution from Model S predictions becomes insignificant in the presence of experimental test data.

(iii) Out-of-sample 10th percentile estimates based on Model S individually, and on Model S plus 5 test points, are close to their true values. B-basis values are also close to nominal values based on actual experiments.

(iv) For a dataset involving a larger variability in test data (generated by Monte Carlo simulations), results suggest that the Model S contribution is approximately equivalent to 4 test data points.

(v) For datasets with large variability in the test data, the number of actual test data points used can be significantly reduced (about 2 times) by using Model S predictions in out-of-sample calculations.

Advantages of the proposed methodology:

- Nonparametric; only mild distributional assumptions are made. Leads naturally to robust estimates.

- Direct estimation of percentiles as a function of the factors.

- Relatively simple approach; results are readily interpretable.

- It is possible to quantify the contribution to the response from different sources of information: physical models, experimental data, expert opinion, etc.

- Allows for pooling of data across different stacking sequences, resulting in only moderate sample size requirements for useful inferences.

- Allows for pooling of data across various experimental setups, resulting in the possibility to reduce the number of complicated (expensive) experiments in lieu of simple (cheap) ones.

- Numerical efficiency: implemented via linear programming, leads to fast and stable numerical procedures.

Recommendations for implementation:

1. Certify a given analytical model (Model S etc.) for some area of application.

2. Build the factor model by splitting the available data into two portions: *training set* and *test set*. Use the first portion for model-building; the second for model cross-validation. Also, validate the factor model with Monte Carlo simulation of possible scenarios.

3. Use the resulting factor model combining modelling and experimental data for calculation of the 10th (1st) percentile and B-Basis (A-Basis) for the new variations of material (out-of-sample applications of the factor model).

9.1 Introduction

A basic problem in engineering design is how to manage uncertainty and risk in the properties of materials. Albeit costly, physical testing of the materials has traditionally been the primary method of quantifying the uncertainty. Recent advances in analytical physics-based engineering models and associated hardware/software technology, have made this an increasingly important contributor. From a monetary perspective, the lower cost of the modelling method makes it more desirable than the test method.

Rather than opt for one method over the other, a more effective risk management strategy might conceivably be the effective integration of the analytical models and the physical test program. The main objective of this study is to develop a sound methodology for such an integration. The resulting procedure will allow inferences in the form of point and interval estimates of percentiles to be made. Special cases include A-Basis and B-Basis values, defined in Appendix A.

Bayesian hierarchical modelling is one possibility when attempting to integrate heterogenous sources of data. Problems with this approach include the parametric assumptions about prior distributions that must be made, and the resulting large estimation errors for small data sets.

Financial engineering has recently made significant progress in working with percentile statistics and related optimization techniques. At the core of these developments is CVaR (Conditional Value-at-Risk), a new risk measure with appealing mathematical properties. The concepts and related methodology, are widely applicable to other engineering areas, and especially the military.

Drawing from our experience with these risk measures, we propose the use of factor models in combination with CVaR regression, as an alternative approach to integrate analytical model and physical test data. This involves treating a subset of the test data as the response variable, with sufficient statistics extracted from the remaining test data and model data as explanatory variables. CVaR regression, which includes the popular *quantile regression* of Koenker and Basset (1978) as a special case, is used to estimate the desired quantile of the response variable. In a case study, we demonstrate that CVaR regression is a viable methodology for combining analytical model and experimental test data.

9.2 Combining Model and Experimental Data: Factor Models and CVaR Regression

As outlined in the Introduction, we will use factor models to integrate model (or several models) predictions and experimental test data. In the conducted case study we integrated Model S predictions and experimental test data (henceforth *model* and *test* data) on the strength of composite materials.

Although beyond the scope of the present work, the approach is immediately gener-alizable to the integration of various other sources of heterogeneity, such as expert opinion.

Model Setup

The idea is to use each of the test data values as the response in turn, with the remaining test data values as well as the model data, as explanatory variables or *factors*. Let Y_{ij} denote the jth test data point corresponding to the ith stacking sequence, $i = 1, \ldots, I$, and $j = 1, \ldots, n_i$.

We suppose that factors can be evaluated with the available model runs and experi-mental data, which can directly predict the percentiles of the failure distribution. For instance, we may have a structural reliability model which generates with a Monte Carlo simulation a histogram of failure load distribution for a specific stacking se-quence and a specific experiment setup. Various statistical characteristics for this distribution can be considered as factors: mean, standard deviation, difference be-tween 90th and 50th percentile, and others. Also a nonlinear transformation of data may be conducted (e.g. the logarithmic transformation for lognormally distributed data) to improve performance of the factor model.

Here, for demonstration purposes, we exemplify the approach with the factor model for the available dataset where for each stacking sequence only upper and lower limits of the failure distribution were generated by the Model S. Let the pair (X_{i1}, X_{i2}) be the Model S data for the ith stacking sequence. In this particular study, to restrict the number of factors, we condense the data from each source, test and model, into a pair of summary statistics: the mean and standard deviation. In statistical parlance, this pair is *sufficient* for the corresponding unknown parameters when sampling from a normal distribution (however, we *do not assume normality* of data; the approach is *distribution-free*). Other data reduction measures of location (e.g. median) and dis-persion (e.g. lower semi-deviation) could also be used for datasets with asymmetrical distributions. Let (m_i, s_i) and (μ_i, σ_i) denote the sample mean and standard deviation for the test and model data, respectively, in stacking sequence i; that is

$$m_i = \frac{1}{n_i} \sum_{j=1}^{n_i} Y_{ij}, \qquad s_i^2 = \frac{1}{n_i - 1} \sum_{j=1}^{n_i} (Y_{ij} - m_i)^2,$$

$$\mu_i = (X_{i2} + X_{i1})/2, \qquad \sigma_i = (X_{i2} - X_{i1})/2.$$

We now fit the regression or factor model:

$$Y_{ij} = c_0 + c_1 \mu_i + c_2 \sigma_i + c_3 m_i + c_4 s_i + \varepsilon_{ij}, \tag{9.1}$$

where $c_0, \ldots, c_4$ are unknown regression coefficients to be estimated from the data, and ε_{ij} is the residual error corresponding to the jth test data point in the ith stacking sequence, $i = 1, \ldots, I, j = 1, \ldots, n_i$. We develop a nonstandard *CVaR regression* technique, estimating percentiles of the response, rather than its mean value.

Factor model (9.1) essentially combines various estimates of strength in a linear way. In this framework, we use the regression model (see [12]) with the *Conditional Value-at-Risk (CVaR)* measure (see, [10],[11]). There are several variants of the methodology; CVaR can either be viewed as an optimality criterion, or as a constraint in the linear regression optimization problem. Here we consider a special case which leads to the same estimates as *quantile regression* due to Koenker and Bassett (1978), which directly estimates the quantiles or percentiles of the response variable Y, as a function of the generic factors μ, σ, m, and s.

The technical details behind the approach considered in the report are outlined in Appendices B and C. Appendix B on "Risk Measures" defines *deviation CVaR* measure of dispersion, which we use as the optimality criterion in this linear regression percentile estimation framework.

To introduce deviation CVaR, let the residual value ε be a continuous random variable with expectation $\mathsf{E}\varepsilon$ (the general case, including discrete distributions, is discussed in Appendix B). For a fixed probability level $0 \leq \tau \leq 1$, the *Value-at-Risk (VaR)* is defined to be the τth quantile of ε. For a given level τ, $\mathrm{VaR}_\tau(\varepsilon)$ is a lower bound on the largest $(1-\tau)$ fraction of values of ε. First, let us consider the upper tail of the distribution, i.e., $\tau > 0.5$. For continuously distributed ε, the deviation CVaR denoted by $\mathrm{CVaR}_\tau^\Delta(\varepsilon)$, is a conditional average of values $\varepsilon - \mathsf{E}\varepsilon$ under condition that ε is in the τ tail of the distribution, i.e.

$$\mathrm{CVaR}_\tau^\Delta(\varepsilon) \equiv \mathsf{E}\left[\varepsilon - \mathsf{E}\varepsilon \mid \varepsilon \geq \mathrm{VaR}_\tau(\varepsilon)\right], \qquad \tau > 0.5.$$

Deviation CVaR at level τ provides an assessment of the $(1-\tau)$ fraction of the largest differences of failure loads and their mean values. It is a dispersion measure satisfying certain desirable axioms, see [11]. From the definition, deviation CVaR is the average distance between the mean value and the values in the τ tail of the distribution, and as such, is a one-sided measure of the width of such distribution. Compared, to [11] we have changed the sign of VaR and CVaR to avoid negative values for the positive failure loads ε, similarly to [10]. Deviation CVaR is convenient in statistical evaluations of percentiles, since linear regression with deviation CVaR (CVaR regression) lends itself naturally to estimation of such percentiles. For the lower tail, deviation CVaR is defined as

$$\mathrm{CVaR}_{1-\tau}^\Delta(-\varepsilon) \equiv \mathsf{E}\left[\mathsf{E}\varepsilon - \varepsilon \mid \varepsilon \leq \mathrm{VaR}_\tau(\varepsilon)\right], \qquad \tau \leq 0.5.$$

Here, deviation CVaR is the average distance between the mean value and the values in the lower τ tail of the distribution.

CVaR regression then entails using the following loss function or optimality criterion in linear regression:

$$\mathcal{P}_\tau^2 = \begin{cases} \mathrm{CVaR}_\tau^\Delta(\varepsilon), & \text{if } \tau \geq 0.5, \\ \mathrm{CVaR}_{1-\tau}^\Delta(-\varepsilon), & \text{if } \tau < 0.5. \end{cases} \tag{9.2}$$

For discrete distributions, the definition of the loss function is somewhat more complicated, but the intent is the same. The loss is approximately (or exactly) equal to the conditional expected value of the tail of the distribution. In the discrete case on hand, the minimization of the optimality criterion can be effectively achieved through linear programming, a robust and highly efficient numerical optimization procedure.

Minimizing deviation measure $\mathcal{P}_\tau^2$ of regression residuals (see, also, equation (9.23) in Appendix C), leads to the following estimated equation for the τth quantile of the failure load ε, as a function of the generic factors $\{\mu, \sigma, m, s\}$:

$$\hat{Q}_Y(\tau) = \hat{c}_0(\tau) + \hat{c}_1(\tau)\mu + \hat{c}_2(\tau)\sigma + \hat{c}_3(\tau)m + \hat{c}_4(\tau)s. \tag{9.3}$$

However, it can be observed that $\mathcal{P}_\tau^2$ in (9.2) depends on parameters c_1, c_2, c_3, c_4, but not on c_0. Therefore, the minimization procedure determines only $\hat{c}_1(\tau), \ldots, \hat{c}_4(\tau)$. The coefficient $\hat{c}_0(\tau) = \text{VaR}_\tau(\hat{\varepsilon}(\tau))$ is chosen so as to ensure $\hat{Q}_Y(\tau)$ is an unbiased quantile estimate, where $\hat{\varepsilon}(\tau)$ is the residual in an optimal point. This procedure leads to the same weights as the quantile regression, see Appendix C. All parameters, $c_0, \ldots, c_4$, can be found in one run of the linear programming problem [Uryasev and Rockafellar 2000, 2002].

One important advantage of quantile/CVaR regression is that only mild distributional assumptions ((i) and (ii) in Appendix C) are made, making the approach essentially nonparametric. In particular, different stacking sequences are allowed to have different standard deviations. Note that:

> The A-basis ($\tau = 0.01$) and B-basis ($\tau = 0.10$) estimates for factor vector $\mathbf{x}' = [1, \mu_i, \sigma_i, m_i, s_i]$, are given by equation (9.17) in Appendix C, with $\alpha = 0.05$.

Appendix C demonstrates the connection between Koenker and Bassett's quantile regression idea, and our *CVaR regression*. Essentially, the objective function used to estimate the regression model parameters in the former, is equivalent to a CVaR objective function in the latter. The equivalence is in the sense that one obtains exactly the same parameter estimates, but the objective functions evaluated at these estimates differ. Since the parameter estimates are the minimizers of these objective functions, the attained minimum values are natural candidates for measures of goodness-of-fit. We argue that the measure of goodness-of-fit in CVaR regression (equation (9.2)), being normalized by sample size, is better suited for making comparisons across fits from models with different sample sizes and confidence levels. Additionally, and of relevance in A and B-bases where lower bounds on estimates are more important than upper bounds, it takes into account the lower tail behavior of the residual error distribution.

Problems and Extensions

Although not done in this study, factor model (9.1) can be easily extended to incorporate other factors contributing to differences in subpopulations. In particular, material/batch indicator variables for open-hole tension and compression, could be included.

Another important issue we have chosen to circumvent, is the potential need to incorporate weighting of coefficients when levels of factors have unequal numbers of replicates. This is the subject of *weighted regression*, and techniques used there could also be applied here. Simple solutions might be to assign weights proportional to the number of replicates in that level; or inversely proportional to the level standard deviation. For demonstration purposes, we consider only the simplest case of equal replication within levels of each material/batch.

Finally, perhaps the most complicated extension concerns the area of optimal design of experiments. Specifically: optimal choice of level combinations of factors with sufficient data to calibrate the factor model; selection of samples from different materials/batches for testing; and allocation of treatments to these experimental units. Again we are not addressing this here, but design of experiment principles should receive careful attention in any future work that might be contemplated.

The suggested CVaR regression technique is based on minimization of the deviation CVaR measure of the regression error. Alternatively, we can consider techniques which estimate percentiles by linear regression with the CVaR constraints, see, [13]. In further studies we can compare the presented in this paper approach versus the approach in [13].

Section 9.4 describes in more detail the issues which are suggested to be addressed in future studies.

9.3 Case Study

In this section we apply the methodology (CVaR regression) outlined in the previous section to the supplied data set, henceforth referred to as the *original dataset*. The original dataset consists of 31 rows corresponding to stacking sequences, each with a lower and upper Model S prediction, but unequal numbers of test data points. A sample of two rows from the dataset is seen if Table 9.1. The original dataset contained one stacking sequence with the very large number of 294 test points. This was omitted from the in-sample analyses, and held in reserve for subsequent out-of-sample cross validation. The test and model failure loads (ksi) for the resulting *full dataset* are plotted in Figure 9.1 against stacking sequence number. There are 186 test points in this dataset.

From the full dataset, a subset consisting of the 15 stacking sequences that had at least 5 test data points was selected. (There are a total of 143 test points in this dataset.) From these, to avoid introduction of coefficients compensating unequal number of experiments for different stacking sequences, exactly 5 test points were (randomly) selected within each stacking sequence. (Issues relating to statistical analyses with

Table 9.1. Two rows of failure loads from the original dataset.

Row Number	Test Data Points	Model S Upper Limit	Model S Average	Model S Lower Limit
1	137.34	115.6		109
	118.96			
	119.41			
	110.64			
	103.67			
mean	118	mean	112.3	
2	48.30	56.64		51.68
	48.36			
	49.54			
	49.88			
	46.02			
mean	48.42	mean	54.16	

unequal numbers of datapoints will be addressed in subsequent studies.) The resulting *reduced dataset*, containing $5 \times 16 = 80$ test points, is seen in Figure 9.2.

In-sample Results

We perform two in-sample analyses with these data. The first will use only Model S model factors; the second will use Model S plus five test points in each stacking sequence.

Analysis 1. In the first analysis, we fit a factor model using only the Model S predictors as factors. That is, we fit the model,

$$Y_{ij} = c_0 + c_1 \mu_i + c_2 \sigma_i + \varepsilon_{ij},$$

to the full dataset, where $i = 1, \ldots, 31$, and the range of j varies for each i. Total sample size (number of test points) here is $n = 186$. The 10th and 90th percentile estimated surfaces are:

$$\hat{Q}_Y(0.10) = -2.427 + 0.974\mu - 1.008\sigma,$$

and

$$\hat{Q}_Y(0.90) = -6.870 + 1.138\mu + 0.789\sigma,$$

with deviation measures $\mathcal{P}_{0.10}^2 = 8.906$ and $\mathcal{P}_{0.90}^2 = 13.337$. The estimated percentiles are shown on Figure 9.3 along with the test data points, as a function of the (ordered) stacking sequence number. Overall, 10.8% (10.8%) of test points fell above (below) the estimated 90th (10th) percentiles in their respective stacking sequence. The fitted values are seen to provide a reasonably good in-sample fit, even in the absence of any test point information.

Analysis 2. In the second analysis, we fit the complete factor model using both Model S and test point data as factors. That is, we fit regression model (9.1) to the reduced dataset, where $i = 1, \ldots, 16$, and $j = 1, \ldots, 5$. The pairs of stacking sequences (2,23) and (3,24) were also merged. Total sample size here is $n = 5 \times 16 = 80$. The 10th and 90th percentile estimated surfaces are:

$$\hat{Q}_Y(0.10) = 0.430 - 0.015\mu - 0.075\sigma + 1.013m - 1.129s,$$

and

$$\hat{Q}_Y(0.90) = -0.167 + 0.058\mu - 0.080\sigma + 0.943m + 1.345s,$$

with deviation measures $\mathcal{P}^2_{0.10} = 4.901$ and $\mathcal{P}^2_{0.90} = 5.225$. A substantial drop in the respective deviation measures from Analysis 1 has occurred, indicative of an improvement in the fit. The estimated percentiles are shown on Figure 9.4 along with all 143 test data points, as a function of the (ordered) stacking sequence number. From (9.16), 95% confidence intervals for the model parameters of $Q_Y(0.10)$, are as follows:

$$c_0 : (-1.58, 2.44), \quad c_1 : (-0.10, 0.07), \quad c_2 : (-3.55, 0.20),$$

$$c_3 : \quad (0.93, 1.10), \quad c_4 : (-1.31, -0.95),$$

so that only c_3 and c_4 are significant. Overall, taking all 143 test points into consideration, 16.8% (15.4%) of test points fell above (below) the estimated 90th (10th) percentiles in their respective stacking sequence. The fitted values are seen to provide a reasonably good in-sample fit, even in the absence of any test point information.

Summary Remarks

- Model S information provides plausible percentile estimates, even in the absence of any experimental test information.

- In the considered dataset measurements only from one lab for each stacking sequence are available. This undoubtedly translates into smaller variability, both within and among stacking sequences, resulting in perhaps over-optimistic fits.

- Small variability in the experimental data may lead to incorrect estimation of the predictive capabilities of the model. The contribution from Model S information becomes insignificant in the presence of a small amount of experimental test information.

- We think that data from several labs needs to be incorporated into the analyses for a more realistic assessment of the predictive capabilities of each source of information.

Out-of-sample Results

The out-of-sample predictive capabilities of the models fitted in Analyses 1 and 2 are now assessed on the omitted stacking sequence with 294 test points. The Model S lower and upper failure load bounds for this stacking sequence are 37.86 and 39.75 respectively, which from Figure 9.2 are seen to be at the lower extremum of all Model S bounds. Also, the actual 10th percentile of all 294 test points is 35.84, which is on the boundary of range of percentile estimates for other stacking sequences.

Analysis 3. Using only the Model S values, the fitted model from Analysis 1 yields an estimate of 34.04 for the 10th percentile of failure loads. The calculated value of the B-Basis is 31.75. As it is supposed to be, the estimated B-Basis value, 31.75, is lower than the actual experimental 10th percentile estimate which equals to 35.84. A histogram of the test failure loads for this out-of-sample stacking sequence is seen in Figure 9.5.

Analysis 4. Using Model S and the five randomly chosen test values 38.12, 37.214, 37.637, 37.707, and 35.63, the fitted model from Analysis 2 yields the estimate of 36.36 for the 10th percentile of failure loads. The calculated value of the B-Basis is 35.21. The estimated B-Basis value, 35.21, is lower than the actual experimental 10th percentile, 35.84.

A histogram of the test failure loads for this out-of-sample stacking sequence is seen in Figure 9.6.

Aiming at a more precise quantification, the procedure of randomly selecting 5 test points and fitting the model of Analysis 2, was repeated 5,000 times. In 78% of cases, the actual 10th percentile estimate of failure loads, 35.84, is above the B-Basis value. This is somewhat lower than the nominal value of 95%, but it must be remembered that this is an out-of-sample inference.

Monte Carlo Results

For our final round of analyses, the Model S predictions for the 31 stacking sequences with at least 3 test points were selected from the full dataset. To simulate a more heterogenous scenario and incorporate between lab variability, 3 sets of 3 points (9 altogether) were then generated by Monte Carlo for each stacking sequence. This was done so as to ensure a greater between lab than within one lab variability, each set of 3 points viewed as originating from a different lab. The resulting *artificial dataset*, containing $9 \times 31 = 279$ test points, is seen in Figure 9.7.

Analysis 5. In the first scenario, and in analogy with Analysis 1, we considered estimation of the 10th and 90th percentiles based on Model S only information. The 10th percentile estimated surface is:

$$\hat{Q}_Y(0.10) = 4.377 + 0.826\mu - 1.186\sigma,$$

with deviation measure $\mathcal{P}^2_{0.10} = 15.077$. The results are presented in Figure 9.8. 95% confidence intervals for the model parameters of $Q_Y(0.10)$, are as follows:

$$c_0 : (-0.96, 9.72), \quad c_1 : (0.76, 0.89), \quad c_2 : (-1.88, -0.49),$$

so that only c_1 and c_2 are significant.

Analysis 6. We now add test point information. We start by selecting 3 test points within each stacking sequence to form m and s; using the remaining 6 points as responses. We do this for each of the $\binom{9}{3} = 84$ possible ways to select 3 points from 9, thus creating a dataset of size $n = 6 \times 84 \times 31 = 15,624$. The 10th percentile estimated surface for this Model S plus 3 points dataset is:

$$\hat{Q}_Y(0.10) = 2.487 + 0.671\mu - 0.717\sigma + 0.202m - 0.377s,$$

with deviation measure $\mathcal{P}^2_{0.10} = 14.715$. 95% confidence intervals for the model parameters are as follows:

$$c_0 : (1.87, 3.11), \quad c_1 : (0.64, 0.70), \quad c_2 : (-0.80, -0.64),$$

$$c_3 : (0.18, 0.23), \quad c_4 : (-0.41, -0.34),$$

Although all coefficients are significant, the addition of 3 test points per stacking sequence does not appreciably change the model based on Model S only information. $\mathcal{P}^2_{0.10}$ has decreased by 2.4% (from 15.077 to 14.715) and the coefficients from the Model S contribution outweigh those from the test point contribution. This demonstrates that the Model S for this particular dataset has a high predictive power and adding several actual measurements do not bring a lot of new information.

Analysis 7. In an attempt to quantify the Model S contribution in terms of test data points, we consider a factor model with only test point information. The calculation of m and s within each stacking sequence is based on selecting 7 test points; the remaining 2 points used as responses. Since there are $\binom{9}{7} = 36$ possible ways to select 7 points from 9, the resulting dataset has size $n = 2 \times 36 \times 31 = 2,232$. The 10th percentile estimated surface for this 7 points dataset is:

$$\hat{Q}_Y(0.10) = 0.725 + 0.933m - 0.967s,$$

with deviation measure $\mathcal{P}^2_{0.10} = 14.484$.

The deviation measure for the 7 test points dataset, 14.484, is approximately the same as the deviation measure for the Model S plus 3 test points dataset, 14.715. Thus the *Model S contribution can be roughly equated to 4 test data points.*

Analysis 8. Lastly, we mimic Analysis 2 by incorporating Model S and all 9 test
points into the construction of the factors. m and s are now calculated from all 9
test points within each stacking sequence; the same 9 points used as responses.
The resulting dataset has size $n = 9 \times 31 = 279$. The 10th percentile estimated
surface for this Model S plus 9 test points dataset is:

$$\hat{Q}_Y(0.10) = 0.000 + 0.041\mu - 0.140\sigma + 0.965m - 1.214s,$$

with deviation measure $\mathcal{P}^2_{0.10} = 11.697$. The results are presented in Figure 9.9.
95% confidence intervals for the model parameters are as follows:

$$c_0 : (-2.63, 2.63), \quad c_1 : (-0.13, 0.22), \quad c_2 : (-0.48, 0.20),$$

$$c_3 : (0.79, 1.14), \quad c_4 : (-1.46, -0.97),$$

so that the Model S contribution is not significant.

9.4 Summary and Recommendations

We have proposed a risk management strategy for integrating various sources of information into one coherent factor model. The sound methodology of CVaR Regression
was developed to enable direct estimation of percentile characteristics of the response,
as a function of explanatory variables. This was applied to the provided dataset in order
to incorporate information from analytical model (Model S) predictions and physical
test data. The results, based on a combination of in-sample analyses, out-of-sample
analyses, and Monte Carlo simulations, suggest the following.

(i) Model S information provides plausible percentile estimates, even in the absence of any experimental test information.

(ii) In the available dataset, only one lab measurement is available for each stacking sequence, which results in a low variability of actual measurements. This
condition makes the contribution from Model S insignificant in the presence of
even small amounts of experimental data.

(iii) For one out-of-sample stacking sequence, 10th percentile estimates based on
Model S as well as on Model S plus 5 test points are close to their true values.
However, coverage probabilities for B-Basis estimates deviated about 20% from
their nominal 95% levels.

(iv) With Monte Carlo simulations variability in measurements between different
labs was introduced. For the simulated dataset, in the presence of Model S plus
experimental test information, Model S contribution is approximately equivalent to 4 test data points.

(v) The number of test data points needed can be significantly reduced by using
Model S predictions; both in-sample and out-of-sample. Experiments showed
that Model S provides acceptable estimates of percentiles and B-Basis even
without any actual measurements.

Advantages of the Proposed Methodology

We list some of the benefits of the proposed factor model with CVaR regression approach.

- Nonparametric; only mild distributional assumptions are made. Leads naturally to robust estimates.

- Leads to direct estimation of percentiles as a function of the factors.

- Relatively simple approach; results are readily interpretable.

- It is possible to quantify the contribution to the response from different sources of information: physical models, experimental data, expert opinion, etc.

- Allows for pooling of data across different stacking sequences, resulting in only moderate sample size requirements for useful inferences.

- Allows for pooling of data across various experimental setups, resulting in the possibility to reduce the number of sophisticated (expensive) experiments in lieu of simple (cheap) ones.

- Numerical efficiency: implemented via linear programming, leads to fast and stable numerical procedures.

Recommendations for Implementation and Future Research

We conclude with some guidelines for applying the proposed methodology to new situations. This report briefly discussed the factor model approach and suggested implementation steps. However, for practical applications, each step should be elaborated in detail, tested and documented in user-friendly format. Firstly, we outline the basic steps in applying the methodology. Secondly, we discuss several issues which need to be addressed in further studies.

Basic Steps of the Factor Analysis Approach

1. Certify a given analytical model (Model S etc.) for some area of application.

2. Build the factor model by splitting the available data into in-sample and out-of-sample portions. Use the former for model-building (the *training set*); the latter for model cross-validation (the *test set*).

3. Validate the choice of selected model on the out-of-sample portion, and by using Monte Carlo simulation of possible scenarios.

4. Use the resulting factor model for calculation of the 1st (10th) percentile and A-Basis (B-Basis).

Issues to be Addressed in Further Studies

1. The model certification process should be investigated in detail. Certification issues have been discussed in the framework of [1]. However, much more research in this area is needed.

2. Model S predictive capabilities have been studied in this report. Other structural reliability models can be investigated and compared with Model S.

3. Guidelines for the planning of the experiments should be developed. How much data is needed for designing training and test sets? How to allocate resources between activities on conducting experiments and model developments? How to take into account existing data which may provide unbalanced information on different variations of material?

4. *Optimality considerations.* We have suggested factor models for estimating percentiles of failure load distribution and CVaR regression methodology for finding the optimal weighting coefficients. Calculations for the considered example show that optimal coefficients are quite sensitive to variability in the actual measurements. We think that weighting coefficients for various factors may not need to be optimal for achieving acceptable precision of predictions. It may be reasonable to identify a set of factors and fix some coefficients for these factors for certain areas of application. These factors and *fixed coefficients* can be tested with various datasets and with extensive Monte Carlo calculations.

5. Procedures for calculating A-basis and B-basis in combination with factor models should be studied. In this report, percentile confidence intervals were calculated under a mild parametric assumption on the kernel bandwidth. Other promising approaches such as bootstrapping (which is nonparametric), also need to be evaluated.

6. Issues relating to robustness of the approach and sensitivity to various factors, e.g. amount of data, variability of data inputs, impact of outliers, etc., should be investigated.

7. The current case study estimated 10th and 90th percentiles of the failure load distribution. The amount of data in the supplied dataset is not sufficient for estimating the 1st and 99th percentiles in the framework of the CVaR regression methodology. Parametric approaches that are more suited for high percentile estimation, such as *extreme value theory*, are worthy of further investigation.

8. Performance of the suggested CVaR regression approach can be compared with other approaches, including the Bayes approach, by testing with several datasets and Monte Carlo simulations.

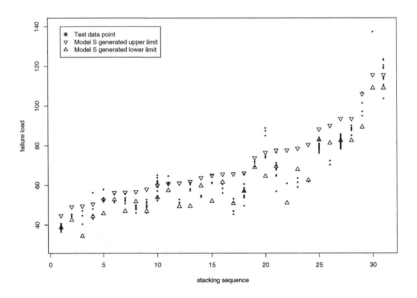

Fig. 9.1. Model S upper and lower failure load bounds for the full dataset.

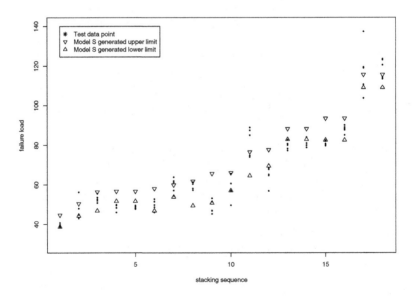

Fig. 9.2. Model S upper and lower failure load bounds for the reduced dataset.

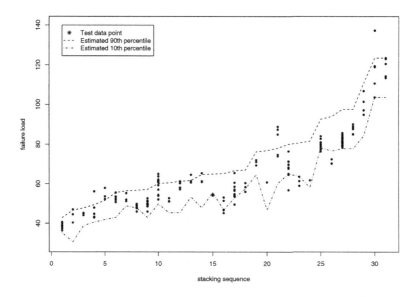

Fig. 9.3. Estimated 10th and 90th percentiles for the full dataset based on Model S only information.

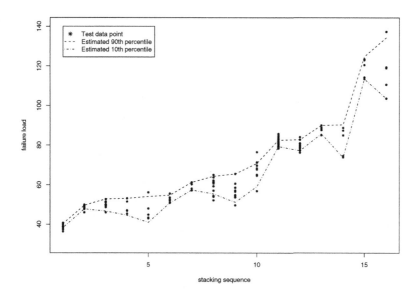

Fig. 9.4. Estimated 10th and 90th percentiles for the reduced dataset based on Model S plus 5 randomly selected test points.

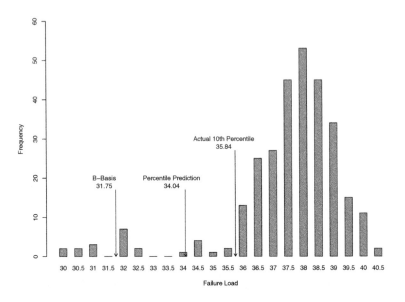

Fig. 9.5. Histogram of the 294 test failure loads for the out-of-sample stacking sequence. The actual and estimated 10th percentiles are shown, along with the B-Basis value. These are based on the fitted model of Analysis 1.

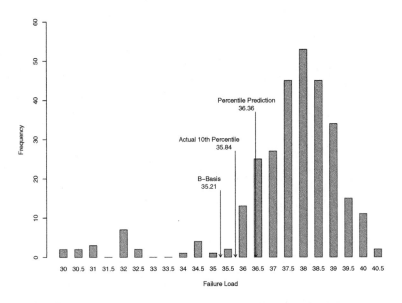

Fig. 9.6. Histogram of the 294 test failure loads for the out-of-sample stacking sequence. The actual and estimated 10th percentiles are shown, along with the B-Basis value. These are based on the fitted model of Analysis 2, with test points 38.12, 37.214, 37.637, 37.707, 35.63.

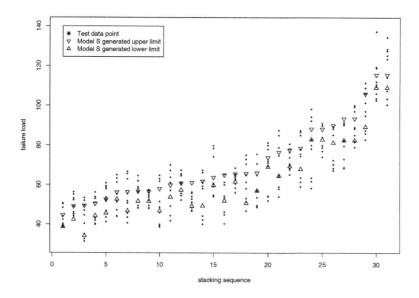

Fig. 9.7. Model S upper and lower failure load bounds for the artificial dataset, along with all 279 test data points.

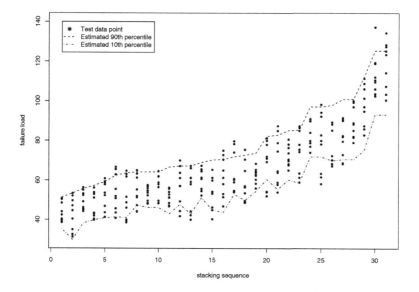

Fig. 9.8. Estimated 10th and 90th percentiles based on Model S only information for the artificial dataset.

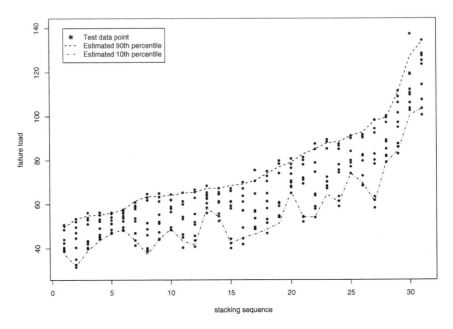

Fig. 9.9. Estimated 10th and 90th percentiles based on Model S plus 9 test points for the artificial dataset.

Acknowledgements

This effort was jointly accomplished by the Boeing led team and the United States Government under the guidance of NAVAIR. This work was funded by DARPA/DSO and administered by NAVAIR under Technology Investment Agreement N00421-01-3-0098. The program would like to acknowledge the guidance and support of Dr. Leo Christodoulou of DARPA/DSO for this effort.

A Definition of A-basis and B-basis values

Let $\{X_1, \dots, X_n\}$ be a random sample of observations from random variable X with distribution function $F(x) \equiv P(X \leq x)$. For probability level $0 \leq \tau \leq 1$, let $\xi_\tau \equiv \xi_\tau(X_1, \dots, X_n)$ be a statistic (a function of the random sample) satisfying,

$$0.95 = P\left[1 - F(\xi_\tau) > 1 - \tau\right] \tag{9.4}$$
$$= P\left[F(\xi_\tau) < \tau\right] \tag{9.5}$$
$$= P\left[\xi_\tau < F^{-1}(\tau)\right].$$

Representation (9.5) means that (ξ_τ, ∞) is one-sided confidence interval for $F^{-1}(\tau)$, so that ξ_τ is a lower 95% *confidence* bound for the τth quantile of X. Equivalently, representation (9.4) means that ξ_τ is a lower 95% *tolerance* bound for the $(1-\tau)$th quantile of X. An *A-basis* value is defined to be $\xi_{0.01}$, and a *B-basis* value, $\xi_{0.10}$.

B Risk Measures

Financial *risk measures* provide a useful class of objective functions with a wide spectrum of applications. To introduce these, let Y be a continuous random variable with $EY < \infty$, and distribution function $F(y)$. For a fixed probability level $0 \le \tau \le 1$, the *Value-at-Risk (VaR)* is defined to be the τth quantile of Y,

$$\text{VaR}_\tau(Y) \equiv F^{-1}(\tau).$$

With Y measuring profit/loss, the conceptual simplicity of VaR has resulted in its widespread incorporation into standard banking regulation, where it is imperative to assess credit risk.

However, note that for a given level τ, $\text{VaR}_\tau(Y)$ is merely a lower bound on the worst $(1-\tau)$ fraction of losses, and thus fails to capture any information about the magnitude of losses beyond that level. This and other undesirable mathematical properties inherent in VaR, are documented by Artzner *et al.* (1997, 1999), who introduce an axiomatic method for the proper construction of risk measures. With $(\nu)^+ = \max\{0, \nu\}$ for any real ν, they propose an alternative measure of risk, *Conditional Value-at-Risk (CVaR)*, defined as the solution of the minimization problem

$$\text{CVaR}_\tau(Y) \equiv \min_{a \in \mathscr{R}} \left\{ a + \frac{1}{1-\tau} E(Y-a)^+ \right\}. \tag{9.6}$$

Rockafellar and Uryasev (2000) shows that the minimizer is actually $\text{VaR}_\tau(Y)$. Also, Rockafellar and Uryasev (2002) show that for continuous Y,

$$\text{CVaR}_\tau(Y) = E[Y \mid Y \ge \text{VaR}_\tau(Y)],$$

i.e., CVaR equals conditional expectation of the τ tail of the distribution. For general distributions, including discrete distributions, CVaR approximately (or exactly) equals the conditional expectation of the τ tail of the distribution. In this sense, CVaR at level τ provides a more realistic assessment of the worst $(1-\tau)$ fraction of losses, by evaluating their expected value. As it has been proved by Pflug (2000), CVaR is a *coherent* measure of risk in the sense of Artzner *et al.* (1997, 1999), and satisfies several desirable properties: translation invariance, sub-additivity, positive homogeneity, and monotonicity with respect to stochastic dominance. Figure B illustrates the relationship between VaR and CVaR. Note that one always has $\text{VaR}_\tau(Y) \le \text{CVaR}_\tau(Y)$.

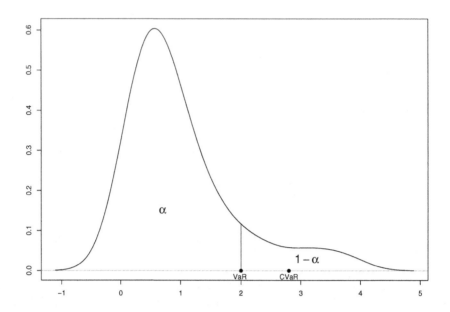

Fig. 9.10. Example of the relationship between VaR and CVaR at quantile α for a continuous distribution.

For discrete Y with probability mass points at $y_1, \ldots, y_n$, and corresponding masses $p_1, \ldots, p_n$, (9.6) becomes,

$$\text{CVaR}_\tau(Y) = \min_{a \in \mathscr{R}} \left\{ a + \frac{1}{1-\tau} \sum_{i=1}^{n} (y_i - a)^+ p_i \right\}. \tag{9.7}$$

In transitioning to an empirical setting when a sample of observations $\{y_1, \ldots, y_n\}$ from Y is available, VaR and CVaR can be estimated via method of moments:

$$\widehat{\text{VaR}}_\tau(Y) = \arg\min_{a \in \mathscr{R}} \left\{ a + \frac{1}{n(1-\tau)} \sum_{i=1}^{n} (y_i - a)^+ \right\} \tag{9.8}$$

$$\widehat{\text{CVaR}}_\tau(Y) = \min_{a \in \mathscr{R}} \left\{ a + \frac{1}{n(1-\tau)} \sum_{i=1}^{n} (y_i - a)^+ \right\}. \tag{9.9}$$

This amounts to assigning the uniform mass $p_i = 1/n$ to each y_i in (9.7).

As already introduced in Section 9.2, for general random variable Y the quantity

$$\text{CVaR}_\tau^\Delta(Y) \equiv \text{CVaR}_\tau(Y - EY), \tag{9.10}$$

is a *deviation measure*, as defined by Rockafellar, Uryasev and Zabarankin (2002). This class of functionals was introduced to generalize the properties of symmetric deviation measures such as standard deviation, thus allowing for the definition of more asymmetric ones. In words, (9.10) equals approximately (or exactly) to the expectation of the right tail of the centered random variable $Y - EY$, and as such can be viewed as a measure of dispersion of its right tail.

When $\tau < 0.5$ (A and B-basis), it is more meaningful to think about minimizing left-tail CVaR, rather than right-tail. We therefore propose the measure:

$$\mathcal{P}_\tau^2 = \begin{cases} \text{CVaR}_\tau^\Delta (Y), & \text{if } \tau \geq 0.5, \\ \text{CVaR}_{1-\tau}^\Delta (-Y), & \text{if } \tau < 0.5, \end{cases} \tag{9.11}$$

as the optimality criterion or measure of goodness-of-fit in quantile regression estimation. This gives rise to what we are calling *CVaR regression*. Since Pflug (2000) shows that for any random variable Z,

$$\text{CVaR}_\tau (Z - EZ) = \frac{\tau}{1 - \tau} \text{CVaR}_{1-\tau} (EZ - Z),$$

we can rewrite (9.11) as,

$$\mathcal{P}_\tau^2 = \begin{cases} \text{CVaR}_\tau^\Delta (Y), & \text{if } \tau \geq 0.5, \\ \frac{1-\tau}{\tau} \text{CVaR}_\tau^\Delta (Y), & \text{if } \tau < 0.5. \end{cases}$$

In this way, the emphasis is on adequately accounting for under-estimates of quantiles, the major source of concern in this study.

In case of discrete distributions, minimization of CVaR with loss functions which linearly depend upon control parameters can be reduced to linear programming, see, Rockafellar and Uryasev, 2000, 2002.

In the next section, we will demonstrate that there is a connection between CVaR regression as we have defined it, and quantile regression as defined by Koenker and Bassett (1978). This will lead naturally to the use of CVaR regression as a generalization of quantile regression.

C Quantile Regression and CVaR Regression

Consider the linear regression model

$$Y = \beta_0 + \beta_1 X_1 + \ldots + \beta_p X_p + \varepsilon = X'\beta + \varepsilon, \tag{9.12}$$

where $\beta = [\beta_0, \ldots, \beta_p]'$, and $X = [1, X_1, \ldots, X_p]'$. The factors $\{X_1, \ldots, X_p\}$ are viewed as independent random variables, and independent from ε which has cdf F and pdf f. A random sample of observations from this model

$$Y_i = \beta_0 + \beta_1 X_{i1} + \ldots + \beta_p X_{ip} + \varepsilon_i = X_i'\beta + \varepsilon, \quad i = 1, \ldots, n,$$

with the $\{\varepsilon_i\}$ i.i.d., can be concisely written as,

$$\mathbf{Y} = \mathrm{X}\beta + \varepsilon, \tag{9.13}$$

where X denotes the n by $p+1$ design matrix whose ith row is $\mathbf{X}'_i$. When transitioning to the empirical setting, we will use lower case, $\{y_i\}$ and $\{x_{ij}\}$, to denote realized values of the corresponding random variables.

Now, note that for random variable Z and constant c, the quantile functions of $Z + c$ and Z are related by $F^{-1}_{Z+c}(\tau) = F^{-1}_Z(\tau) + c$. Since

$$P(Y \leq y | \mathbf{X} = \mathbf{x}) = F\left(y - \mathbf{x}'\beta\right),$$

it follows immediately that the *conditional quantile function* of Y given $\mathbf{x} = [1, x_1, \ldots, x_p]'$, $Q_Y(\tau|\mathbf{x})$, and the quantile function of ε, are related according to,

$$Q_Y(\tau|\mathbf{x}) = \mathbf{x}'\beta + F^{-1}(\tau) \equiv \mathbf{x}'\beta(\tau), \tag{9.14}$$

where $\beta_0(\tau) = \beta_0 + F^{-1}(\tau)$, and $\beta_j(\tau) = \beta_j$, for $j = 1, \ldots, p$. Viewed as a function of the design point $\mathbf{x}$, $Q_Y(\tau|\mathbf{x})$ describes how the τth quantile surface of Y varies as a function of the factors.

Quantile Regression

If $\mathbf{y}' = [y_1, \ldots, y_n]$ is the observed vector of responses, Koenker and Bassett (1978) proposed the following estimator of $\beta(\tau)$:

$$\hat{\boldsymbol{\beta}}(\tau) = \arg \min_{\mathbf{b} \in \mathscr{R}^{p+1}} \sum_{i=1}^{n} \left\{ \tau(y_i - \mathbf{x}'_i\mathbf{b})^+ - (1 - \tau)(y_i - \mathbf{x}'_i\mathbf{b})^- \right\} \tag{9.15}$$

$$\equiv \arg \min_{\mathbf{b} \in \mathscr{R}^{p+1}} \mathcal{P}^0_\tau(\mathbf{b}),$$

where $(\nu)^+ = \max\{0, \nu\}$ and $(\nu)^- = \min\{0, \nu\}$. This is a natural generalization to linear regression, of the consistent estimator of the τth quantile

$$\hat{b} = \arg \min_{b \in \mathscr{R}} \sum_{i=1}^{n} \left\{ \tau(y_i - b)^+ - (1 - \tau)(y_i - b)^- \right\},$$

in the special case of (9.12) when $\beta_j = 0$, $j = 1, \ldots, p$ (the *location model*). For any design point $\mathbf{x}$, the empirical conditional quantile function is then defined by Koenker and Bassett (1982) as

$$\hat{Q}_Y(\tau|\mathbf{x}) = \mathbf{x}'\hat{\boldsymbol{\beta}}(\tau).$$

In case (9.15) does not have a unique solution, the minimum value of $\hat{Q}_Y(\tau|\mathbf{x})$ over all such $\hat{\boldsymbol{\beta}}(\tau)$ is taken.

Koenker and Bassett (1978) also show that under the mild design conditions,

(i) $\{\varepsilon_i\}$ i.i.d. with continuous cdf F, and continuous and positive density, f, at $F^{-1}(\tau)$,

(ii) $\lim\limits_{n\to\infty} n^{-1}X'X \equiv D$, is a positive definite matrix,

with $\Omega \equiv \tau(1-\tau)/f^2 \left(F^{-1}(\tau) \right) D^{-1}$, we have the asymptotic result:

$$\sqrt{n} \left(\hat{\beta}(\tau) - \beta(\tau) \right) \xrightarrow{d} N\left(0, \Omega \right).$$

Defining $\Omega_n \equiv n^{-1}\Omega$, this means $\hat{\beta}(\tau)$ and $x'\hat{\beta}(\tau)$ are asymptotically normal

$$\hat{\beta}(\tau) \sim AN(\beta(\tau), \Omega_n), \quad \text{and} \quad x'\hat{\beta}(\tau) \sim AN(x'\beta(\tau), x'\Omega_n x).$$

The only obstacle to immediate construction of asymptotics-based confidence intervals for regression quantiles therefore, is the estimation of the *sparsity function*,

$$s(\tau) = \left[f\left(F^{-1}(\tau) \right) \right]^{-1}.$$

Koenker (1994), argues that a natural candidate is

$$\hat{s}_n(\tau) = \left[\hat{F}_n^{-1}(\tau + h_n) - \hat{F}_n^{-1}(\tau - h_n) \right] /(2h_n),$$

with $\hat{F}_n^{-1}(\cdot)$ the usual empirical quantile function based on the quantile regression residuals. The most difficult question of course is selection of the bandwidth h_n. The heuristic argument and simulations in Koenker (1994), indicate that the normal model version of the Hall and Sheather (1988) method is promising. For construction of $(1-\alpha)100\%$ confidence intervals, this method identifies

$$h_n = \left[\frac{3z_{1-\alpha/2}^2 e^{-z_\tau^2}}{4n\pi(2z_\tau^2 + 1)} \right]^{1/3},$$

as the optimal bandwidth. (For probability r, z_r denotes the rth quantile of a standard normal random variable.) With the above choices, Ω_n can be estimated by,

$$\hat{\Omega}_n = \tau(1-\tau)(X'X)^{-1}\hat{s}_n(\tau)^2.$$

This leads immediately to the following $(1-\alpha)100\%$ two-sided confidence interval bounds for $x'\beta(\tau)$:

$$x'\hat{\beta}(\tau) \pm z_{1-\alpha/2}\sqrt{x'\hat{\Omega}_n x}, \tag{9.16}$$

and the lower $(1-\alpha)100\%$ one-sided confidence interval bound for $x'\beta(\tau)$:

$$x'\hat{\beta}(\tau) + z_\alpha \sqrt{x'\hat{\Omega}_n x}. \tag{9.17}$$

(Confidence bounds for the kth element of $\beta(\tau)$, are obtained by setting the kth element of x to 1, zeroes elsewhere.) Note that the A-basis and B-basis values for Y at x are given by the last formula with $\tau = 0.01$ and $\tau = 0.10$, respectively, when $\alpha = 0.05$.

Connection with CVaR Regression

It can be shown that for any real numbers τ and ν,

$$\tau(\nu)^+ - (1-\tau)(\nu)^- = (1-\tau)\left[\frac{1}{1-\tau}(\nu)^+ - \nu\right].$$

If ν is viewed as a random variable, this equality will continue to hold when the expectation of both sides is taken, i.e.

$$\mathsf{E}\left[\tau(\nu)^+ - (1-\tau)(\nu)^-\right] = (1-\tau)\left[\frac{1}{1-\tau}\mathsf{E}(\nu)^+ - \mathsf{E}(\nu)\right].$$

Defining $\bar{\mathbf{b}} = [b_1, \ldots, b_p]'$ and $\bar{\mathbf{x}}_i = [x_{i1}, \ldots, x_{ip}]'$, the empirical version of the above result applied to (9.15) gives,

$$
\begin{aligned}
P_\tau^0(\mathbf{b}) &= n(1-\tau)\left[\frac{1}{n(1-\tau)}\sum_{i=1}^n (y_i - \mathbf{x}_i'\mathbf{b})^+ - \frac{1}{n}\sum_{i=1}^n (y_i - \mathbf{x}_i'\mathbf{b})\right] \\
&= n(1-\tau)\left[\frac{1}{n(1-\tau)}\sum_{i=1}^n (y_i - \bar{\mathbf{x}}_i'\bar{\mathbf{b}} - b_0)^+ - \frac{1}{n}\sum_{i=1}^n (y_i - \bar{\mathbf{x}}_i'\bar{\mathbf{b}} - b_0)\right] \\
&= n(1-\tau)\left[b_0 + \frac{1}{n(1-\tau)}\sum_{i=1}^n (y_i - \bar{\mathbf{x}}_i'\bar{\mathbf{b}} - b_0)^+ - \frac{1}{n}\sum_{i=1}^n (y_i - \bar{\mathbf{x}}_i'\bar{\mathbf{b}})\right].
\end{aligned}
$$

Let us denote

$$P_\tau^1(\bar{\mathbf{b}}) \equiv \left\{\widehat{\mathrm{CVaR}}_\tau\left(Y - \sum_{j=1}^p X_j b_j\right) - \hat{\mathsf{E}}\left(Y - \sum_{j=1}^p X_j b_j\right)\right\},$$

where $\hat{\mathsf{E}}(Y - \sum_{j=1}^p X_j b_j) = \frac{1}{n}\sum_{i=1}^n (y_i - \bar{\mathbf{x}}_i'\bar{\mathbf{b}})$. Then,

$$
\begin{aligned}
P_\tau^0(\mathbf{b}) &= n(1-\tau)\left[b_0 + \frac{1}{n(1-\tau)}\sum_{i=1}^n (y_i - \bar{\mathbf{x}}_i'\bar{\mathbf{b}} - b_0)^+ \right. \\
&\qquad\qquad\left. - \hat{\mathsf{E}}\left(Y - \sum_{j=1}^p X_j b_j\right)\right] \\
&= n(1-\tau)\left[b_0 + \frac{1}{n(1-\tau)}\sum_{i=1}^n (y_i - \bar{\mathbf{x}}_i'\bar{\mathbf{b}} - b_0)^+ \right. \\
&\qquad\qquad\left. - \widehat{\mathrm{CVaR}}_\tau\left(Y - \sum_{j=1}^p X_j b_j\right) + P_\tau^1(\bar{\mathbf{b}})\right].
\end{aligned}
$$

Now, since

$$\min_{b_0 \in \mathscr{R}} \left\{ b_0 + \frac{1}{n(1-\tau)} \sum_{i=1}^{n} (y_i - \bar{\mathbf{x}}_i' \mathbf{b} - b_0)^+ \right\} = \widehat{\mathrm{CVaR}}_\tau \left(Y - \sum_{j=1}^{p} X_j b_j \right),$$

with minimizer $b_0 = \widehat{\mathrm{VaR}}_\tau (Y - \sum_{j=1}^{p} X_j b_j)$, the minimizers of (9.15) are

$$\hat{\boldsymbol{\beta}}(\tau) = \left[\hat{\beta}_1(\tau), \ldots, \hat{\beta}_p(\tau) \right]' = \arg\min_{\mathbf{b} \in \mathscr{R}^p} \mathcal{P}_\tau^1(\mathbf{b}), \tag{9.18}$$

$$\hat{\beta}_0(\tau) = \widehat{\mathrm{VaR}}_\tau \left(Y - \sum_{j=1}^{p} X_j \hat{\beta}_j(\tau) \right). \tag{9.19}$$

The attained minimum is

$$\mathcal{P}_\tau^0 \left(\hat{\boldsymbol{\beta}}(\tau) \right) = n(1-\tau) \mathcal{P}_\tau^1 \left(\hat{\boldsymbol{\beta}}(\tau) \right), \tag{9.20}$$

which can be viewed as a measure of goodness-of-fit. However, we propose that a more natural measure is the normalized quantity

$$\mathcal{P}_\tau^1 \left(\hat{\boldsymbol{\beta}}(\tau) \right) = \frac{1}{n(1-\tau)} \mathcal{P}_\tau^0 \left(\hat{\boldsymbol{\beta}}(\tau) \right). \tag{9.21}$$

If $\hat{\varepsilon}(\tau) \equiv Y - \hat{\beta}_0(\tau) - \sum_{j=1}^{p} X_j \hat{\beta}_j(\tau)$ denotes the fitted residual random variable, then by the translation invariance of CVaR,

$$\begin{aligned} \mathcal{P}_\tau^1 \left(\hat{\boldsymbol{\beta}}(\tau) \right) &= \widehat{\mathrm{CVaR}}_\tau \left(\hat{\beta}_0 + \hat{\varepsilon}(\tau) \right) - \hat{\mathsf{E}} \left(\hat{\beta}_0 + \hat{\varepsilon}(\tau) \right) \\ &= \widehat{\mathrm{CVaR}}_\tau \left(\hat{\beta}_0 + \hat{\varepsilon}(\tau) \right) - \hat{\beta}_0 - \hat{\mathsf{E}} \left(\hat{\varepsilon}(\tau) \right) \\ &= \widehat{\mathrm{CVaR}}_\tau \left(\hat{\varepsilon}(\tau) - \hat{\mathsf{E}} \hat{\varepsilon}(\tau) \right) \\ &= \widehat{\mathrm{CVaR}}_\tau^\Delta (\hat{\varepsilon}(\tau)). \end{aligned} \tag{9.22}$$

In CVaR regression, as discussed in Appendix B, we use the measure of goodness-of-fit:

$$\mathcal{P}_\tau^2 \left(\hat{\boldsymbol{\beta}}(\tau) \right) = \begin{cases} \widehat{\mathrm{CVaR}}_\tau^\Delta (\hat{\varepsilon}(\tau)), & \text{if } \tau \geq 0.5, \\ \widehat{\mathrm{CVaR}}_{1-\tau}^\Delta (-\hat{\varepsilon}(\tau)), & \text{if } \tau < 0.5. \end{cases}$$

$$= \begin{cases} \widehat{\mathrm{CVaR}}_\tau^\Delta (\hat{\varepsilon}(\tau)), & \text{if } \tau \geq 0.5, \\ \frac{1-\tau}{\tau} \widehat{\mathrm{CVaR}}_\tau^\Delta (\hat{\varepsilon}(\tau)), & \text{if } \tau < 0.5. \end{cases} \tag{9.23}$$

In this manner, the emphasis is on adequately accounting for under-estimates of $Q_Y(\tau|\mathbf{x})$, the major source of concern. The minimizer, $\hat{\boldsymbol{\beta}}(\tau)$, continues to be the same, but the new minimum (measure of goodness-of-fit) is

$$\mathcal{P}_\tau^2 \left(\hat{\boldsymbol{\beta}}(\tau) \right) = \begin{cases} \frac{1}{n(1-\tau)} \mathcal{P}_\tau^0(\hat{\boldsymbol{\beta}}(\tau)), & \text{if } \tau \geq 0.5, \\ \frac{1}{n\tau} \mathcal{P}_\tau^0(\hat{\boldsymbol{\beta}}(\tau)), & \text{if } \tau < 0.5. \end{cases} \tag{9.24}$$

Remark 1. CVaR regression in its most general form doesn't determine an estimator for $\beta_0(\tau)$. The value in (9.19) was chosen so as to agree with the estimate from quantile regression. This is an unbiased estimate as we will now show. From the linear regression model formulation, we know

$$Y = \sum_{j=1}^{p} X_j \beta_j + \beta_0 + \varepsilon,$$

which by (9.14) can be written as,

$$Y = \sum_{j=1}^{p} X_j \beta_j(\tau) + \beta_0(\tau) - F_\varepsilon^{-1}(\tau) + \varepsilon.$$

This means that

$$Y - \sum_{j=1}^{p} X_j \beta_j(\tau) = \beta_0(\tau) - F_\varepsilon^{-1}(\tau) + \varepsilon \equiv \bar{\varepsilon}(\tau).$$

Applying the same argument used to obtain (9.14) with $Z = \varepsilon$ and $c = \beta_0(\tau) - F_\varepsilon^{-1}(\tau)$, gives

$$F_{\bar{\varepsilon}(\tau)}^{-1}(\tau) = \beta_0(\tau) - F_\varepsilon^{-1}(\tau) + F_\varepsilon^{-1}(\tau) = \beta_0(\tau).$$

Thus any unbiased estimator for $F_{\bar{\varepsilon}(\tau)}^{-1}(\cdot)$ will result in an unbiased $\hat{\beta}_0(\tau)$.

References

[1] "Accelerated Insertion of Materials - Composites (AIM-C): Methodology", Boeing Phantom Works Report Number 2004P0020, V 1.2.0, 12 May 2004.
[2] Artzner P., Delbaen, F., Eber, J., and Heath, D. (1997), "Thinking Coherently", *Risk*, 10, 68-71.
[3] Artzner P., Delbaen, F., Eber, J., and Heath, D. (1999), "Coherent measures of risk", *Mathematical Finance*, 9, 203-228.
[4] Hall, P. and Sheather, S. (1988), "On the distribution of a studentized quantile", *J. Royal Statist. Soc. B*, 50, 381-391.
[5] Koenker, R. and Bassett, G. (1978), "Regression quantiles", *Econometrica*, 46, 33-50.
[6] Koenker, R., and Bassett, G. (1982), "An empirical quantile function for linear models with iid errors", *Journal of the American Statistical Association*, 77, 407-415.
[7] Koenker, R. (1994), "Confidence intervals for quantile regression", *Proceedings of the 5th Prague Symposium on Asymptotic Statistics*, P. Mandl and M. Huskova (eds), Heidelberg: Physica-Verlag.

[8] Pflug, G. (2000), "Some remarks on the Value-at-Risk and the Conditional Value-at-Risk", in *Probabilistic Constrained Optimization: Methodology and Applications*, S. Uryasev (ed), Kluwer Academic Publishers.

[9] Rockafellar, R., and Uryasev, S., (2000), "Optimization of conditional value-at-risk", *Journal of Risk*, 2, 21-41.

[10] Rockafellar R.T. and S. Uryasev. (2002), "Conditional Value-at-Risk for General Loss Distributions", *Journal of Banking and Finance*, 26-7, 1443-1471.

[11] Rockafellar, R.T., Uryasev S., and Zabarankin, M. (2002), "Deviation Measures in Risk Analysis and Optimization.", *Research Report 2002-7*, ISE Dept., University of Florida.

[12] Rockafellar, R.T., Uryasev S., and Zabarankin, M. (2002), "Deviation Measures in Generalized Linear Regression", *Research Report 2002-9*, ISE Dept., University of Florida.

[13] Trindade, A.A., Uryasev S., and Zrazhevsky, G. (2003), "Controlling Risk via Asymmetric Rezidual Error Tail Constraints with an Application to Financial Returns ", *Research Report 2003*, ISE Dept., University of Florida, to appear.

Combining Model and Test Data for Optimal Determination of Percentiles and Allowables: CVaR Regression Approach, Part II

Stan Uryasev[1] and A. Alexandre Trindade[2]

[1] American Optimal Decisions, Inc.
and Department of Industrial and Systems Engineering
University of Florida
uryasev@ufl.edu

[2] Department of Statistics
University of Florida

Summary. This report makes a more detailed assessment of the CVaR regression method proposed in [3] for determining A-Basis and B-Basis allowables, and quantifying the impact of test data and different analytical models on failure load predictions. Although the method can in principle be applied to any desired quantile, we will focus only on the 10th (B-Basis) in this study. We consider failure data arising from two sources: (1) a controlled environment where data is simulated from different Weibull distributions; (2) a supplied dataset similar to that of [3] augmented with failure load predictions from two additional analytical models (Model S2 and Model S3). Using absolute deviation between true and estimated 10th percentiles and the CVaR regression goodness-of-fit measure introduced in [3] as accuracy-assessment criteria, the key findings are as follows. The accuracy of CVaR regression is relatively insensitive to the number of batches present, but fairly sensitive to the number of test points per batch[3]. There are diminishing benefits in using more than 10 batches, or more than 10 test points per batch, in any one application of CVaR regression. The estimates of A-basis and B-basis are fairly robust, in the sense that they are not severely affected by miscalibrations (biases or errors) in the analytical models. Among the analytical models used as the sole input with no (input) test data, the best performer is Model S, followed by Model S2. Model S3 is the worst performer. The models contribute substantially to percentile prediction when up to 3 test points are used as input. When 4 test points are used as input, the 3 models can be roughly equated to the input information provided by one additional test point.

Key words: information integration, quantile regression, conditional value-at-risk, B-basis, tolerance limit,Weibull distribution

---

[3] A "batch" is defined as any one single source of variability affecting the test point failure data.

Executive Summary

In [3], we proposed a coherent methodology for integrating various sources of variability on properties of materials in order to accurately predict percentiles of their failure load distribution. The approach, **CVaR regression**, involved the linear combination of factors that are associated with failure load, into a statistical regression or *factor model*. The method can be used for determining A-Basis and B-Basis allowables, and quantifying the impact of experimental (or test) data and different analytical models on failure load predictions. The present report builds on this work by considering failure data arising from two sources: (1) a controlled environment where data is simulated from different Weibull distributions with parameters in ranges plausibly mimicking failure data from a supplied dataset similar to that of [3]; (2) a supplied dataset with failure load predictions from three analytical models.

Specific Objectives

1. Pooling of various sources of information of possibly different origins: models, experiments and expert opinions.

2. Develop a methodology for estimating percentiles of failure distributions and allowables.

3. Minimize amount of data needed for certification process.

4. Take into account various sources of uncertainty.

5. Validate the approach in a controlled statistical environment.

6. Demonstrate the efficacy of the approach with case studies.

Summary of Accomplished Tasks and Findings

1. Developed factor model for direct estimation of percentiles using:

   a) various sources of information (models, experiments, expert opinions); it is possible to quantify value of different sources of information.

   b) statistical characteristics: mean, st.dev., deviation CVaR, etc.

2. Simple, clear, computationally effective methodology enables the pooling of data across:

   a) many individual materials: relatively small requirements on size of datasets.

   b) various experimental setups: crediting simple experiments to more sophisticated (expensive) ones.

3. Developed CVaR statistical techniques for optimal estimation (weighting) of coefficients in the factor model, and corresponding confidence intervals for unknown parameters (A-basis and B-basis).

   a) Technique is new; developed in the framework of the AIM-C project.

b) Approach was especially designed for estimating percentiles and constructing confidence intervals (A and B-basis).

c) No distributional assumptions (such as normality) are made; method is non-parametric.

d) CVaR deviation goodness-of-fit measure has exceptional mathematical and computational properties, allowing easy and efficient implementation of the methodology via linear programming: high speed of calculations; analysis of large datasets feasible; stable results.

4. Case studies were performed with simulated data.

a) Minimal number of batches needed to calibrate the CVaR regression model was determined; sensitivity to this number.

b) Minimal number of experimental test points per batch was determined; sensitivity to this number.

c) Approximate relationship established between CVaR deviation error (observed) and true error (unobserved) in percentile estimation.

d) Sensitivity of the approach to errors in analytical model information was assessed; methodology is robust to biases.

e) 10th percentile estimates based on Model S individually, and on Model S plus 5 test points, are close to true values. B-basis values are also close to nominal values based on actual experiments.

5. Two case studies carried out for open-hole coupon dataset: estimation of 10th percentile and B-basis (failure data plus 3 analytical models).

a) CVaR regression calculations provided plausible estimates of percentiles of failure load distribution.

b) CVaR regression with analytical models only as predictor variables, provide plausible percentile and B-basis estimates, even in the absence of any experimental test data (used as predictors).

c) Benefits of combining models for predicting percentiles were quantified.

d) Benefits of combining models and experimental data were evaluated.

6. Our investigations provide compelling evidence that the methodology can effectively integrate modeling and experimental data, and reduce overall testing cost.

10.1 Introduction

In [3], we proposed a coherent methodology for integrating various sources of variability on properties of materials in order to accurately predict percentiles of their failure load distribution. The approach, **CVaR regression**, involved the linear combination of factors that are associated with failure load, such as might arise from analytical model and experimental test data, into a statistical regression or factor model. The method can be used for determining A-Basis and B-Basis allowables, and quantifying the impact of experimental test data and different analytical models on failure load characteristics. The purpose of the present report is to perform a more detailed evaluation of this CVaR regression method. The term **batch** is used here in the broad sense to denote any one single source of variability affecting the experimental test data. For example: stacking sequence of laminates, type of test, batch number, etc.

Using the terminology of [3], let Y_{ij} denote the j-th failure load value for the i-th batch, obtained from test data, $i = 1, \ldots, I$, $j = 1, \ldots, N_i$. Let (m_i, s_i) and (μ_i, σ_i) denote the sample mean and standard deviation for the test and model data, respectively, in batch i. There are several ways to form these summary statistics; two of these are outlined in Appendix A. Armed with these summary statistics, we can fit the regression or factor model:

$$Y_{ij} = c_0 + c_1\mu_i + c_2\sigma_i + c_3 m_i + c_4 s_i + \varepsilon_{ij}, \tag{10.1}$$

where the unknown regression coefficients $c_0, \ldots, c_4$ are to be estimated from the data by minimizing the deviation measure

$$\mathcal{P}_\tau^2 = \begin{cases} \mathrm{CVaR}_\tau^\Delta(\varepsilon), & \text{if } \tau \geq 0.5, \\ \mathrm{CVaR}_{1-\tau}^\Delta(-\varepsilon), & \text{if } \tau < 0.5. \end{cases} \tag{10.2}$$

Additional terms can be added to (10.1) in order to accommodate more factors, e.g. different analytical models. When the failure data is skewed, a better fit may be attained by symmetrizing it, e.g. taking logarithms. This will make the mean and variance more representative measures of location and dispersion within batches. More robust summary statistic measures, such as quantiles and tail means (CVaR), could also be used.

$\mathcal{P}_\tau^2$ can be used as an absolute measure of goodness-of-fit in CVaR regression, much like the Mean Square Error (MSE) of ordinary regression. For comparing two fitted models, M_2 to M_1 say, the quantity

$$\mathcal{P}_\tau^2(M_2, M_1) = \left[1 - \frac{\mathcal{P}_\tau^2(M_2)}{\mathcal{P}_\tau^2(M_1)}\right] 100\%,$$

measures the percentage improvement in fit obtained by using M_2 over M_1 (negative values meaning that the fit has worsened). This is exactly analogous to the *partial R^2* of ordinary regression, which measures the percentage change in MSE(M_2) over MSE(M_1). If M_0 denotes the model with just an intercept term, then $\mathcal{P}_\tau^2(M_2, M_0)$ is exactly equivalent to the R^2 for model M_2.

A direct result of the CVaR regression fitting process is the following estimated equation for the τth quantile of the failure load, as a function of factors $\{\mu, \sigma, m, s\}$:

$$\hat{Q}_Y(\tau) = \hat{c}_0(\tau) + \hat{c}_1(\tau)\mu + \hat{c}_2(\tau)\sigma + \hat{c}_3(\tau)m + \hat{c}_4(\tau)s. \tag{10.3}$$

The calculation of estimates of the A and B-bases, requires additional computations; see Appendix C.1 of [3] for details.

Note that in contrast with the Bayesian approaches that have hitherto been suggested in this context, CVaR regression makes minimal assumptions about underlying distributions, and is therefore essentially nonparametric. Also, CVaR regression naturally pools together information from all other batches when predicting/fitting for any one particular batch; another feature absent from proposed Bayesian approaches.

The report is divided into three sections. Sections 10.2 and 10.3 are concerned with assessing the performance of CVaR regression in a controlled environment consisting of data simulated from known distributions. In Section 10.2 the model data is drawn from the same distribution as the test data, whilst in Section 10.3 it is drawn from a distribution different from the test data. Section 10.4 examines the performance of the methodology in the context of a supplied dataset similar to that analyzed in [3], augmented with failure data on two additional analytical models. Details on the implementation of the CVaR Regression methodology with an accompanying flowchart, are provided in the Appendix.

10.2 Assessing the Performance of CVaR Regression in a Controlled Environment: Model Data Drawn from Correct Distribution

In this section we investigate the performance of CVaR regression in a controlled environment consisting of data simulated from known Weibull distributions. The density function $f(x)$ of a two-parameter Weibull distribution with shape and scale parameters α and β respectively, is given by the formula,

$$f(x) = \frac{\alpha}{\beta} \left(\frac{x}{\beta} \right)^{\alpha-1} \exp\left\{ -\left(\frac{x}{\beta} \right)^{\alpha} \right\}, \qquad \alpha > 0, \ \beta > 0.$$

Specifically, we consider the following problems.

- Calibration of the CVaR regression method by investigating the sensitivity to number of batches (m), and number of test points within each batch (n). As already stated, the term **batch** is used here in the broad sense to denote any one single source of variability affecting the test point data. For example: stacking sequence of laminates, type of test, batch number, etc. Any given batch is characterized by a particular choice of the shape and scale parameters of a Weibull distribution from which the failure load data for that batch is drawn.

- Relating CVaR deviation as defined by (10.2), to true error in 10th quantile estimation. The significance of this is immediately understood when we note that the former is observed while the latter is not.

- Quantifying the sensitivity of CVaR regression to errors/biases in the analytical model data.

We assess goodness-of-fit by measuring Mean Absolute Deviation (MAD) between true and estimated 10th quantiles across batches, and Mean CVaR Deviation (MCD) for CVaR regression fits with $\tau = 0.1$. The CVaR deviation is that given by equation (10.2).

In this Section we consider only simulations where the model data is drawn from the same Weibull distribution as the test data; the latter viewed as the *correct* distribution. This "perfect information" setup allows us to benchmark the performance of the methodology in an idealized setting. By considering simulated model data drawn from a distribution that differs from that of the test data, Section 10.3 will address the problem discussed in the last point above.

Choice of Weibull Model Parameters

In order to simulate failure data that realistically corresponds to the values typically encountered in the supplied dataset (discussed in Section 10.4), we chose the range of the Weibull distribution shape and scale parameters as follows. The 19 batches that had at least 5 test points were selected. A Weibull distribution was then fitted via maximum likelihood to the failure values of all test points within each batch. Figure 10.1 shows the resulting estimated parameters. Plots of the corresponding density functions appear in Figure 10.2. In all ensuing simulations in this section, we draw failure loads randomly from Weibulls with shape and scale parameters in these ranges. Specifically, the ranges are:

$$10 \leq \alpha \leq 80, \quad \text{and} \quad 40 \leq \beta \leq 120.$$

Calibrating CVaR Regression Using Only Model Data

We consider first CVaR regression using only model data as covariates (e.g. Model S). Plotted in the ensuing figures are the MAD and the MCD. The suffix "1" is appended to form MAD1 and MCD1, signaling that the corresponding deviations are based on CVaR regression fits using only **model** data as covariates. Other numbers will be appended later for other cases. The deviations are plotted as functions of number of batches (m) used in the CVaR regression fit, as well as number of test points (n) present in each batch.

Henceforth, and until further notice, a particular value of MAD or MCD is based on 100 draws from the same Weibull distribution. The choice of shape and scale parameters, constituting a batch, is made randomly from the region plotted in Figure 10.1. Each draw is of sample size $n + 100$. The first n points are then used as test data, and the remaining 100 as model data. A new set of shape and scale parameters is then randomly selected to produce data on the next batch, and so on. The formation of the CVaR regression inputs is done by using all test data as both response and covariates,

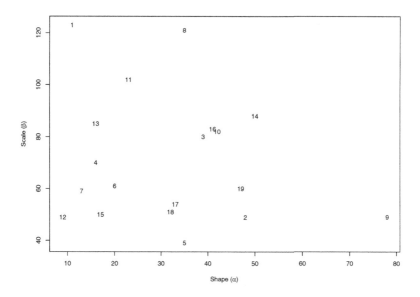

Fig. 10.1. Estimated shape and scale parameters of a Weibull fit to the test data from the 19 batches of the supplied dataset that had at least 5 test points per batch. Plots of the corresponding density functions appear in Figure 10.2.

as detailed in the Appendix. The deviations are plotted as functions of number of batches (m) used in the CVaR regression fit, as well as number of test points (n) within each batch.

Figure 10.3, shows MAD1 error between true and estimated 10th quantiles for CVaR regression fits based on model data and different numbers (n) of test points, as a function of number of batches (m). Figure 10.4, shows the corresponding MCD1 error. Figure 10.5 shows part of the data from Figures 10.4 and 10.3 in the same plot. Generally, we see that MAD1 decreases while MCD1 increases, with increasing m. Also, MAD1 is lower while MCD1 is higher, at higher values of n, for any given m.

Figure 10.6, shows the MAD1 error between true and estimated 10th quantiles for CVaR regression fits based on model data and different numbers of test points (n) and batches (m), as a function of n. Figure 10.7, shows the corresponding MCD1 error. Figure 10.8 shows part of the data from Figures 10.7 and 10.6 in the same plot. Generally, we see that MAD1 decreases while MCD1 increases, with increasing n. Also, MAD1 is lower while MCD1 is higher, at higher values of m, for any given n. Figures 10.9 and 10.10 provide a concise summary of the behavior of the MAD1 and MCD1 errors in Figures 10.3-10.8, via a three-dimensional perspective plot, as a function of both n and m. This offers a broader view than the two-dimensional plots above. Generally, MAD decreases with both m and n, levelling off at moderate values of these. MCD behaves inversely, increasing with m and n.

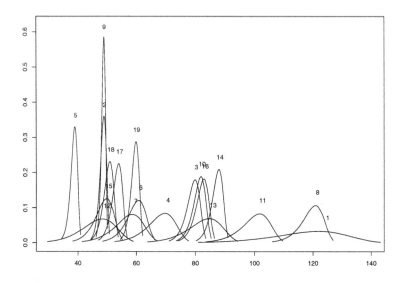

Fig. 10.2. Plots of the Weibull density functions corresponding to the shape and scale parameters appearing in Figure 10.1.

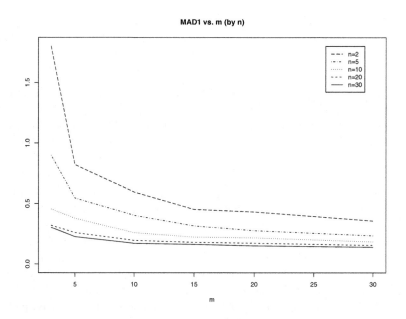

Fig. 10.3. Mean Absolute Deviation (MAD1) error between true and estimated 10th quantiles for CVaR regression fits based on model data and different numbers of test points (n), as a function of number of batches (m). Only model data was used as covariates. MAD1 is based on 100 draws from the same Weibull distribution within each batch, each draw of sample size $n + 100$ (n used as test data; 100 as model data).

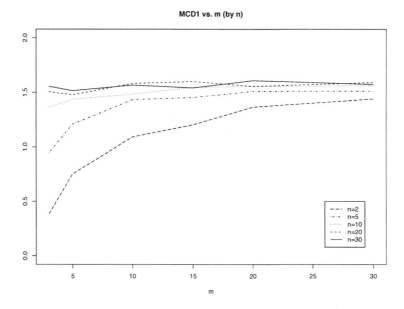

Fig. 10.4. Mean CVaR Deviation (MCD1) error for CVaR regression fits based on model data and different numbers (n) of test points, as a function of number of batches (m). Only model data was used as covariates. MCD1 is based on 100 draws from the same Weibull distribution within each batch, each draw of sample size $n + 100$ (n used as test data; 100 as model data.

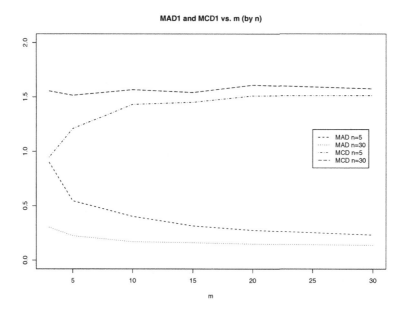

Fig. 10.5. MAD1 and MCD1 error from Figures 10.4 and 10.3 combined in the same plot.

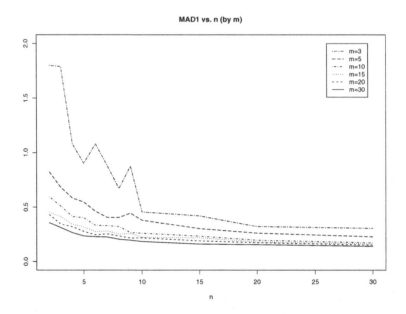

Fig. 10.6. Mean Absolute Deviation (MAD1) error between true and estimated 10th quantiles for CVaR regression fits based on model data and different numbers of test points (n) and batches (m), as a function of n.

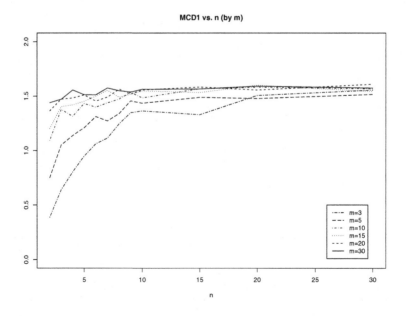

Fig. 10.7. Mean CVaR Deviation (MCD1) error between true and estimated 10th quantiles for CVaR regression fits based on model data and different numbers of test points (n) and batches (m), as a function of n.

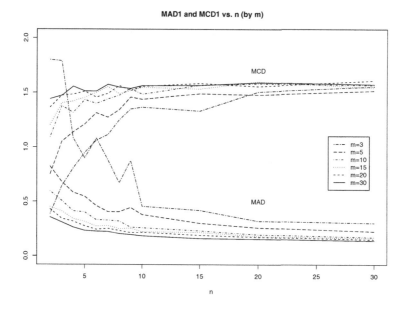

Fig. 10.8. MAD1 and MCD1 error from Figures 10.7 and 10.6 combined in the same plot.

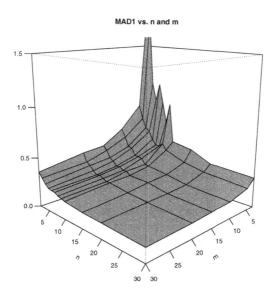

Fig. 10.9. MAD1 error perspective plot of the data in Figures 10.3-10.7, as a function of both number of test points (n) and number of batches (m).

MCD1 vs. n and m

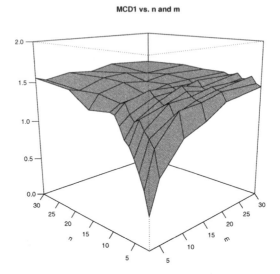

Fig. 10.10. MCD1 error perspective plot of the data in Figures 10.3-10.7, as a function of both number of test points (n) and number of batches (m).

Summary Remarks:

- MAD1 decreases with increasing m (n), plateauing off at about $m = 10$ ($n = 10$). The rate of decrease is less pronounced at larger values of n (m).

- MCD1 increases with m (n), plateauing off at about $m = 10$ ($n = 10$). The rate of increase is less pronounced at larger values of n (m).

- For large m and n, MCD is between 4 to 7 times larger than MAD.

Calibrating CVaR Regression Using Both Model and Test Data

We now consider the case of model and test point data as covariates, and compute the mean absolute deviation (MAD3) between true and estimated 10th quantiles, and the mean CVaR deviation (MCD3) for CVaR regression fits. The "3" signals that the corresponding deviation is based on CVaR regression fits using both **model** and **test** point data as covariates.

Figure 10.11, shows MAD3 error between true and estimated 10th quantiles for CVaR regression fits based on model data and different numbers (n) of test points, as a function of number of batches (m). Figure 10.12, shows the corresponding MCD3 error. Figure 10.13 shows part of the data from Figures 10.12 and 10.11 in the same

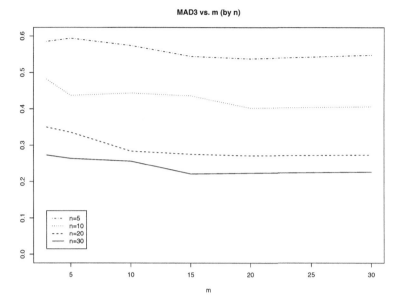

Fig. 10.11. Mean Absolute Deviation (MAD3) error between true and estimated 10th quantiles for CVaR regression fits based on model data and different numbers of test points (n), as a function of number of batches (m). Both model and test data were used as covariates. MAD3 is based on 100 draws from the same Weibull distribution within each batch, each draw of sample size $n + 100$ (n used as test data; 100 as model data).

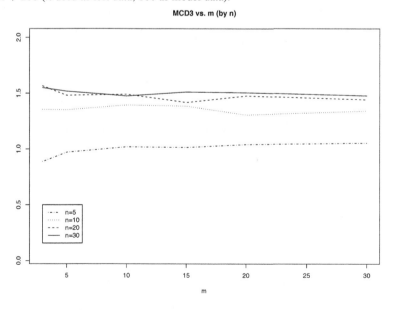

Fig. 10.12. Mean CVaR Deviation (MCD3) error for CVaR regression fits based on model data and different numbers (n) of test points, as a function of number of batches (m). Both model and test data were used as covariates. MCD3 is based on 100 draws from the same Weibull distribution within each batch, each draw of sample size $n + 100$ (n used as test data; 100 as model data.

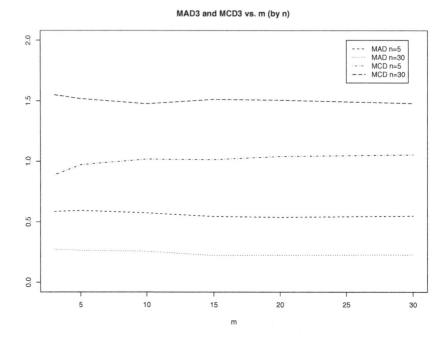

Fig. 10.13. MAD3 and MCD3 error from Figures 10.12 and 10.11 combined in the same plot.

plot. These graphs show that both MAD3 and MCD3 are relatively constant with m, the former (latter) being higher (lower) at lower values of n. As before, MCD3 is uniformly higher than MAD3, thus providing an upper bound on the MAD error.

Figure 10.14, shows the MAD3 error between true and estimated 10th quantiles for CVaR regression fits based on model data and different numbers of test points (n) and batches (m), as a function of n. Figure 10.15, shows the corresponding MCD3 error. Figure 10.16 shows part of the data from Figures 10.15 and 10.14 in the same plot. These graphs show that MAD3 (MCD3) decreases (increases) with n. Beyond $n = 5$, MCD3 is uniformly higher than MAD3, thus providing an upper bound on the MAD error.

Figures 10.17 and 10.18 provide a concise summary of the behavior of the MAD3 and MCD3 errors in Figures 10.11-10.16, via a three-dimensional perspective plot, as a function of both n and m. This offers a broader view than the two-dimensional plots above. Generally, MAD decreases with n, but is constant with m. MCD is also constant with m, but increases with n.

MAD3 vs. n (by m)

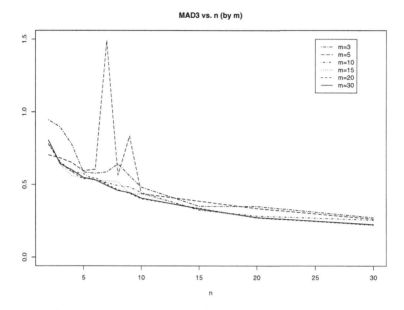

Fig. 10.14. Mean Absolute Deviation (MAD3) error between true and estimated 10th quantiles for CVaR regression fits based on model data and different numbers of test points (n) and batches (m), as a function of n.

MCD3 vs. n (by m)

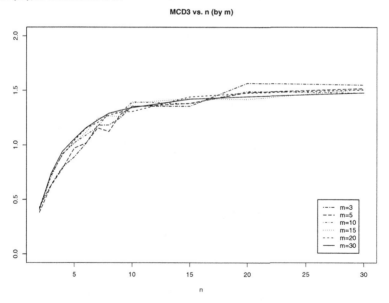

Fig. 10.15. Mean CVaR Deviation (MCD3) error between true and estimated 10th quantiles for CVaR regression fits based on model data and different numbers of test points (n) and batches (m), as a function of n.

MAD3 and MCD3 vs. n (by m)

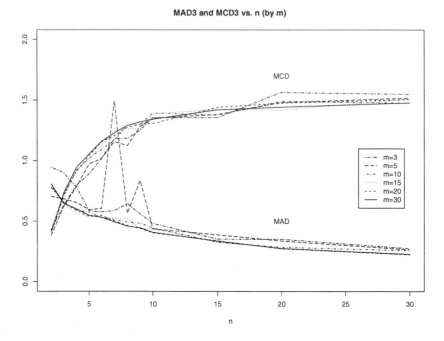

Fig. 10.16. MAD3 and MCD3 error from Figures 10.15 and 10.14 combined in the same plot.

MAD3 vs. n and m

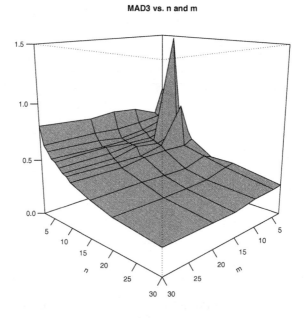

Fig. 10.17. MAD3 error perspective plot of the data in Figures 10.11-10.15, as a function of both number of test points (n) and number of batches (m).

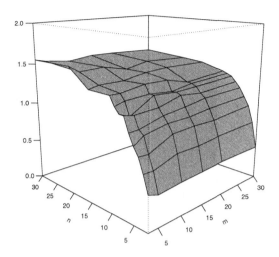

Fig. 10.18. MCD3 error perspective plot of the data in Figures 10.11-10.15, as a function of both number of test points (n) and number of batches (m).

Summary Remarks:

- As a function of m, both MAD3 and MCD3 error are relatively constant, with lower (higher) values of MAD3 (MCD3) at higher (lower) values of n.

- MAD3 decreases from about 0.6 to 0.3 as n increases from 5 to 30.

- MCD3 increases from about 1 to 1.5 as n increases from 5 to 30.

- The decrease in MAD3 with increasing n, plateaus off at about $n = 10$.

- The increase in MCD3 with increasing n, plateaus off at about $n = 10$.

- For large m and n, MCD is between 2 to 6 times larger than MAD.

10.3 Assessing the Performance of CVaR Regression in a Controlled Environment: Model Data Drawn from Incorrect Distribution

To quantify the sensitivity of CVaR regression to errors in the model data, we consider deterministic and stochastic perturbations, with respect to test data (viewed as the *correct* distribution), in both the shape and scale parameters of the Weibull distribution

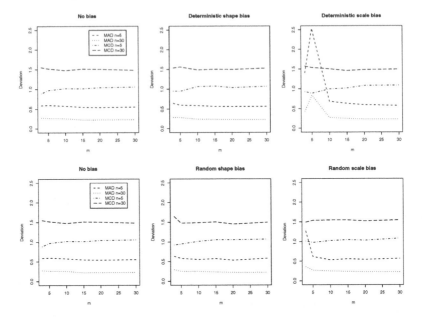

Fig. 10.19. MAD3 and MCD3 errors similar to Figure 10.13 (no bias, column 1), but with deterministic perturbations in the shape ($\alpha \rightarrow \alpha + 20$, column 2) and scale ($\beta \rightarrow \beta + 20$, column 3) parameters of the Weibull distribution from which the model data is drawn.

from which the model data is drawn. The deterministic perturbations involve the changes

$$\alpha \rightarrow \alpha + 20, \quad \text{and} \quad \beta \rightarrow \beta + 20,$$

and the random perturbations the changes

$$\alpha \rightarrow \alpha + Z, \quad \text{and} \quad \beta \rightarrow \beta + Z.$$

(Z denotes a random value drawn from a Normal distribution with mean 0 and standard deviation 10.)

These perturbations can be thought of as introducing systematic and random biases in the model data.

Figure 10.19 shows MAD3 and MCD3 error, by number of test points (n), as a function of number of batches (m), as in Figure 10.13. Column 2 (3) shows the effect of deterministic and random perturbations in the shape (scale) parameter. Figure 10.13 (no bias) is reproduced in the first column for comparison. These graphs show that MCD3 is virtually unaffected by both shape and scale biases. MAD3 is moderately affected by scale biases, increasing dramatically at lower values of m and n.

Figure 10.20 shows MAD3 and MCD3 error, by number of batches (m), as a function of number of test points (n), as in Figure 10.16. Column 2 (3) shows the effect of deterministic and random perturbations in the shape (scale) parameter. Figure 10.16

(no bias) is reproduced in the first column for comparison purposes. Again, MCD3 is virtually unaffected by both shape and scale biases. MAD3 is severely affected by scale biases, increasing dramatically at lower values of m and n.

Figures 10.21–10.28 show MAD3 and MCD3 error perspective plots similar to Figures 10.17 and 10.18, but with deterministic and random shape and scale perturbations. Figure 10.29 (Figure 10.30) combines all the MAD3 (MCD3) plots into one, with Figure 10.17 (Figure 10.18) reproduced in the first column for comparison. MAD3 decreases with m and n as already noted, with a dramatic increase at low values of m and n with respect to the unbiased plots. The plots also clearly show that MCD3 is relatively unaffected by both shape and scale biases.

Summary Remarks:

- MAD is virtually unaffected by shape biases.

- MAD is moderately affected by scale biases; dramatically at lower values of m and n.

- MCD is virtually unaffected by both shape and scale biases.

The fact that the location of a Weibull depends more on its scale parameter β than on its shape parameter α, is undoubtedly partly responsible for the larger impact of scale biases. Another way to assess this is to consider the Kullback-Leibler distance between the perturbed Weibull and the true one. The Kullback-Leibler distance of distribution g from distribution f, is a measure of the *information lost when g is used to approximate f*; Kullback and Leibler (1951). Figure 10.31 plots the Kullback-Leibler distance of various Weibulls from a Weibull(40,80), as a function of the approximating Weibull's shape and scale parameters. As we can see, this distance varies more with scale than it does with shape, and rises dramatically when the approximating Weibull's scale parameter is too low.

10.4 Assessing the Predictive Capabilities of CVaR Regression with Data from Several Analytical Models

In this section we apply the CVaR regression methodology to the supplied dataset similar to the *full dataset* of [3], augmented with failure load information from two additional analytical models. From a total of 28 batches, only the 19 batches with at least 5 test points per batch were selected. As in [3], the dataset was symmetrized by selecting exactly 5 test points per batch. In addition to the Model S failure load prediction model data within each batch already present, data on two additional models, Model S2 and Model S3, was added. The objective is to extend the results in [3], thus quantifying the benefits of introducing other models.

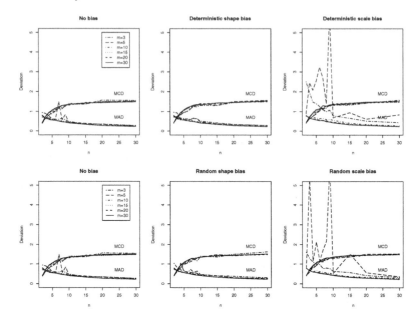

Fig. 10.20. MAD3 and MCD3 errors similar to Figure 10.16 (no bias, column 1), but with random perturbations in the shape ($\alpha \to \alpha + Z$, column 2) and scale ($\beta \to \beta + Z$, column 3) parameters of the Weibull distribution from which the model data is drawn. (Z denotes a random value drawn from a Normal distribution with mean 0 and standard deviation 10.)

MAD3: deterministic shape bias

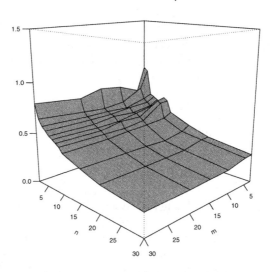

Fig. 10.21. MAD3 error perspective plot similar to Figure 10.17, but with a deterministic perturbation in the shape ($\alpha \to \alpha + 20$) parameter of the Weibull distribution from which the model data is drawn.

MCD3: deterministic shape bias

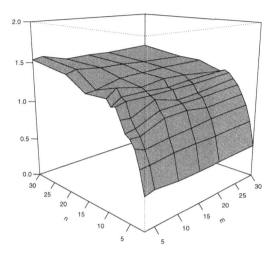

Fig. 10.22. MCD3 error perspective plot similar to Figure 10.18, but with a deterministic perturbation in the shape ($\alpha \rightarrow \alpha + 20$) parameter of the Weibull distribution from which the model data is drawn.

MAD3: deterministic scale bias

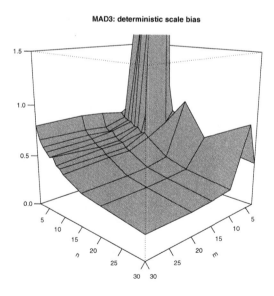

Fig. 10.23. MAD3 error perspective plot similar to Figure 10.17, but with a deterministic perturbation in the scale ($\beta \rightarrow \beta + 20$) parameter of the Weibull distribution from which the model data is drawn.

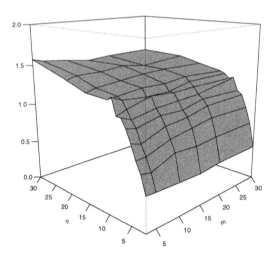

Fig. 10.24. MCD3 error perspective plot similar to Figure 10.18, but with a deterministic perturbation in the scale ($\beta \to \beta + 20$) parameter of the Weibull distribution from which the model data is drawn.

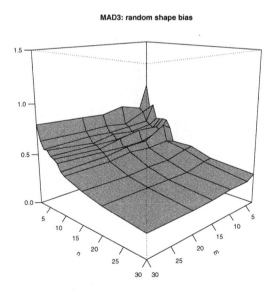

Fig. 10.25. MAD3 error perspective plot similar to Figure 10.17, but with a random perturbation in the shape ($\alpha \to \alpha + 20$) parameter of the Weibull distribution from which the model data is drawn.

MCD3: random shape bias

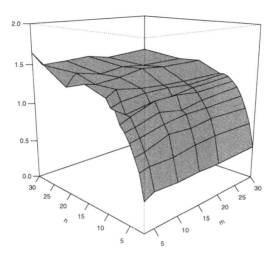

Fig. 10.26. MCD3 error perspective plot similar to Figure 10.18, but with a random perturbation in the shape ($\alpha \to \alpha + 20$) parameter of the Weibull distribution from which the model data is drawn.

MAD3: random scale bias

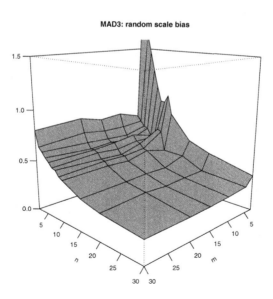

Fig. 10.27. MAD3 error perspective plot similar to Figure 10.17, but with a random perturbation in the scale ($\beta \to \beta + 20$) parameter of the Weibull distribution from which the model data is drawn.

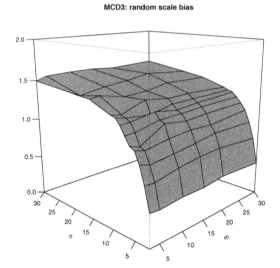

Fig. 10.28. MCD3 error perspective plot similar to Figure 10.18, but with a random perturbation in the scale ($\beta \to \beta + 20$) parameter of the Weibull distribution from which the model data is drawn.

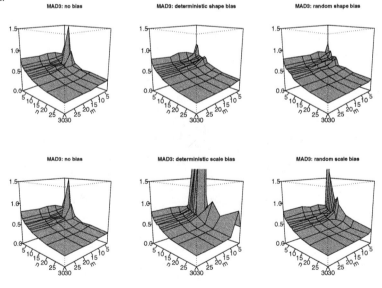

Fig. 10.29. Composite perspective plots of MAD3 error in Figures 10.21-10.28. Figure 10.17 (no bias) is reproduced in column 1 for comparison with deterministic (column 2) and random (column 3) perturbations in the shape and scale parameters of the Weibull distribution from which the model data is drawn.

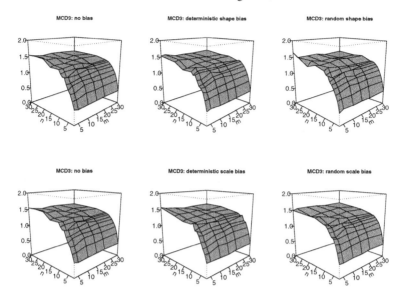

Fig. 10.30. Composite perspective plots of MCD3 error in Figures 10.21-10.28. Figure 10.18 (no bias) is reproduced in column 1 for comparison with deterministic (column 2) and random (column 3) perturbations in the shape and scale parameters of the Weibull distribution from which the model data is drawn.

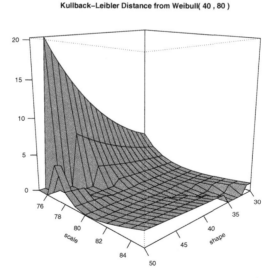

Fig. 10.31. Kullback-Leibler distance of various Weibull distributions from a Weibull(40,80), as a function of the approximating Weibull's shape and scale parameters.

Table 10.1. Results of CVaR regression applied to the supplied dataset with test points $(\hat{c}_1,\hat{c}_2)$ and the three Models: Model S $(\hat{c}_3,\hat{c}_4)$, Model S2 $(\hat{c}_5,\hat{c}_6)$ and Model S3 $(\hat{c}_7,\hat{c}_8)$. Within each batch, all 5 test points were used as output, but only a subset was used as input (column 1).

Number of Test Points Used as Input	Const. $\hat{c}_0$	Test Data $\hat{c}_1$	$\hat{c}_2$	Model S $\hat{c}_3$	$\hat{c}_4$	Model S2 $\hat{c}_5$	$\hat{c}_6$	Model S3 $\hat{c}_7$	$\hat{c}_8$	Goodness-of-fit CVaR Measure $(\mathcal{P}^2_{0.10})$
0	36.73							0.59	-0.31	27.16
0	39.26					0.57	0.01			24.66
0	14.27			1.10	-4.30					16.86
0	13.32			0.51	-6.01	0.66	-1.24	0.04	0.04	13.82
1	14.29	0.30		0.11	-5.13	0.83	-1.06	-0.08	0.07	12.61
2	15.75	0.44	0.22	-0.26	-5.71	1.16	-1.03	-0.18	0.09	12.36
3	12.75	0.62	-0.27	-0.14	-3.88	0.71	-0.72	-0.09	0.05	11.82
4	11.08	0.88	-0.88	-0.10	-1.64	0.33	-0.37	-0.06	0.03	10.79
5	10.30	1.00	-1.44							10.29
5	10.88	0.94	-1.46	0.07	-0.26					10.21
5	11.30	1.04	-1.44					-0.04	0.03	10.14
5	11.00	1.01	-1.43			0.01	-0.22			10.01
5	9.27	0.97	-1.43	0.16	0.16	-0.11	-0.18	-0.00	0.04	9.72

We repeat Analyses 1 and 2 of [3]. The 10th percentile estimated surface for a generic factor model under consideration, takes on the form:

$$\hat{Q}_Y(0.10) = \hat{c}_0 + \hat{c}_1 m + \hat{c}_2 s + \hat{c}_3 \mu^{(1)} + \hat{c}_4 \sigma^{(1)} + \hat{c}_5 \mu^{(2)} + \hat{c}_6 \sigma^{(2)} + \hat{c}_7 \mu^{(3)} + \hat{c}_8 \sigma^{(3)},$$

where the superscript (i), $i = 1,2,3$, identifies the Model used to compute the summary statistic corresponding to that particular coefficient. The method of using complementary subsets for response and covariate CVaR regression input formation, outlined in the Appendix, is used here.

The results are summarized in Table 10.1. The form of the table allows us to easily identify how many test points and which Models were used as input in which case. For example, in row 2 no test points and only Model S2 were used as input. The results are also ranked according to the CVaR goodness-of-fit measure, $\mathcal{P}^2_{0.10}$. According to this criterion, we can conclude the following concerning these various sources of input to CVaR regression:

- Among the Models used as the sole input with no test data, the best performer is Model S, followed by Model S2. Model S3 is the worst.
- With no test data as input, the contribution from all 3 analytical models results in an improvement of approximately:

  - 50% over Model S3 by itself;

  - 44% over Model S2 by itself;

  - 18% over Model S by itself.

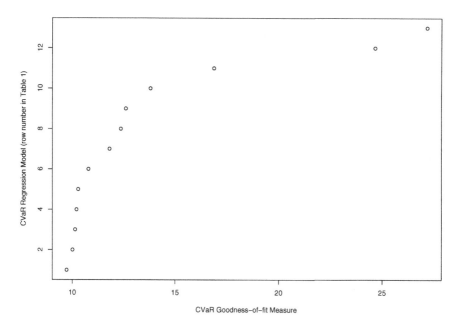

Fig. 10.32. CVaR goodness-of-fit measure plotted against CVaR regression model number (row number in Table 10.1).

The combining of model information is therefore advantageous.

■ When all 5 test points are used as input, each of the 3 Models by themselves provides practically no additional improvement (less than 3%). The addition of all 3 Models results in only about a 6% improvement.

■ The models contribute substantially to percentile prediction when up to 3 test points are used as input. This can be seen in Table 10.1 as the magnitude of the estimated coefficients shifts from the model to the test data as the number of test points used as input increases.

■ The contribution from 4 test points plus all 3 Models, is approximately equivalent to the contribution from 5 test points by themselves. This implies the 3 Models can be roughly equated to one test point.

Figure 10.32 gives a visual summary of the rate of decrease in $\mathcal{P}^2_{0.10}$ as a function of the fitted model (row number in Table 10.1).

We have noticed that the reason for the under-performance of Models 2 and (especially) 3 relative to Model 1, is that they have more outliers with respect to the test data (the range of failure loads predicted by these Models within batches often fails to capture any test data). This suggests the need for an additional calibration procedure before using a given analytical model. We performed one such crude calibration, by removing from the data batches whose range of values predicted by Model 1 failed to capture any test point data. Only 12 batches remained. The same was done for Models 2 and 3, but the criterion for batch exclusion there was that there should be no test points within one standard deviation of the corresponding Model mean. Ten (10) and 5 batches remained, respectively. CVaR regression was run on only test points plus each model. The results are summarized in Table 10.2. Note that the values of the CVaR

Table 10.2. Results of CVaR regression applied to the supplied dataset with outlying batches removed. The coefficients correspond to test points ($\hat{c}_1,\hat{c}_2$) and the three Models: Model S ($\hat{c}_3,\hat{c}_4$), Model S2 ($\hat{c}_5,\hat{c}_6$) and Model S3 ($\hat{c}_7,\hat{c}_8$). Within each batch, all 5 test points were used as output and their summary statistics used as input. Batches were excluded if the Model failed to capture any of its test points (Model S), or if there were no test points within one standard deviation of the Model mean (Model S2 and Model S3).

Number of Batches	\multicolumn{9}{c}{Estimated Coefficients of $Q_Y(0.10)$}	$\mathcal{P}^2_{0.10}$ Table 10.2 (Table 10.1)
	Const. $\hat{c}_0$	Test Data $\hat{c}_1$	$\hat{c}_2$	Model S $\hat{c}_3$	$\hat{c}_4$	Model S2 $\hat{c}_5$	$\hat{c}_6$	Model S3 $\hat{c}_7$	$\hat{c}_8$	
12	0.24	0.98	-1.55	0.02	0.02					10.04 (10.21)
10	2.27	1.16	-1.25			-0.18	-0.27			9.94 (10.01)
5	-3.41	1.26	-1.32					-0.22	0.19	7.24 (10.14)

goodness-of-fit measure $\mathcal{P}^2_{0.10}$ have all decreased from their values in Table 10.1 (in parentheses in last column). The reason for this is in part due to an improved fit, and in part due to a reduction in the number of batches used in model fitting.

10.5 Suggestions for Future Research

The simulations and case studies presented thus far show that ours is a promising methodology for integrating various sources of variability on properties of materials. However, only the simplest of scenarios have hitherto been tackled; the methodology clearly warrants further and deeper investigation. In this section we outline some possible directions for future research.

1. We have hitherto assumed that the experimental test data is observed without error, i.e. it comes from the true failure distribution. The extent to which CVaR regression estimates are impacted when the test data is contaminated with errors, deserves careful attention.

2. Coefficient estimation in the current form of CVaR regression is wholly data driven, in the sense that there is no prior weighting of factors to reflect the suspected precision of a particular factor relative to the others. In ordinary regression,

this is achieved by choosing appropriate weights for least squares parameter estimation, giving rise to *weighted regression*. Analogous ideas could be applied to CVaR regression.

3. The performance of the methodology in its present form, has only been assessed on 10th percentile inference (B-basis). Inference on the 1st percentile (A-basis), may require quite a large number of test data points per batch in order to be effective. Specialized variants of CVaR regression may need to be developed to cope with this situation.

4. As a follow-up to the previous point, it may be possible to adapt some form of Extreme Value Theory (EVT) to provide an alternative to, or be used in conjunction with, CVaR regression. EVT methods were designed for inference on the tails of distributions where data is usually sparse. It may be possible to adapt some of its key ideas in order to alleviate the problem of insufficient test data when A-basis (or even B-basis) estimates are desired.

5. Assess the capabilities of the methodology for pooling experimental data from various sources, and compare this with Bayes and empirical Bayes methods.

6. Use a Bayesian hierarchical approach instead of CVaR regression to obtain the predictive distribution of the failure load. The contribution from each analytical model could be assessed/incorporated via a Bayesian hyperprior of weights.

7. The suggested CVaR methodology is nonparametric. Assess the impact of parametric assumptions on distributions on the Bayesian approach, in order to obtain estimates of similar quality to CVaR regression.

Acknowledgements

This effort was jointly accomplished by the Boeing led team and the United States Government under the guidance of NAVAIR. This work was funded by DARPA/DSO and administered by NAVAIR under Technology Investment Agreement N00421-01-3-0098. The program would like to acknowledge the guidance and support of Dr. Leo Christodoulou of DARPA/DSO for this effort.

A CVaR Regression Implementation Details and Flowchart

Mathematical Notation

- Vectors appear in boldface.

- Matrices appear in Roman capitals.

- The superscript $'$ denotes array transposition.

- $\min(a, b)$ and $\max(a, b)$ denote respectively the minimum and maximum of a and b.

- τ denotes any real number between 0 and 1.

- For positive integers N and n with $n \leq N$,

$$\binom{N}{n} = \frac{N!}{n!(N-n)!},$$

denotes the number of distinct combinations of n items from N.

- z_τ denotes the τth quantile of a Normal distribution with mean 0 and variance 1, i.e.

$$z_\tau = \Phi^{-1}(\tau), \quad \text{where} \quad \Phi(t) = \int_{-\infty}^{t} \frac{1}{\sqrt{2\pi}} \exp\{-x^2/2\} dx.$$

Introduction and Formation of Input Data Arrays

This report provides a detailed explanation of the CVaR regression method proposed in [3] for determination of A-Basis and B-Basis allowables. The details are laid out in a manner that will enable a programmer with only elementary mathematical knowledge to implement the methodology in the form of a computer program. CVaR regression provides a method for integrating different sources of information on failure loads, such as might arise from analytical model and experimental test data. The term **batch** is used here in the broad sense to denote any one single source of variability affecting the test point data. For example: stacking sequence of laminates, type of test, batch number, etc.

Suppose we wish to build a CVaR regression model to integrate information from K analytical models with the physical test data. The available data spans I batches. First define the following:

- Let Y_{ij} be jth failure load value for the ith batch, obtained from test data, $i = 1, \ldots, I, j = 1, \ldots, N_i$.

- Let N be the total number of test data points

$$N = \sum_{i=1}^{I} N_i.$$

We then create the data arrays used as the input into the CVaR Regression Module of the next section, as outlined below. The test data can be utilized for covariate and response formation in several ways. We detail two ways, both of which have been used in this report.

Formation of Input Data Arrays: All Test Data Used as Both Response and Covariates

1. Form $\mathbf{y}$, the response vector of length $n = N$ consisting of all failure loads Y_{ij} grouped by batches. The batches and test points within batches, can be arranged in any order, but once decided upon, this order must be maintained throughout.

$$\mathbf{y} = [Y_{11}, Y_{12}, \ldots, Y_{1N_1}, Y_{21}, \ldots, Y_{2N_2}, \ldots, Y_{I1}, \ldots, Y_{IN_I}]'$$
$$= [y_1, \ldots, y_n]'.$$

2. Form, m_i, the sample mean of failure loads for the test data in batch i, for all $i = 1, \ldots, I$.

$$m_i = \frac{1}{N_i} \sum_{j=1}^{N_i} Y_{ij}.$$

3. Form, s_i, the sample standard deviation of failure loads for the test data in batch i, for all $i = 1, \ldots, I$.

$$s_i = \sqrt{\frac{1}{N_i - 1} \sum_{j=1}^{N_i} (Y_{ij} - m_i)^2}.$$

4. Similarly to m_i, form $\mu_i^{(k)}$, the sample mean of failure loads for the data from the kth model in batch i, for all $k = 1, \ldots, K$, and $i = 1, \ldots, I$.

5. Similarly to s_i, form $\sigma_i^{(k)}$, the sample standard deviation of failure loads for the data from the kth model in batch i, for all $k = 1, \ldots, K$, and $i = 1, \ldots, I$.

6. Form the design matrix, X, of CVaR regression. This consists of the test and model sample summary statistics arranged in a matrix of n rows and $l + 1 = 2K + 3$ columns (the first column consists of the number 1). The rows correspond to those in $\mathbf{y}$.

$$X = \begin{bmatrix} 1 & m_1 & s_1 & \mu_1^{(1)} & \sigma_1^{(1)} & \mu_1^{(2)} & \sigma_1^{(2)} & \cdots & \mu_1^{(K)} & \sigma_1^{(K)} \\ \vdots & \vdots & \vdots & \vdots & \vdots & \vdots & \vdots & \vdots & \vdots & \vdots \\ 1 & m_1 & s_1 & \mu_1^{(1)} & \sigma_1^{(1)} & \mu_1^{(2)} & \sigma_1^{(2)} & \cdots & \mu_1^{(K)} & \sigma_1^{(K)} \\ 1 & m_2 & s_2 & \mu_2^{(1)} & \sigma_2^{(1)} & \mu_2^{(2)} & \sigma_2^{(2)} & \cdots & \mu_2^{(K)} & \sigma_2^{(K)} \\ \vdots & \vdots & \vdots & \vdots & \vdots & \vdots & \vdots & \vdots & \vdots & \vdots \\ 1 & m_2 & s_2 & \mu_2^{(1)} & \sigma_2^{(1)} & \mu_2^{(2)} & \sigma_2^{(2)} & \cdots & \mu_2^{(K)} & \sigma_2^{(K)} \\ \vdots & \vdots & \vdots & \vdots & \vdots & \vdots & \vdots & \vdots & \vdots & \vdots \\ \vdots & \vdots & \vdots & \vdots & \vdots & \vdots & \vdots & \vdots & \vdots & \vdots \\ 1 & m_I & s_I & \mu_I^{(1)} & \sigma_I^{(1)} & \mu_I^{(2)} & \sigma_I^{(2)} & \cdots & \mu_I^{(K)} & \sigma_I^{(K)} \\ \vdots & \vdots & \vdots & \vdots & \vdots & \vdots & \vdots & \vdots & \vdots & \vdots \\ 1 & m_I & s_I & \mu_I^{(1)} & \sigma_I^{(1)} & \mu_I^{(2)} & \sigma_I^{(2)} & \cdots & \mu_I^{(K)} & \sigma_I^{(K)} \end{bmatrix} \begin{matrix} \uparrow \\ \\ N_1 \text{ rows} \\ \downarrow \\ \uparrow \\ N_2 \text{ rows} \\ \downarrow \\ \\ \vdots \\ \\ \uparrow \\ N_I \text{ rows} \\ \downarrow \end{matrix}$$

Formation of Input Data Arrays: Complementary Subsets of Test Data Used as Response and Covariates

In this procedure, the test data points in each batch are partitioned into two disjoint sets. The points in the one set are used as the response, while those in the other are used to form the covariates. This is repeated for all remaining possible ways to partition the test points so as to result in different response and covariate sets. The details are as follows:

1. Repeat the following loop for each batch i, where $i = 1, \ldots, I$:

   a) Form all $p_i = \binom{N_i}{w}$ possible partitions of the N_i test points into two disjoint sets consisting of w and $n_i = N_i - w$ points, respectively, where $0 \leq w \leq N_i$, is constant for all i. The points in the first set will be used to form the covariates (the *covariate set*); those in the second will be used as responses (the *response set*).

   b) Let $Y_{i(1)}^{(q)}, \ldots, Y_{i(n_i)}^{(q)}$ denote the test points in the qth response set, $q = 1, \ldots, p_i$. Form $\mathbf{y}_i$, the vector of length $n_i p_i$ consisting of the responses ordered as follows,

   $$\mathbf{y}_i = [Y_{i(1)}^{(1)}, \ldots, Y_{i(n_i)}^{(1)}, Y_{i(1)}^{(2)}, \ldots, Y_{i(n_i)}^{(2)}, \ldots, Y_{i(1)}^{(p_i)}, \ldots, Y_{i(n_i)}^{(p_i)}]'.$$

   c) Let $Y_{i(n_i+1)}^{(q)}, \ldots, Y_{i(N_i)}^{(q)}$ denote the test points in the qth covariate set, $q = 1, \ldots, p_i$. Form $m_i^{(q)}$ and $s_i^{(q)}$, the sample mean and standard deviation of the test data in the qth covariate set,

   $$m_i^{(q)} = \frac{1}{w} \sum_{j=n_i+1}^{N_i} Y_{i(j)}^{(q)},$$

   $$s_i^{(q)} = \sqrt{\frac{1}{w-1} \sum_{j=n_i+1}^{N_i} \left(Y_{i(j)}^{(q)} - m_i^{(q)} \right)^2}.$$

   If $w = 0$, the test data is not used as covariates (only the model). If $w = 1$, then $s_i^{(q)} = 0$. If $w = N_i$, we use all the test data as both response and covariates as in the preceding subsection.

   d) Form $\mu_i^{(k)}$ and $\sigma_i^{(k)}$, the sample mean and standard deviation of failure loads for the data from the kth model in batch i, for all $k = 1, \ldots, K$.

   e) Form the design matrix, $\mathbf{X}_i$, for batch i. This consists of the test and model sample summary statistics arranged in a matrix of $n_i p_i$ rows and $2K + 3$ columns (the first column consists of the number 1). The rows correspond to those in $\mathbf{y}_i$.

$$X_i = \begin{bmatrix} 1 & m_i^{(1)} & s_i^{(1)} & \mu_i^{(1)} & \sigma_i^{(1)} & \mu_i^{(2)} & \sigma_i^{(2)} & \cdots & \mu_i^{(K)} & \sigma_i^{(K)} \\ \vdots & \vdots & \vdots & \vdots & \vdots & \vdots & \vdots & \vdots & \vdots & \vdots \\ 1 & m_i^{(1)} & s_i^{(1)} & \vdots & \vdots & \vdots & \vdots & \vdots & \vdots & \vdots \\ 1 & m_i^{(2)} & s_i^{(2)} & \vdots & \vdots & \vdots & \vdots & \vdots & \vdots & \vdots \\ \vdots & \vdots & \vdots & \vdots & \vdots & \vdots & \vdots & \vdots & \vdots & \vdots \\ 1 & m_i^{(2)} & s_i^{(2)} & \vdots & \vdots & \vdots & \vdots & \vdots & \vdots & \vdots \\ \vdots & \vdots & \vdots & \vdots & \vdots & \vdots & \vdots & \vdots & \vdots & \vdots \\ \vdots & \vdots & \vdots & \vdots & \vdots & \vdots & \vdots & \vdots & \vdots & \vdots \\ 1 & m_i^{(p_i)} & s_i^{(p_i)} & \vdots & \vdots & \vdots & \vdots & \vdots & \vdots & \vdots \\ \vdots & \vdots & \vdots & \vdots & \vdots & \vdots & \vdots & \vdots & \vdots & \vdots \\ 1 & m_i^{(p_i)} & s_i^{(p_i)} & \mu_i^{(1)} & \sigma_i^{(1)} & \mu_i^{(2)} & \sigma_i^{(2)} & \cdots & \mu_i^{(K)} & \sigma_i^{(K)} \end{bmatrix} \begin{matrix} \uparrow \\ n_i \text{ rows} \\ \downarrow \\ \uparrow \\ n_i \text{ rows} \\ \downarrow \\ \\ \vdots \\ \\ \uparrow \\ n_i \text{ rows} \\ \downarrow \end{matrix}$$

2. Form $\mathbf{y}$, the response vector of length $n = \sum_{i=1}^{I} n_i p_i$ comprised of the vertical concatenation of the batch response vectors,

$$\mathbf{y} = \begin{bmatrix} \mathbf{y}_1 \\ \vdots \\ \mathbf{y}_I \end{bmatrix}.$$

3. Form the design matrix, X, of CVaR regression. This is a matrix of n rows and $l + 1 = 2K + 3$ columns, comprised of the vertical concatenation of the batch design matrices X_i.

$$X = \begin{bmatrix} X_1 \\ \vdots \\ X_I \end{bmatrix}.$$

Example
Suppose batch i has $\{10.1, 12.3, 14.5, 16.7\}$ as test data, and $\{12.4, 16.8\}$ as model data. We will illustrate the calculations in item 1(a). Here $K = 1$, $N_i = 4$, and suppose we choose $w = 2$. Then $p_i = \binom{4}{2} = 6$, and X_i will be of dimension (12×5). We obtain,

$$\mu_i^{(1)} = (12.4 + 16.8)/2 = 14.6, \quad \text{and} \quad \sigma_i^{(1)} = 16.8 - 12.4 = 4.4.$$

For the first covariate set, $\{14.5, 16.7\}$, we obtain

$$m_i^{(1)} = (14.5 + 16.7)/2 = 15.6, \quad \text{and} \quad s_i^{(1)} = 16.7 - 14.5 = 2.2.$$

Computing $m_i^{(2)}, \ldots, m_i^{(6)}$ and the corresponding $s_i^{(2)}, \ldots, s_i^{(6)}$ similarly, gives eventually,

$$
\mathbf{y}_i =
\begin{bmatrix}
10.1 \\
12.3 \\
10.1 \\
14.5 \\
10.1 \\
16.7 \\
12.3 \\
14.5 \\
12.3 \\
16.7 \\
14.5 \\
16.7
\end{bmatrix}, \qquad
\mathbf{X}_i =
\begin{bmatrix}
1\ 15.6\ 2.2\ 14.6\ 4.4 \\
1\ 15.6\ 2.2\ 14.6\ 4.4 \\
1\ 14.5\ 4.4\ 14.6\ 4.4 \\
1\ 14.5\ 4.4\ 14.6\ 4.4 \\
1\ 13.4\ 2.2\ 14.6\ 4.4 \\
1\ 13.4\ 2.2\ 14.6\ 4.4 \\
1\ 13.4\ 6.5\ 14.6\ 4.4 \\
1\ 13.4\ 6.5\ 14.6\ 4.4 \\
1\ 12.3\ 4.4\ 14.6\ 4.4 \\
1\ 12.3\ 4.4\ 14.6\ 4.4 \\
1\ 11.2\ 2.2\ 14.6\ 4.4 \\
1\ 11.2\ 2.2\ 14.6\ 4.4
\end{bmatrix}.
$$

The CVaR Regression Module

This module takes on the following inputs from the previous section:

Inputs:

- $\mathbf{y}$: the test data (vector of length n);
- $\mathbf{X}$: the design matrix (n by $l + 1$);
- τ: the probability level of the desired quantile ($0 \leq \tau \leq 0.5$).

Here for simplicity we describe only the case when $0 \leq \tau \leq 0.5$. The $\tau > 0.5$ case can be treated symmetrically; see Appendix C in [3]. From these inputs, the following are computed:

1. Let $\mathbf{x}_r$ designate the vector of length $l + 1$ consisting of the rth row of $\mathbf{X}$. Defining $\overline{\mathbf{X}}$ to be $\mathbf{X}$ without the first column, Let $\bar{\mathbf{x}}_r$ designate the vector of length l consisting of the rth row of $\overline{\mathbf{X}}$. With this notation, we can write

$$
\mathbf{X} =
\begin{bmatrix}
\mathbf{x}'_1 \\
\vdots \\
\mathbf{x}'_n
\end{bmatrix},
\qquad \text{and} \qquad
\overline{\mathbf{X}} =
\begin{bmatrix}
\bar{\mathbf{x}}'_1 \\
\vdots \\
\bar{\mathbf{x}}'_n
\end{bmatrix}.
$$

2. Let $\mathbf{c}$ denote the vector of length $l + 1$ of CVaR regression coefficients

$$
\mathbf{c} = [c_0, c_1, \ldots, c_l]'.
$$

In addition, let $\bar{\mathbf{c}}$ be $\mathbf{c}$ without c_0. Let $\mathcal{P}_\tau^2(\bar{\mathbf{c}})$ designate the goodness-of-fit criterion (or objective function) for the fitted τth quantile CVaR regression surface, evaluated at $\bar{\mathbf{c}}$ defined as,

$$
\mathcal{P}_\tau^2(\bar{\mathbf{c}}) = \widehat{\mathrm{CVaR}}_{1-\tau}^{\Delta}(-\bar{\varepsilon}) = \frac{1-\tau}{\tau}\widehat{\mathrm{CVaR}}_\tau^{\Delta}(\bar{\varepsilon}).
$$

(See formulas (22) and (23) in [3].) $\widehat{\mathrm{CVaR}}_{1-\tau}^{\Delta}\left(-\bar{\varepsilon}\right)$ is the average of the largest $n\tau$ elements in the list of centered negative residuals,

$$-\bar{\varepsilon}_i = -\left[y_i - \bar{\mathbf{x}}_i'\bar{\mathbf{c}} - \frac{1}{n}\sum_{j=1}^{n}(y_j - \bar{\mathbf{x}}_j'\bar{\mathbf{c}})\right], \quad i = 1, \ldots, n.$$

As such, $\widehat{\mathrm{CVaR}}_{1-\tau}^{\Delta}\left(-\bar{\varepsilon}\right)$ is an estimate of the distance between the mean and the mean of the right tail beyond the $(1-\tau)$th quantile of the CVaR regression residual distribution.

If $n\tau$ is not an integer, we must count a fraction of an element in the list of residuals. For example, suppose that $n = 5$, $\tau = 0.3$, and the list of centered negative residuals is $\{6, 4, 12, 1, 3\}$. In this case, $\widehat{\mathrm{CVaR}}_{1-\tau}^{\Delta}\left(-\bar{\varepsilon}\right)$ equals the average of the largest $5(0.3) = 1.5$ elements. Since the two largest elements are 6 and 12, we have

$$\widehat{\mathrm{CVaR}}_{1-\tau}^{\Delta}\left(-\bar{\varepsilon}\right) = \frac{1(12) + 0.5(6)}{1.5} = \frac{15}{1.5} = 10.$$

3. Let $\hat{\bar{\mathbf{c}}}(\tau)$ denote the minimizer of $\mathcal{P}_\tau^2(\bar{\mathbf{c}})$ over all $\bar{\mathbf{c}} \in \mathscr{R}^l$, at quantile τ, i.e.

$$\hat{\bar{\mathbf{c}}}(\tau) = \arg\min_{\bar{\mathbf{c}} \in \mathscr{R}^l} \mathcal{P}_\tau^2(\bar{\mathbf{c}}). \tag{10.4}$$

For details on how this minimization can be performed efficiently, see Section A.

4. Define

$$\hat{c}_0(\tau) = \widehat{\mathrm{VaR}}_{1-\tau}\left(-\hat{\bar{\varepsilon}}\right) - \frac{1}{n}\sum_{j=1}^{n}(y_j - \bar{\mathbf{x}}_j'\hat{\bar{\mathbf{c}}}(\tau)).$$

If $\{-\bar{\varepsilon}_{(1)} \leq -\bar{\varepsilon}_{(2)} \leq \cdots \leq -\bar{\varepsilon}_{(n)}\}$ denote the ordered centered negative residuals, we define

$$\widehat{\mathrm{VaR}}_{1-\tau}\left(-\bar{\varepsilon}\right) = -\bar{\varepsilon}_{(r)},$$

where r is the unique integer satisfying $r - 1 \leq n(1-\tau) < r$. In the above example with $n = 5$ and $\tau = 0.3$, $(1-\tau)n = 3.5$, and therefore $\widehat{\mathrm{VaR}}_{1-\tau}\left(-\bar{\varepsilon}\right)$ is the 4th ordered element, i.e. 6.

5. Denote by $\hat{\mathbf{c}}(\tau) = [\hat{c}_0(\tau), \hat{\bar{\mathbf{c}}}(\tau)']'$, the estimator of $\mathbf{c}$ in CVaR regression fitted at quantile level τ.

6. Let $\hat{\varepsilon}_i^*$ denote the ith residual from the CVaR regression fit, i.e.

$$\hat{\varepsilon}_i^* = y_i - \mathbf{x}_i'\hat{\mathbf{c}}(\tau),$$

and let $\hat{\varepsilon}_{(1)}^* \leq \hat{\varepsilon}_{(2)}^* \leq \cdots \leq \hat{\varepsilon}_{(n)}^*$ denote these n ordered residuals. Define

$$\hat{F}_n^{-1}(\tau) = \widehat{\mathrm{VaR}}_\tau\left(\hat{\varepsilon}^*\right) = \hat{\varepsilon}_{(r)}^*,$$

where r is the unique integer satisfying $r - 1 \leq n\tau < r$.

7. Let $h_n(\tau)$ be the bandwidth at quantile τ defined by,

$$h_n(\tau) = \left[\frac{3z_{0.975}^2 \exp\{-z_\tau^2\}}{4n\pi(2z_\tau^2 + 1)} \right]^{1/3}.$$

8. Let $\hat{s}_n(\tau)$ be the sparsity function at quantile τ defined by,

$$\hat{s}_n(\tau) = \frac{\hat{F}_n^{-1}(\tau + h_n(\tau)) - \hat{F}_n^{-1}(\tau - h_n(\tau))}{2h_n(\tau)}.$$

9. Let $\hat{\Omega}_n(\tau)$ be the asymptotic covariance matrix at quantile τ defined by,

$$\hat{\Omega}_n(\tau) = \tau(1 - \tau)(X'X)^{-1}\hat{s}_n(\tau)^2.$$

It has $l + 1$ columns and $l + 1$ rows.

The following is the desired output from the CVaR Regression Module:

Outputs:

- $\hat{c}(\tau)$: the optimal coefficients;
- $\mathcal{P}_\tau^2\left(\hat{\bar{c}}(\tau)\right)$: the CVaR goodness-of-fit measure;
- $\hat{\Omega}_n(\tau)$: the asymptotic covariance matrix.

Final Quantities of Interest

Using the outputs from the CVaR Regression Module, we can calculate several quantities. Suppose that the summary statistics for a particular batch of interest are available. We can arrange them into a vector as follows,

$$\mathbf{x} = [1, m_1, s_1, \mu_1^{(1)}, \sigma_1^{(1)}, \mu_1^{(2)}, \sigma_1^{(2)}, \ldots, \mu_1^{(K)}, \sigma_1^{(K)}]'.$$

From this we can compute the following.

τth quantile of failure load at $\mathbf{x}$. The estimated value for the τth quantile of the failure load when the batch values are as in $\mathbf{x}$, is given by

$$\begin{aligned}
\hat{Q}_Y(\tau) = {} & \hat{c}_0(\tau) + \hat{c}_1(\tau)m_1 + \hat{c}_2(\tau)s_1 + \hat{c}_3(\tau)\mu_1^{(1)} + \hat{c}_4(\tau)\sigma_1^{(1)} \\
& + \cdots + \hat{c}_{2K+1}(\tau)\mu_1^{(K)} + \hat{c}_{2K+2}(\tau)\sigma_1^{(K)}. \quad (10.5)
\end{aligned}$$

For example, the estimated 10th quantile of the failure load is $\hat{Q}_Y(0.10)$.

Goodness-of-fit of τ-CVaR regression. The overall goodness-of-fit measure of the τth quantile CVaR regression factor model fitted to the data, is simply $\mathcal{P}_\tau^2(\hat{c}(\tau))$. For example, the goodness-of-fit of a 10th quantile CVaR regression fit is $\mathcal{P}_\tau^2(\hat{c}(0.10))$.

A-Basis. The A-Basis estimate when the batch values are as in $\mathbf{x}$, is given by

$$\mathbf{x}'\hat{\mathbf{c}}(0.01) + z_{0.05}\sqrt{\mathbf{x}'\hat{\Omega}_n(0.01)\mathbf{x}}. \qquad (10.6)$$

B-Basis. The B-Basis estimate when the batch values are as in $\mathbf{x}$, is given by

$$\mathbf{x}'\hat{\mathbf{c}}(0.10) + z_{0.05}\sqrt{\mathbf{x}'\hat{\Omega}_n(0.10)\mathbf{x}}. \qquad (10.7)$$

CVaR Regression Flowchart Diagram

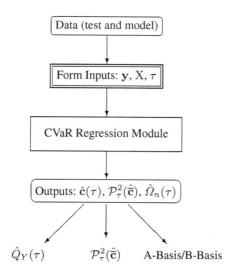

Fig. 10.33. CVaR regression flowchart diagram.

Minimization of CVaR: Reduction to Linear Programming

The minimization problem

$$\min_{\overline{\mathbf{c}}\in\mathscr{R}^l} \mathcal{P}_\tau^2(\overline{\mathbf{c}}) = \min_{\overline{\mathbf{c}}\in\mathscr{R}^l} \widehat{\mathrm{CVaR}}_{1-\tau}^{\Delta}(-\overline{\varepsilon}), \qquad (10.8)$$

can be reduced to the following linear programming problem; see [2]:

$$\min\left\{\eta + \frac{1}{n\tau}\sum_{i=1}^{n} z_i\right\},$$

where

$$z_i \geq -y_i + \frac{1}{n}\sum_{i=1}^n y_i + \left(\bar{\mathbf{x}}_i' - \frac{1}{n}\sum_{i=1}^n \bar{\mathbf{x}}_i'\right)\bar{\mathbf{c}} - \eta, \quad i = 1,\ldots,n,$$

$$z_i \geq 0, \quad i = 1,\ldots,n,$$

and

$$\eta \in \mathscr{R}, \quad \bar{\mathbf{c}} \in \mathscr{R}^l.$$

The minimization is conducted with respect to $\bar{\mathbf{c}} = [\bar{c}_1,\ldots,\bar{c}_l]' \in \mathscr{R}^l$; $\eta \in \mathscr{R}$; and $z_i \in \mathscr{R}, i = 1,\ldots,n$.

Let $\hat{\bar{\mathbf{c}}} = [\hat{\bar{c}}_1,\ldots,\hat{\bar{c}}_l]'$, $\hat{\eta}$, and $\hat{z}_1,\ldots,\hat{z}_n$ denote the minimizers. The minimization simultaneously minimizes $\widehat{\mathrm{CVaR}}_{1-\tau}^{\Delta}(-\bar{\varepsilon})$ and finds $\widehat{\mathrm{VaR}}_{1-\tau}(-\bar{\varepsilon})$ at the optimal $-\bar{\varepsilon}$. In other words,

$$\min_{\bar{\mathbf{c}} \in \mathscr{R}^l} \widehat{\mathrm{CVaR}}_{1-\tau}^{\Delta}(-\bar{\varepsilon}) = \widehat{\mathrm{CVaR}}_{1-\tau}^{\Delta}\left(-\hat{\bar{\varepsilon}}\right) = \mathcal{P}_\tau^2(\hat{\bar{\mathbf{c}}}) = \hat{\eta} + \frac{1}{n\tau}\sum_{i=1}^n \hat{z}_i,$$

and

$$\widehat{\mathrm{VaR}}_{1-\tau}\left(-\hat{\bar{\varepsilon}}\right) = \hat{\eta}.$$

References

[1] Kullback, S., and Leibler, R.A. (1951), "On information and sufficiency", *Annals of Mathematical Statistics* 22, 79-86.
[2] Rockafellar R.T. and S. Uryasev. (2002), "Conditional Value-at-Risk for General Loss Distributions", *Journal of Banking and Finance*, 26-7, 1443-1471.
[3] Uryasev S., Trindade A., and Uryasev Jr., S. (2003), "Combining Model and Test Data for Optimal Determination of Percentiles and Allowables: CVaR Regression Approach, Part I". American Optimal Decisions, Inc., report conducted in the framework of AIM-C program.

11

Semidefinite Programming for Sensor Network and Graph Localization

Yinyu Ye

Management Science and Engineering and, by courtesy, Electrical Engineering
Stanford University
Stanford, CA 94305
yinyu-ye@stanford.edu

Summary. We survey recent developments of using semidefinite programming for solving the sensor network and graph localization problem. The semidefinite programming (SDP) relaxation based method was initially proposed by [15; 16], theoretically analyzed in [44], and further improved by a gradient-based local search method from [32]. An optimization problem is set up so as to minimize the error in sensor positions to fit distance measures. Observable criteria are developed to certify the quality of the point estimation of sensors or to detect erroneous sensors. The performance of this technique is highly satisfactory compared to other techniques. Very few anchor nodes are required to accurately estimate the position of all the unknown nodes in a network. Also the estimation errors are minimal even when the anchor nodes are not suitably placed within the network or the distance measurements are noisy.

Key words: semidefinite programming, sensor network

11.1 Introduction

One of the most studied problems in distance geometry is the *Graph Realization or Localization* problem, in which one is given a graph $G = (V, E)$ and a set of non–negative weights $\{d_{ij} : (i,j) \in E\}$ on its edges, and the goal is to compute a realization of G in the Euclidean space $\mathscr{R}^d$ for a given dimension d, i.e. to place the vertices of G in $\mathscr{R}^d$ such that the Euclidean distance between every pair of adjacent vertices v_i, v_j equals to the prescribed weight d_{ij}. This problem and its variants arise from applications in various areas, such as molecular conformation, dimensionality reduction, Euclidean ball packing, and more recently, wireless sensor network localization [3; 15; 19; 26; 40; 43]. In the sensor networks setting, the vertices of G correspond to sensors, the edges of G correspond to communication links, and the weights correspond to distances. Furthermore, the vertices are partitioned into two sets – one is the *anchors*, whose exact positions are known (via GPS, for example);

and the other is the *sensors*, whose positions are unknown. The goal is to determine the positions of all the sensors. We shall refer to this problem as the *Sensor Network Localization* problem. Note that we can view the Sensor Network Localization problem as a variant of the Graph Realization problem in which a subset of the vertices are constrained to be in certain positions.

In many sensor networks applications, sensors collect data that are location–dependent. Thus, another related question is whether the given instance has a unique realization in the required dimension (say, in $\mathscr{R}^2$). Indeed, most of the previous works on the Sensor Network Localization problem fall into two categories – one deals with computing a realization of a given instance [15; 19; 20; 26; 39; 40; 41; 43], and the other deals with determining whether a given instance has a unique realization in $\mathscr{R}^d$ using graph rigidity [20; 23]. It is interesting to note that from an algorithmic viewpoint, the two problems above have very different characteristics. Under certain non–degeneracy assumptions, the question of whether a given instance has a unique realization on the plane can be decided efficiently [29], while the problem of computing a realization on the plane is NP–complete in general, even if the given instance has a unique realization on the plane [7]. Thus, it is not surprising that all the aforementioned heuristics for computing a realization of a given instance do not guarantee to find it in the required dimension. On another front, there has been attempts to characterize families of graphs that admit polynomial time algorithms for computing a realization in the required dimension. For instance, Eren et. al. [20] have shown that the family of *trilateration graphs* has such property. (A graph is a trilateration graph in dimension d if there exists an ordering of the vertices $1, \ldots, d+1, d+2, \ldots, n$ such that (i) the first $d + 1$ vertices form a complete graph, and (ii) each vertex $j > d + 1$ has at least $d + 1$ edges to vertices earlier in the sequence.) However, the question of whether there exist larger families of graphs with such property is open.

Recently, an SDP relaxation based method has been developed for the *Sensor Network Localization* problem. The decision and optimization problems are set up so as to minimize the error in sensor positions for fitting the distance measures. Observable criteria are developed to certify the quality of the point estimation. The basic idea behind the technique is to convert the nonconvex quadratic distance constraints into linear constraints and matrix inequalities by introducing an SDP relaxation to remove the quadratic term in the formulation.

The method has been proved to be a polynomial time algorithm for solving a family of *uniquely localizable* graphs. Informally, a graph is uniquely localizable in dimension d if (i) it has a unique realization in $\mathscr{R}^d$, and (ii) it does not have any nontrivial realization whose affine span is $\mathscr{R}^h$, where $h > d$. Specifically, we present an SDP model that guarantees to find the unique realization in polynomial time when the input graph is uniquely localizable. The proof employs SDP duality theory and properties of interior–point algorithms for SDP. To the best of our knowledge, this is the first time such a theoretical guarantee is proven for a general localization algorithm. Moreover, in view of the hardness result of Aspnes et. al. [7], our result is close to be the best possible in terms of identifying the largest family of efficiently realizable graphs. Furthermore, we introduce the concept of strong localizability. Informally, a graph is strongly localizable if it is uniquely localizable and remains so under slight perturbations. We show that the SDP model will identify all the strongly localizable subgraphs in the input graph.

We should mention here that various researchers have used SDP to study the Sensor Network Localization problem (or its variants) before. For instance, Schoenberg G. Young and A. S. Householder [42; 50] have explored the fundamental connection of the Euclidean matrix and positive semidefinite matrix. Their work was the basis for what is now called classical multidimensional scaling, originally developed by Torgerson and by Gower [46; 22]; also see Mardia et al. [36]. Unfortunately, the multidimensional scaling method typically requires complete distance information in order to make a localization. Barvinok [10; 11] has studied this problem in the context of quadratic maps and used SDP theory to analyze the possible dimensions of the realization. Alfakih/Wolkowicz, Laurent, and Trosset [4; 5; 31; 47; 48] have related this problem to the *Euclidean Distance Matrix Completion* problem, and obtained an SDP formulation and/or a variable reduction for the former. Moreover, Alfakih has obtained a characterization of rigid graphs in [1] using Euclidean distance matrices and has studied some of the computational aspects of such characterization in [2] using SDP. However, these papers mostly addressed the question of realizability of the input graph, and the analysis of their SDP models only guarantee that they will find a realization whose dimension lies within a certain range. In fact, SDP has also been used to compute and analyze distance geometry problems where the realization is allowed to have a certain amount of distortion in the distances [12; 33]. Again, these methods can only guarantee to find a realization that lies in a range of dimensions. (For some related work in this direction, see, e.g., [8].) Thus, these models are not quite suitable for our application where the dimension of the realization space is fixed. In contrast, our SDP method takes advantage of the presence of anchors and gives a condition which guarantees that the SDP model will find a realization in the required dimension.

The computational performance of the method is highly satisfactory compared to other techniques. Very few anchor nodes are required to accurately estimate the position of all the unknown nodes in a network. Also the estimation errors are minimal even when the anchor nodes are not suitably placed within the network. It is demonstrated that the method produces a solution that is very close to the optimal one even the graph is not *uniquely localizable* or the input distance data are noisy.

The paper is organized as follows. We introduce the localization decision problem and its SDP formulation and relaxation first. Then we show few theories to support our SDP feasibility model. Subsequently, we develop an SDP optimization model to handle incomplete and/or inaccurate distance information and an SDP rounding method to produce solution to the original problem. Finally, we present a distributed SDP method and the preliminary computational results. Sensor network measurements are bound to contain noise, and throughout the paper we will consider how well various techniques will respond to varying distributions and levels of noise.

11.2 Decision Problems and SDP Relaxations

We begin with some notations. The trace of a matrix A is denoted by $\text{Trace}(A)$. We use I and $\mathbf{0}$ to denote the identity matrix and the matrix of all zeros, respectively, whose dimensions will be clear from the context. The inner product of two matrices P and Q is denoted by $P \bullet Q = \text{Trace}(P^T Q)$. The 2–norm of a vector x, denoted by $\|x\|$, is given by $\sqrt{x \bullet x}$. A positive semidefinite matrix X is denoted by $X \succeq 0$.

We use the *Sensor Network Localization* problem, which is defined as follows, as our target graph localization problem. We are given m *anchor* points $a_1, \ldots, a_m \in \mathscr{R}^d$ whose locations are known, and n *sensor* points $x_1, \ldots, x_n \in \mathscr{R}^d$ whose locations we wish to determine. Furthermore, we are given the Euclidean distance values d_{kj} between a_k and x_j for some k, j, and d_{ij} between x_i and x_j for some $i < j$. Specifically, let $N_a = \{(k, j) : d_{kj} \text{ is specified}\}$ and $N_x = \{(i, j) : i < j, d_{ij} \text{ is specified}\}$. The Sensor Network Localization problem is then to find a realization of $x_1, \ldots, x_n \in \mathscr{R}^d$ such that:

$$
\begin{aligned}
\|a_k - x_j\|^2 &= d_{kj}^2, \quad &\forall\, (k, j) \in N_a, \\
\|x_i - x_j\|^2 &= d_{ij}^2, \quad &\forall\, (i, j) \in N_x.
\end{aligned}
\tag{11.1}
$$

We would like to develop fast algorithms to answer questions like: Does the network have a realization of x_j's in $\mathscr{R}^d$? Is the realization unique? As we shall see in subsequent sections, these questions can be answered efficiently.

In general, problem (11.1) is a non–convex optimization problem and difficult to solve. In fact, most previous approaches adopt global optimization techniques such as nonlinear least square methods, or geometric methods such as trilateration. An alternate approach, called the semidefinite programming method, is recently developed in [15] and related earlier work [3; 31]. We shall summarize this approach below.

Let $X = [x_1\ x_2\ \ldots\ x_n]$ be the $d \times n$ matrix that needs to be determined. Then, for all $(i, j) \in N_x$, we have:

$$
\|x_i - x_j\|^2 = e_{ij}^T X^T X e_{ij},
$$

and for all $(k, j) \in N_a$, we have:

$$
\|a_k - x_j\|^2 = (a_k; e_j)^T [I_d; X]^T [I_d; X](a_k; e_j).
$$

Here, $e_{ij} \in \mathscr{R}^n$ is the vector with 1 at the i–th position, -1 at the j–th position and zero everywhere else; $e_j \in \mathscr{R}^n$ is the vector of all zeros except an -1 at the j–th position; $(a_k; e_j) \in \mathscr{R}^{d+n}$ is the vector of a_k on top of e_j; and I_d is the d–dimensional identity matrix. Thus, problem (11.1) becomes: find a symmetric matrix $Y \in \mathscr{R}^{n \times n}$ and a matrix $X \in \mathscr{R}^{d \times n}$ such that:

$$
\begin{aligned}
e_{ij}^T Y e_{ij} &= d_{ij}^2, &\forall\, (i, j) \in N_x, \\
(a_k; e_j)^T \begin{pmatrix} I_d & X \\ X^T & Y \end{pmatrix} (a_k; e_j) &= d_{kj}^2, &\forall\, (k, j) \in N_a, \\
Y &= X^T X.
\end{aligned}
$$

The SDP method is to relax the constraint $Y = X^T X$ to $Y \succeq X^T X$, where $Y \succeq X^T X$ means that $Y - X^T X \succeq 0$. It is well–known [17] that the condition $Y \succeq X^T X$ is equivalent to:

$$Z = \begin{pmatrix} I_d & X \\ X^T & Y \end{pmatrix} \succeq 0. \tag{11.2}$$

Thus, we can write the relaxed problem as a standard SDP problem, namely, find a symmetric matrix $Z \in \mathcal{R}^{(d+n) \times (d+n)}$ to:

$$\max \quad 0$$

$$\text{s.t.} \quad Z_{1:d,1:d} = I_d,$$

$$(\mathbf{0}; e_{ij})(\mathbf{0}; e_{ij})^T \bullet Z = d_{ij}^2, \quad \forall \, (i,j) \in N_x, \tag{11.3}$$

$$(a_k; e_j)(a_k; e_j)^T \bullet Z = d_{kj}^2, \quad \forall \, (k,j) \in N_a,$$

$$Z \succeq 0,$$

where $Z_{1:d,1:d}$ is the $d \times d$ principal submatrix of Z. Note that this formulation forces any possible feasible solution matrix to have rank at least d.

The dual of the SDP relaxation is given by:

$$\min \quad I_d \bullet V + \sum_{(i,j) \in N_x} y_{ij} d_{ij}^2 + \sum_{(k,j) \in N_a} w_{kj} d_{kj}^2$$

$$\text{s.t.} \quad \begin{pmatrix} V & 0 \\ 0 & 0 \end{pmatrix} + \sum_{(i,j) \in N_x} y_{ij}(\mathbf{0}; e_{ij})(\mathbf{0}; e_{ij})^T \tag{11.4}$$

$$+ \sum_{(k,j) \in N_a} w_{kj}(a_k; e_j)(a_k; e_j)^T \succeq 0.$$

Note that the dual is always feasible, as $V = 0$, $y_{ij} = 0$ for all $(i,j) \in N_x$ and $w_{kj} = 0$ for all $(k,j) \in N_a$ is a feasible solution.

Analysis of the SDP Relaxation

We now investigate when will the SDP (11.3) have an exact relaxation, i.e. when will the solution matrix Z have rank d. Suppose that problem (11.3) is feasible. This occurs when, for instance, d_{kj} and d_{ij} represent exact distance values for the positions $\bar{X} = [\bar{x}_1 \ \bar{x}_2 \ \ldots \ \bar{x}_n]$. Then, the matrix $\bar{Z} = (I_d; \bar{X})^T (I_d; \bar{X})$ is a feasible solution for (11.3). Now, since the primal is feasible, the minimal value of the dual must be 0, i.e. there is no duality gap between the primal and dual.

Let U be the $(d + n)$–dimensional dual slack matrix, i.e.:

$$U = \begin{pmatrix} V & 0 \\ 0 & 0 \end{pmatrix} + \sum_{(i,j) \in N_x} y_{ij}(0; e_{ij})(0; e_{ij})^T$$

$$+ \sum_{(k,j) \in N_a} w_{kj}(a_k; e_j)(a_k; e_j)^T.$$

Then, from the duality theorem for SDP (see, e.g., [6]), we have:

Theorem 1. *Let $\bar{Z}$ be a feasible solution for (11.3) and $\bar{U}$ be an optimal slack matrix of (11.4). Then,*

1. *complementarity condition holds: $\bar{Z} \bullet \bar{U} = 0$ or $\bar{Z}\bar{U} = 0$.*
2. *$Rank(\bar{Z}) + Rank(\bar{U}) \leq d + n$.*
3. *$Rank(\bar{Z}) \geq d$ and $Rank(\bar{U}) \leq n$.*

An immediate result from the theorem is the following:

Corollary 1. *If an optimal dual slack matrix has rank n, then every solution of (11.3) has rank d. That is, problems (11.1) and (11.3) are equivalent and (11.1) can be solved as an SDP in polynomial time.*

Another technical result is the following:

Proposition 1. *If every sensor point is connected, directly or indirectly, to an anchor point in (11.1), then any solution to (11.3) must be bounded, that is, Y_{jj} is bounded for all $j = 1, \dots, n$.*

In general, a primal (dual) max–rank solution is a solution that has the highest rank among all solutions for primal (11.3) (dual (11.4)). It is known [24; 21; 34] that various path–following interior–point algorithms compute the max–rank solutions for both the primal and dual in polynomial time. This motivates the following definition.

Definition 1. *Problem (11.1) is uniquely localizable if there is a unique localization $\bar{X} \in \mathscr{R}^{d \times n}$ and there is no $x_j \in \mathscr{R}^h$, $j = 1, \dots, n$, where $h > d$, such that:*

$$\begin{aligned}
\|(a_k; 0) - x_j\|^2 &= d_{kj}^2, & \forall\, (k,j) \in N_a, \\
\|x_i - x_j\|^2 &= d_{ij}^2, & \forall\, (i,j) \in N_x, \\
x_j &\neq (\bar{x}_j; 0), & \text{for some } j \in \{1, \dots, n\}.
\end{aligned}$$

The latter says that the problem cannot have a non–trivial localization in some higher dimensional space $\mathscr{R}^h$ (i.e. a localization different from the one obtained by setting $x_j = (\bar{x}_j; 0)$ for $j = 1, \dots, n$), where anchor points are augmented to $(a_k; 0) \in \mathscr{R}^h$, for $k = 1, \dots, m$.

We now develop the following theorem:

Theorem 2. *Suppose that the network is connected. Then, the following statements are equivalent:*

1. *Problem (11.1) is uniquely localizable.*
2. *The max–rank solution matrix of (11.3) has rank d.*
3. *The solution matrix of (11.3), represented by (11.2), satisfies $Y = X^T X$.*

Although unique localizability is an useful notion in determining the solvability of the Sensor Network Localization problem, it is not stable under perturbation. As we shall see below, there exist networks which are uniquely localizable, but may no longer be so after small perturbation of the sensor points. This motivates us to define another notion called strong localizability.

Definition 2. *We say problem (11.1) is strongly localizable if the dual of its SDP relaxation (11.4) has an optimal dual slack matrix with rank n.*

Note that if a network is strongly localizable, then it is uniquely localizable from Theorems 1 and 2, since the rank of all feasible solution of the primal is d.

We now develop the next theorem.

Theorem 3. *If a problem (graph) contains a subproblem (subgraph) that is strongly localizable, then the submatrix solution corresponding to the subproblem in the SDP solution has rank d. That is, the SDP relaxation computes a solution that localizes all possibly localizable unknown points.*

The above theorem establishes an error management of the original problem data. Let $\bar{Z}$ be a feasible solution for (11.3). Then,

$$\text{Trace}(\bar{Y} - \bar{X}^T \bar{X}) = \sum_{j=1}^{n}(\bar{Y}_{jj} - \|\bar{x}_j\|^2),$$

the total trace of the difference matrix measures the quality of the localization solution $\bar{X}$. In particular, individual trace

$$\bar{Y}_{jj} - \|\bar{x}_j\|^2,$$

which can be interpreted as the variance of $\bar{x}_j$, helps us to detect possible distance measure errors, and outlier or defect sensors. These errors often occur in real applications either due to the lack of data information or noisy measurement, and are often difficult to detect since the true location of sensors is unknown.

A Comparison of Notions

We now demonstrate the notions of unique localizability, strong localizability and rigidity. In the following graphs, the three squares are three anchors whose positions

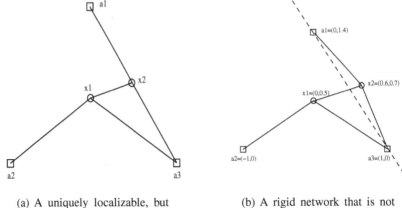

(a) A uniquely localizable, but
not strongly localizable network

(b) A rigid network that is not
uniquely localizable

Fig. 11.1. A comparison of graph notions.

are fixed, two circles are two sensors whose positions need to be determined, and
pair-wise distance values only with edges are known to the SDP method.

It would be nice to have a simple characterization on those graphs which are rigid in
the plane but have higher dimensional realizations. However, finding such a characterization remains a challenging task, as such characterization would necessarily be
non–combinatorial, and would depend heavily on the geometry of the network. For
instance, the networks shown in Figure 11.2, while having the same combinatorial
property as the one shown in Figure 1(b), are uniquely localizable (in fact, they are
both strongly localizable):

11.3 Optimization Problems to Handle Distance Noises

For the localization problem with measurement noises, the story can be quite different. In general there is no solution to satisfy the constraints in (11.1). Therefore,
an optimization problem is set up to minimize the difference between the measured
distance values and the distances formulated from the estimated solutions. In [32]
we assume that distance noises are randomly generated according to the following
formula,

$$d_{ij} = \bar{d}_{ij}(1 + \rho), \tag{11.5}$$

where $\bar{d}_{ij}$ represents the true distance between nodes (either sensors or anchors) i and
j, $\rho \in \mathcal{N}(0, nf^2)$ is a random variable, and nf, called *noisy factor*, is used to adjust
the variance of the distance uncertainty.

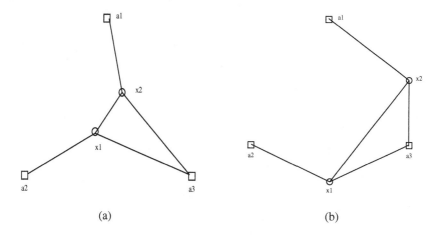

(a)   (b)

Fig. 11.2. Strongly localizable networks.

Maximum Likelihood Estimation and its SDP relaxation

One approach, called maximum likelihood estimation, is introduced below. Let $d :$
$\mathscr{R}^{2\times 2} \to \mathscr{R}$ be the distance function between a sensor/anchor or sensor/sensor pair.
Suppose there are some measurement errors between a_k/x_j and x_i/x_j denoted by
ω_{kj} and ω_{ij}, respectively,

$$d_{kj} = d(x_j, a_k) + \omega_{kj}, \quad \forall (k,j) \in N_a,$$
$$d_{ij} = d(x_j, x_i) + \omega_{ij}, \quad \forall (i,j) \in N_x,$$

where we assume each $\omega_{kj} \sim \mathcal{N}(0, \sigma_{kj}^2)$ and $\omega_{ij} \sim \mathcal{N}(0, \sigma_{ij}^2)$ and they are indepen-

dent.

Let the maximum likelihood function p to estimate X, using all distance measure
information, be

$$p\left(d_{kj}, (k,j) \in N_a; \ d_{ij}, (i,j) \in N_x\right), X)$$

$$= \prod_{j,k;(k,j)\in N_a} \frac{1}{2\pi^{\frac{1}{2}}\sigma_{kj}} \exp\left(-\frac{1}{2\sigma_{kj}^2}(d_{kj} - d(x_j, a_k))^2\right)$$

$$\times \prod_{j,i;(i,j)\in N_x} \frac{1}{2\pi^{\frac{1}{2}}\sigma_{ij}} \exp\left(-\frac{1}{2\sigma_{ij}^2}(d_{ij} - d(x_j, x_i))^2\right),$$

and the maximum likelihood estimation be

$$X_{ml} = \arg\max_X p((d_{kj}, (k,j) \in N_a; d_{ij}, (i,j) \in N_x), X).$$

Then, X_{ml} can be written explicitly as

$$X_{ml} = \arg\min_X \left(\begin{array}{l} \sum_{j,k;(k,j)\in N_a} \frac{1}{\sigma_{kj}^2} \left(d_{kj} - d(x_j, a_k)\right)^2 \\ + \sum_{j,i;(i,j)\in N_x} \frac{1}{\sigma_{ij}^2} \left(d_{ij} - d(x_j, x_i)\right)^2 \end{array} \right). \qquad (11.6)$$

Hence the following optimization problem solves the maximum likelihood estimation problem

$$\begin{aligned} \min \quad & \sum_{j,k;(k,j)\in N_a} \frac{1}{\sigma_{kj}^2}\epsilon_{kj} + \sum_{j,i;(i,j)\in N_x} \frac{1}{\sigma_{ij}^2}\epsilon_{ij} \\ \text{s.t.} \quad & \left(\|x_j - x_i\| - d_{ij}\right)^2 = \epsilon_{ij}, \qquad \forall(i,j)\in N_x, \\ & \left(\|x_j - a_k\| - d_{kj}\right)^2 = \epsilon_{kj}, \qquad \forall(k,j)\in N_a. \end{aligned} \qquad (11.7)$$

If the variance of distance measurements are not known, any reasonable assumption can be applied. Problem (11.7) is not a convex optimization problem, but we can construct its SDP relaxation problem:

$$\begin{aligned} \min \quad & \sum_{j,k;(k,j)\in N_a} \frac{1}{\sigma_{kj}^2}\epsilon_{kj} + \sum_{j,i;(i,j)\in N_x} \frac{1}{\sigma_{ij}^2}\epsilon_{ij} \\ \text{s.t.} \quad & (-d_{ij};1)^T D_{ij} (-d_{ij};1) = \epsilon_{ij}, \qquad \forall(i,j)\in N_x, \\ & (-d_{kj};1)^T D_{kj} (-d_{kj};1) = \epsilon_{kj}, \qquad \forall(k,j)\in N_a, \\ & (0;e_j - e_i)^T Z (0;e_j - e_i) = v_{ij}, \qquad \forall(i,j)\in N_x, \\ & (a_k;-e_j)^T Z (a_k;-e_j) = v_{kj}, \qquad \forall(k,j)\in N_a, \\ & D_{ij} \succeq 0, \qquad\qquad\qquad\qquad\quad \forall(i,j)\in N_x, \\ & D_{kj} \succeq 0, \qquad\qquad\qquad\qquad\quad \forall(k,j)\in N_a, \\ & Z \succeq 0, \end{aligned} \qquad (11.8)$$

where $Z = \begin{pmatrix} I_2 & X \\ X^T & Y \end{pmatrix}$ and $D_{ij} = \begin{pmatrix} 1 & u_{ij} \\ u_{ij} & v_{ij} \end{pmatrix}, \forall(i,j)\in N_x, D_{kj} = \begin{pmatrix} 1 & u_{kj} \\ u_{kj} & v_{kj} \end{pmatrix},$ $\forall(k,j)\in N_a.$

Weighted Maximum Likelihood Estimation and its SDP relaxation

Another formulation to minimize the estimation error is

$$\begin{aligned} \min \quad & \sum_{j,i;(i,j)\in N_x} \varepsilon_{ij}^2 + \sum_{j,k;(k,j)\in N_a} \varepsilon_{kj}^2 \\ \text{s.t.} \quad & \|x_j - x_i\|^2 - d_{ij}^2 = \varepsilon_{ij}, \qquad \forall(i,j)\in N_x, \\ & \|x_j - a_k\|^2 - d_{kj}^2 = \varepsilon_{kj}, \qquad \forall(k,j)\in N_a. \end{aligned} \qquad (11.9)$$

The objective function can be written as,

$$\sum_{j,i;(i,j)\in N_x} (\|x_j - x_i\| + d_{ij})^2 (\|x_j - x_i\| - d_{ij})^2$$
$$+ \sum_{j,k;(k,j)\in N_a} (\|x_j - a_k\| + d_{kj})^2 (\|x_j - a_k\| - d_{kj})^2 . \tag{11.10}$$

If we assume the measurement errors are not too large; $\|x_j - x_i\| \approx d_{ij}$ and $\|x_j - a_k\| \approx d_{kj}$, the objective is approximately equivalent to minimize

$$\sum_{j,i;(i,j)\in N_x} d_{ij}^2(\|x_j - x_i\| - d_{ij})^2 + \sum_{j;k;(k,j)\in N_a} d_{kj}^2(\|x_j - a_k\| - d_{kj})^2 .$$

Comparing this with the objective function of (11.7), one can immediately see that if $\sigma_{ij} = \frac{1}{d_{ij}}$ and $\sigma_{kj} = \frac{1}{d_{kj}}$, the solution of (11.9) is actually the maximum likelihood estimation solution. We call it weighted maximum likelihood estimation.

The SDP relaxation of (11.9) can be written as

$$
\begin{aligned}
\min \quad & \alpha \\
\text{s.t.} \quad & (0; e_j - e_i)^T Z (0; e_j - e_i) - d_{ij}^2 = \varepsilon_{ij}, \quad \forall (i,j) \in N_x, \\
& (a_k; -e_j)^T Z (a_k; -e_j) - d_{kj}^2 = \varepsilon_{kj}, \quad \forall (k,j) \in N_a, \\
& \left(\sum_{(i,j)\in N_x} \varepsilon_{ij}^2 + \sum_{(i,j)\in N_a} \varepsilon_{kj}^2 \right)^{1/2} \leq \alpha, \\
& Z \succeq 0,
\end{aligned}
\tag{11.11}
$$

where $Z = \begin{pmatrix} I_2 & X \\ X^T & Y \end{pmatrix}$. Note that this problem is almost as simple as the SDP relaxation (11.3), and it has merely one more second order cone constraint on α and εs.

If the distance measurements are exactly correct and the sensor network is uniquely localizable, then all three formulations (11.3), (11.8), and (11.11) solves the true sensor locations. Moreover, both (11.8) and (11.11) can deal with the localization problem with distance noises. One may notice the assumption that the noise variance is inversely proportional to the measured distance may not be true in general. However, the reason we still use (11.11) will be clear when the gradient search method is introduced later.

The High-Rank Property of SDP Relaxations

When the distance measurements have errors, one may expect (11.8) or (11.11) to be good approximations of (11.7). However, due to the "max-rank" property of the SDP relaxation described in [44], the solution of (11.8) and (11.11) may not be satisfactory. This property is illustrated in the following example,

Consider an ad hoc sensor network in $\mathscr{R}^2$ with 3 anchors and 1 sensor. The locations of sensors are at $a_1 = (1,0)^T, a_2 = (0,1)^T, a_3 = (-1,-1)^T$. The measured distances between a_1, a_2, a_3 and x are $1.020, 1.041, 1.412$ respectively. By solving the SDP relaxation (11.8), the objective function value is zero and the optimal Z^* is

$$Z^* = \begin{pmatrix} I_2 & x^* \\ x^{*T} & y^* \end{pmatrix}$$

$$= \begin{pmatrix} 1 & 0 & 0.0061 \\ 0 & 1 & -0.0571 \\ 0.0061 & -0.0571 & 0.0958 \end{pmatrix}.$$

Sine it is not a rank 2 solution, we know that a higher dimensional localization exists. One way to interpret this solution is: for this only one unknown sensor, the solution space is lifted to $\mathscr{R}^3$, and its projection on $\mathscr{R}^2$ is $x^* = [0.0061, -0.0571]^T$. The third coordinate of x^* is equal to $\sqrt{y^* - x^{*T} x^*} = 0.0925$. This says a point $\bar{x} = [0.0061, -0.0571, 0.0925]^T \in \mathscr{R}^3$ satisfies the three distance constraints.

In this example we see that when the three distance measurements contradict to each other and there is no localization in $\mathscr{R}^2$, it is still possible to locate the sensor in a higher dimensional space and to make the objective function value zero. The optimal solution in a higher dimensional space always results in a smaller objective function value than the one constrained in $\mathscr{R}^2$ even if both of them are nonzero. Hence, because of the relaxation of the rank requirement, the solution is "lifted" to a higher dimensional space.

A main research topic is how to round the higher-dimension (higher rank) SDP solution into a lower-dimension (rank-2) solution. One way is to ignore the augmented dimensions and use the projection x^* as a suboptimal solution, which is the case in [15]. In [32] we have shown how to use a gradient-based local search method to round and improve the SDP relaxation solution. The method turns out to be extremely effective and efficient.

11.4 SDP Rounding Method

A typical solution using the SDP relaxation (11.8) for an ad hoc sensor network with 5 anchors and 45 sensors ($m = 5, n = 45$), radio range 0.35, are shown in Figure 11.3. (The radio range indicates that the distance values between any two nodes are known to the solver if they are below the range; otherwise they are unknown.) The diamonds and the circles indicate the true positions of the anchors and the sensors, respectively. The red stars show the SDP localizations (The same notations will be used in all examples of this article). This problem is uniquely localizable so that all the sensors are located correctly. Now for the same problem, some distance noises are added. Each d_{ij} or d_{kj} below the radio range is collapsed by a Gaussian noise with mean zero and variance equal to 10% of the actual distance (noise factor $nf = 0.10$). The new result is shown in Figure 11.4. By comparing with the true sensor locations plotted in circles, it is easy to see that most localizations are off from their actual locations.

Since this network is uniquely localizable, the localization errors are completely due to the distance measurement errors. A higher dimensional localization is then found by the SDP relaxation solved by SEDUMI [45] or DSDP2 [13], and apparently it is not quite acceptable by its projection on the 2-dimensional space. Even worse, the localization errors propagate from the sensors near some anchors to the sensors far from any anchors.

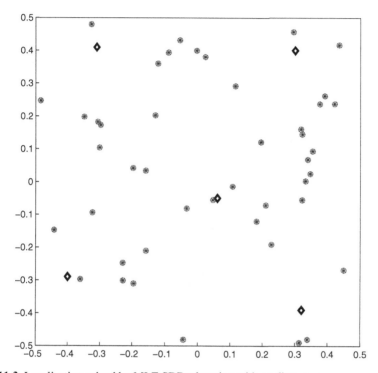

Fig. 11.3. Localization solved by MLE SDP relaxation with no distance measurement error.

The method we suggest to improve the SDP solution is moving every sensor location along the opposite of its gradient direction of the sum of error square function, which will for sure reduce the error function value. The detail and theoretical background of this approach will be given in [32]. We here demonstrate the basic idea of it. Let us begin from (11.6) and for simplification all σ_{ij} and σ_{kj} are assumed to be the same. The maximum likelihood estimation is an unconstrained optimization problem if all constraints are substituted into the objective function,

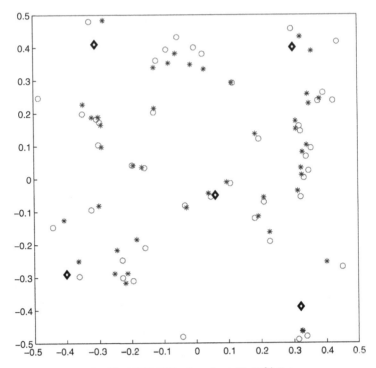

Fig. 11.4. Localization solved by MLE SDP relaxation with 10% distance measurement errors.

$$\min_{X} \ f(X) := \sum_{j,k,(k,j)\in N_a} \left(d_{kj} - \sqrt{(x_j - a_k)^T(x_j - a_k)} \right)^2$$

$$+ \sum_{j,i,(i,j)\in N_x} \left(d_{ij} - \sqrt{(x_j - x_i)^T(x_j - x_i)} \right)^2. \tag{11.12}$$

Differentiate f with respect to x_j, one can find the gradient ∂f_{x_j} for a certain sensor x_j,

$$\partial f_{x_j} \equiv \frac{\partial f}{\partial x_j}$$

$$= \sum_{k,(k,j)\in N_a} \left(d_{kj} - \sqrt{(x_j - a_k)^T(x_j - a_k)} \right) \frac{(x_j - a_k)^T}{\sqrt{(x_j - a_k)^T(x_j - a_k)}}$$

$$+ \frac{1}{2} \sum_{i,(i,j)\in N_x} \left(d_{ij} - \sqrt{(x_j - x_i)^T(x_j - x_i)} \right) \frac{(x_j - x_i)^T}{\sqrt{(x_j - x_i)^T(x_j - x_i)}}. \tag{11.13}$$

It is important to notice that ∂f_{x_j} only relates to the sensors and anchors that are connected to (within the radio range) x_j, and they are local information, so that ∂f_{x_j} for every sensor x_j can be solved *distributedly*.

The following update rule is applied to improve the iterative solution.

$$x_j \longleftarrow x_j - \alpha \cdot \partial f_{x_j}^T \text{ for } j = 1 \text{ to } n, \tag{11.14}$$

where α is the step size, chosen to be 0.05 for this example. In one gradient step, the method calculates the gradient of each sensor and updates its location by this rule. The SDP localization shown in Figure 11.4 is used as the initial solution. Figure 11.5 shows the update trajectories in 50 steps. The red points indicate the new positions of each sensor after each gradient step; contiguous points are connected by blue lines. It can be observed clearly that most sensors are moving toward their actual locations marked by the circles. The final localizations after 50 gradient steps are plotted in Figure 11.6. Comparing Figure 11.4 with Figure 11.6, we see that a much more accurate localization is found.

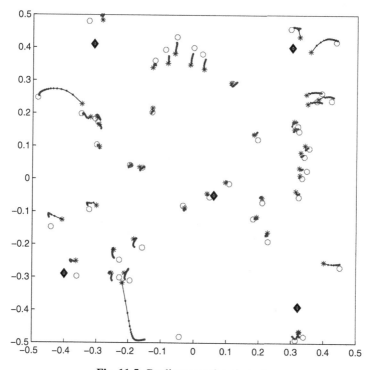

Fig. 11.5. Gradient search trajectories.

To demonstrate that the new localization is indeed better, we can substitute the sensor positions after every gradient step into the objective function to see if the sum of error

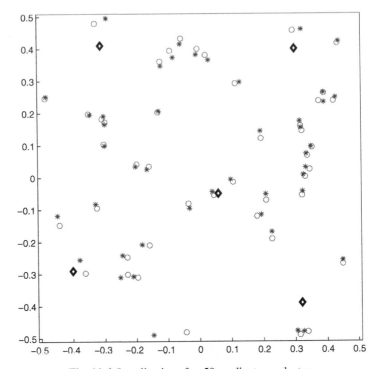

Fig. 11.6. Localization after 50 gradient search steps.

squares is reduced. In Figure 11.7 the objective function values vs number of gradient steps curve are plotted in a blue line. One can see that in the first 20 steps the objective function value drops rapidly, and then the trajectory tends to be flat. This demonstrated that the gradient search method does improve the overall localization result. A natural question is how good the new localization is. To answer this question we need a lower bound of the objective function value. One trivial lower bound of the objective function value is 0; but a better one is the SDP relaxation objective value, since the SDP problem is a relaxation of the original 2-dimensional localization problem. In this case, the SDP objective value is about 0.094, plotted in Figure 11.7 in a red dashed line, and the gradient search method finds a 2-dimensional localization with objective function value about 0.12. Thus an error gap 0.026 of the suboptimality is obtained, which is less than 30% of the error lower bound value.

The gradient-based descent method is a local search method and can be proved to find the global optimal solution only for convex optimization. The ad hoc sensor network localization problem is NOT a convex optimization problem. Hence a pure gradient search method should not work. To see this, another experiment is performed. We use the origin as the initial sensor locations and update them by the same rule (11.14). The updated trajectories are shown in Figure 11.8. Most of these sensors do not converge

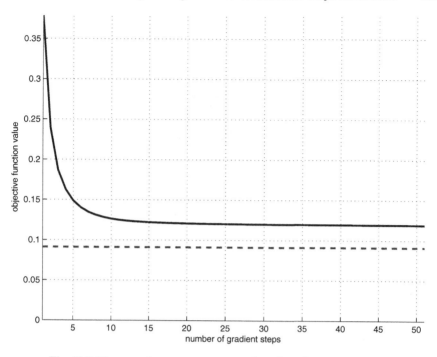

Fig. 11.7. The sum of error squares vs number of gradient search steps.

to their actual positions. Comparing it with the result of Figure 11.6, we can see that
the use of the SDP solution as the initial localization makes all the difference.

11.5 Computational Results

In this section, several examples are used to demonstrate the effectiveness of the SDP
relaxation and the gradient rounding method in solving the localization problem for
ad hoc sensor networks. Simulations were performed on a network of nodes randomly
placed in a square region of size 1×1 centered at the origin. The distances between
the nodes was calculated. Recall that if the distance between two nodes was less than
a specified $radiorange$, a random error was added to it.

$$d_{ij} = \bar{d}_{ij} \cdot (1 + randn(1) * nf),$$

where $\bar{d}_{ij}$ represented the true distance value, noisy factor nf was a given number
between $[0, 1]$ and $randn(1)$ was a standard normal random variable. If the distance
was beyond the given $radiorange$, only the lower bound constraint, $\geq radiorange$,
was applied if necessary. The selection of the lower bounding constraint was based on
an iterative active-constraint generation technique. Because most of these "bounding
away" constraints, i.e., the constraints between two remote nodes, would be inactive

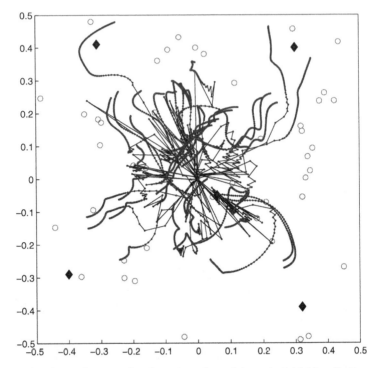

Fig. 11.8. Gradient search trajectories using origin as the initial localization.

or redundant at an optimal solution, very few of these constraints were needed in computation.

The computational results presented here were generated using the interior-point algorithm SDP solvers SeDuMi of [45] and DSDP2.0 of [13] with their interfaces to Matlab on a Pentium 1.2 GHz and 500 MB RAM PC. DSDP is faster due to the fact that the data structure of the problem is more suitable for DSDP. However SeDuMi is often more accurate and robust.

Example 1. In this example, 50 point are generated and three of them chosen as anchors. Distance values are given for various radio ranges. The average estimation error and average trace were defined by

$$\frac{1}{n} \cdot \sum_{j=1}^{n} \|\bar{x}_j - a_j\| \quad \text{and} \quad \frac{1}{n} \cdot \text{Trace}(\bar{Y} - \bar{X}^T \bar{X}),$$

respectively, where $\bar{x}_j$ comes from the SDP solution and a_j is the true position of the jth node. Connectivity indicates how many of the nodes, on average, are within the radio range of a node. The original and the estimated sensors were plotted. The (blue) diamond nodes refer to the positions of the anchors; (green) circle nodes to the true locations of the unknown sensors; and (red) asterisk nodes to their estimated positions from SDP. The discrepancies in the positions can be estimated by the offsets between the true and the estimated points as indicated by the solid lines.

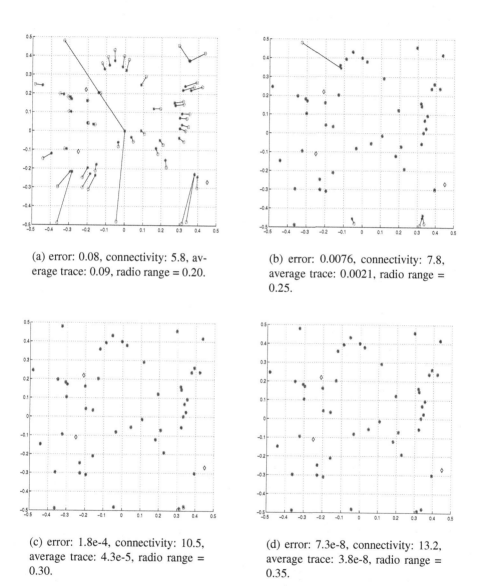

(a) error: 0.08, connectivity: 5.8, average trace: 0.09, radio range = 0.20.

(b) error: 0.0076, connectivity: 7.8, average trace: 0.0021, radio range = 0.25.

(c) error: 1.8e-4, connectivity: 10.5, average trace: 4.3e-5, radio range = 0.30.

(d) error: 7.3e-8, connectivity: 13.2, average trace: 3.8e-8, radio range = 0.35.

Fig. 11.9. Position estimations with 3 anchors, nf = 0, and various radio ranges.

The effect of variable radio ranges and as a result, connectivity, was observed in Figure 11.9. As the radio range varied from 0.2 to 0.35, the localizable subgraph was grew to include all sensors. We also see how the individual trace to predict the quality of the SDP localization solution. In Figure 11.9(b), for the four sensors with large error estimation, their individual traces made up most of the total traces, which match where the real errors are exactly, see Figure 11.10 for the correlation between individual error and trace for each unknown sensor for cases in Figure 11.9(a) and 11.9(b).

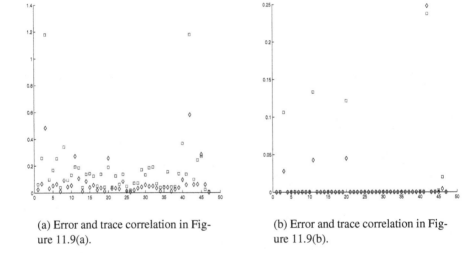

(a) Error and trace correlation in Figure 11.9(a).

(b) Error and trace correlation in Figure 11.9(b).

Fig. 11.10. Diamond: the offset distance between estimated and true positions, Box: the square root of individual trace $\bar{Y}_{jj} - \|\bar{x}_j\|^2$.

In comparison, for the same case as Figure 11.9(c), we computed the results from the Doherty et al. [19] method with the number of anchors 10 and 25, and depicted their pictures in Figure 11.11. As we expected, the estimated positions were all in the convex hull of the anchors.

We have shown in Section 11.4 that the gradient search method can dramatically improve the localization accuracy when the distance measurements are noisy. It would be interesting to know how this method helps when the distance measurements are exact but the network is not uniquely localizable. We select a simple example from [44] to demonstrate this point. We will also use this example to compare the difference between the exact line-search and the constant step size gradient method.

Example 2. Three anchors and two sensors are located at $a_1 = [-\sqrt{3}/2, -1/2]^T$, $a_2 = [-\sqrt{3}/2, 1/2]^T$, $a_3 = [0, 1]^T$, $x_1 = [0, 0]^T$, and $x_2 = [0.77, 0.2]^T$, respec-

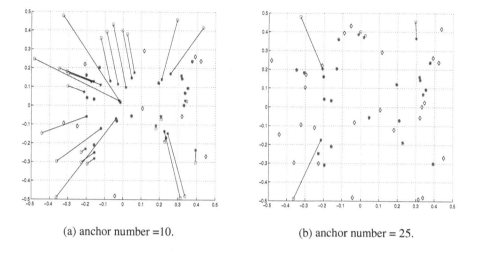

(a) anchor number =10. (b) anchor number = 25.

Fig. 11.11. Position estimations by Doherty et al., radio range = 0.30, nf = 0, and various number of anchors.

tively. The connection of sensor/sensor and sensor/anchor are shown in Figure 11.12 by solid lines. Exact distance measurements of the connections are available. This example is interesting because it is not uniquely localizable but has a unique localization in the original 2-dimensional space. We use both MLE SDP relaxation with fixed step size gradient method and WMLE SDP relaxation with exact line-search gradient method to solve it. The localization and gradient update trajectories are shown in Figure 11.12.

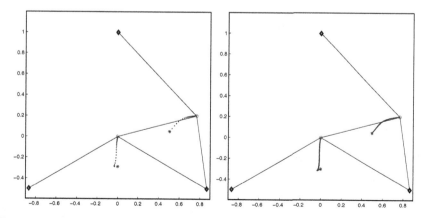

Fig. 11.12. Left: WMLE SDP localization and update trajectory with exact line-search. Right: MLE SDP localization and update trajectory with fixed step size.

Both WMLE and MLE SDP relaxations in this example give wrong solutions. However, after applying gradient search methods, they both converge to the correct ones. This indicates that the gradient rounding can even apply to some of the distance geometry problems which are not uniquely localizable.

We have shown that the gradient method can reduce the objective function value and find a suboptimal solution. In Section 11.4 we also demonstrate that a lower bound of the objective function value can be obtained by the SDP relaxation. The gap between these two values indicates how good the suboptimal solution is. In the following we would like to illustrate how this gap changes as the noise level varies.

Example 3. An ad hoc sensor network with the same five anchors of the inspiring example and 45 sensors being randomly deployed each time in $[-0.5, 0.5] \times [-0.5, 0.5] \in \mathscr{R}^2$. Radio range is set to be 0.35. This problem is repeatedly solved 100 times with noisy factors randomly chosen from $[0, 0.3]$. The solver we use is MLE SDP relaxation combing with gradient search method. In each solution, The SDP objective function value, i.e. the lower bound, and the best objective function value found by the gradient search method are recorded. Figure 11.13 shows the result.

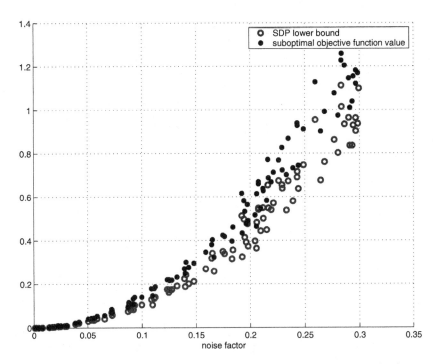

Fig. 11.13. SDP lower bound and suboptimal objective function value vs noisy factor.

In this figure we see that when the noisy factor is larger, both the lower bound and the suboptimal objective function value are growing but keeping close—the gap is roughly 30% of the lower bound value at all different noisy levels. This implies that not only the lower bound is a tight one but also the suboptimal solution is very close to optimal.

11.6 A Distributed SDP Method

Unfortunately, the existing SDP solvers have very poor scalability. They can only handle SDP problems with the dimension and the number of constraints up to few thousands, where in the SDP sensor localization model the number of constraints is in the order of $O(n^2)$, where n is the number of sensors. The difficulty is that each iteration of interior-point algorithm SDP solvers needs to factorize and solve a dense matrix linear system whose dimension is the number of constraints. While we could solve localization problems with 50 sensors in few seconds, we have tried to use several off-the-shell codes to solve localization problems with 200 sensors and often these codes quit either due to memory shortage or having reached the maximum computation time.

We describe an iterative distributed SDP computation scheme, first demonstrated in [16] to overcome this difficulty. We first partition the anchors into many clusters according to their physical positions, and assign some sensors into these clusters if a sensor has a direct connection to one of the anchors. We then solve semidefinite programs *independently* at each cluster, and fix those sensors' positions which have high accuracy measures according the SDP computation. These positioned sensors become 'ghost anchors' and are used to decide the remaining un-positioned sensors. The distributed scheme then repeats.

The distributed scheme is highly scalable and we have solved randomly generated sensor networks of 4, 000 sensors in few minutes for a sequential implementation (that is, the cluster SDP problems are solved sequentially on a single processor), while the solution quality remains as good as that of using the centralized method for solving small networks. We remark that our distributed or decomposed computation scheme should be applicable to solving other Euclidean geometry problems where points are locally connected.

A round of the distributed computation method is straightforward and intuitive:

1. Partition the anchors into a number of clusters according to their geographical positions. In our implementation, we partition the entire sensor area into a number of equal-sized squares and those anchors in a same square form a regional cluster.

2. Each (unpositioned) sensor sees if it has a direct connection to an anchor (within the communication range to an anchor). If it does, it becomes an unknown sensor point in the cluster to which the anchor belongs. Note that a sensor may be assigned into multiple clusters and some sensors are not assigned into any cluster.

3. For each cluster of anchors and unknown sensors, formulate the error minimization problem for that cluster, and solve the resulting SDP model if the number of anchors is more than 2. Typically, each cluster has less than 100 sensors and the model can be solved efficiently.

4. After solving each SDP model, check the individual trace for each unknown sensor in the model. If it is below a predetermined small tolerance, label the sensor as *positioned* and its estimation $\bar{x}_j$ becomes an " anchor". If a sensor is assigned in multiple clusters, we choose the $\bar{x}_j$ that has the smallest individual trace. This is done so as to choose the best estimation of the particular sensor from the estimations provided by solving the different clusters.

5. Consider positioned sensors as anchors and return to Step 1 to start the next round of estimation.

Note that the solution of the SDP problem in each cluster can be carried out at the cluster level so that the computation is highly distributive. The only information that needs to be passed among the neighboring clusters is which of the unknown sensors become positioned after a round of SDP solutions.

In solving the SDP model for each cluster, even if the number of sensors is below 100, the total number of constraints could be in the range of thousands. However, many of those "bounding away" constraints, i.e., the constraints between two remote points, are inactive or redundant at the optimal solution. Therefore, we adapt an iterative active constraint generation method. First, we solve the problem including only partial equality constraints and completely ignoring the bounding-away inequality constraints to obtain a solution. Secondly we verify the equality and inequality constraints and add those violated at the current solution into the model, and then resolve it with a "warm-start" solution. We can repeat this process until all of the constraints are satisfied. Typically, only about $O(n+m)$ constraints are active at the final solution so that the total number of constraints in the model can be controlled at $O(n+m)$.

It has been shown that when the distance measurements are accurate, the distributed method works as well as the original one. However, in the case with measurement noises, the method needs more iterations and much longer time to converge and the result is not quite acceptable. We would like to re-examine the same example, and show that the SDP solution combining with the gradient search method will give us a much better localization within a much shorter time. In this approach, we may stop the SDP iteration earlier. Then, the solution produced by the decomposed SDP strategy is served as the starting solution for the gradient-based search method to minimize the objective function of the entire sensor network. Since the computation complexity of the gradient vector is very low so that the gradient-based method, although applied to the entire network without decomposition, can be completed extremely fast.

Example 4. Consider the same problem in [16] with 1800 sensors, 200 anchors, radio range 0.05 and a larger noisy factor 0.1, where the sensor network is decomposed into 36 equal-sized domains. Figure 11.14 shows the SDP solution after 3 iterations.

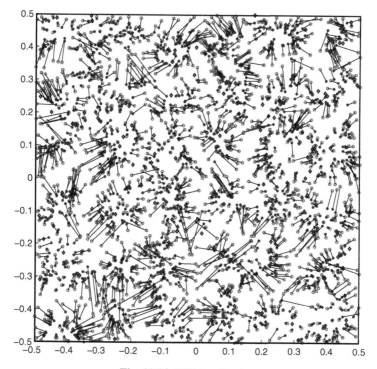

Fig. 11.14. SDP localization.

One can see that the SDP algorithm fails to find accurate solution in most small areas. But after 50 gradient-based search steps, the final localization, shown in Figure 11.15, is greatly improved.

Our program is implemented with Matlab and it uses SEDUMI [45] or DSDP2 [13] as the SDP solver. It costs 165 seconds of SEDUMI or 63 seconds of DSDP2 to get the SDP localization in Figure 11.14 on a PC. Then, after 50 gradient search steps in 50 seconds, the objective function is reduced from 12.81 to 0.230. It can be seen from Figure 11.15 that most of the sensors are located very close to their true positions, although few of them, most of which are close to the boundary of the network, are solved inaccurately. We see, from this example, the dramatic solution improvement and the cost efficiency of the combination of the SDP relaxation and the gradient search methods.

11.7 Conclusions and Further Research

We have surveyed recent SDP methods for the Sensor Network Localization problem, which is a variant of the Graph Realization problem. It has been shown for the

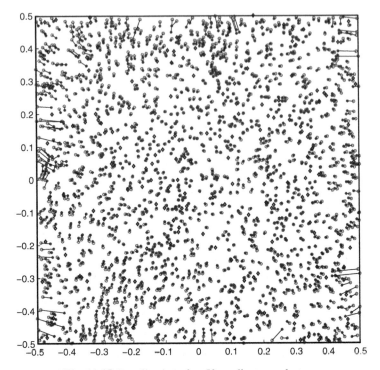

Fig. 11.15. Localization after 50 gradient search steps.

first time that the SDP method yields an algorithm that guarantees to find the solution if the input graph is uniquely localizable. Moreover, we have defined various notions of localizability and demonstrated their relationship with classical rigidity theory. However, this work has still left many interesting open questions unanswered. First, for those instances that are not uniquely localizable, it would be interesting to investigate how many anchors are needed and how should they be placed in order to make the instance uniquely localizable. Secondly, the SDP model assumes that the input data are noise–free. However, sensor measurements are often noisy, and it is important to have a more robust model that can handle noisy data and has good theoretical properties. Thirdly, besides the distance measurements, there may be extra information available to help us determine the desired realization. For instance, we may have angle estimates between a pair of sensors that are within communication range. It would be desirable to develop a model that incorporates and exploits these information. For some results in this direction, see, e.g., [9]. Finally, if anchors are not present, we need to develop effective anchor-free SDP localization models and to extend the distributed computation in the anchor-free environment.

References

[1] A. Y. Alfakih. Graph Rigidity via Euclidean Distance Matrices. *Linear Algebra and Its Applications* 310:149–165, 2000.

[2] A. Y. Alfakih. On Rigidity and Realizability of Weighted Graphs. *Linear Algebra and Its Applications* 325:57–70, 2001.

[3] A. Y. Alfakih, A. Khandani, H. Wolkowicz. Solving Euclidean Distance Matrix Completion Problems Via Semidefinite Programming. *Comput. Opt. and Appl.* 12:13–30, 1999.

[4] A. Y. Alfakih, H. Wolkowicz. On the Embeddability of Weighted Graphs in Euclidean Spaces. *Research Report CORR 98–12, University of Waterloo, Dept. of Combinatorics and Optimization*, 1998.

[5] A. Y. Alfakih, H. Wolkowicz. Euclidean Distance Matrices and the Molecular Conformation Problem. *Research Report CORR 2002–17, University of Waterloo, Dept. of Combinatorics and Optimization*, 2002.

[6] Farid Alizadeh. Interior Point Methods in Semidefinite Programming with Applications to Combinatorial Optimization. *SIAM J. Opt.* 5:13–51, 1995.

[7] James Aspnes, David Goldenberg, Yang Richard Yang. On the Computational Complexity of Sensor Network Localization. *ALGOSENSORS 2004*, in *LNCS* 3121:32–44, 2004.

[8] M. Bǎdoiu. Approximation Algorithm for Embedding Metrics into a Two–Dimensional Space. *Proc. 14th SODA* 434-443, 2003.

[9] M. Bǎdoiu, E. D. Demaine, M. T. Hajiaghayi, P. Indyk. Low–Dimensional Embedding with Extra Information. *Proc. 20th SoCG* 320–329, 2004.

[10] A. Barvinok. Problems of Distance Geometry and Convex Properties of Quadratic Maps. *Disc. Comp. Geom.* 13:189–202, 1995.

[11] A. Barvinok. A Remark on the Rank of Positive Semidefinite Matrices Subject to Affine Constraints. *Disc. Comp. Geom.* 25(1):23–31, 2001.

[12] A. Barvinok. A Course in Convexity. *AMS*, 2002.

[13] S. J. Benson, Y. Ye, and X. Zhang. Solving large-scale sparse semidefinite programs for combinatorial optimization. *SIAM Journal on Optimization*, 10(2):443–461, 2000.

[14] J. Beutel. Geolocation in a PicoRadio Environment, MS Thesis, ETH Zurich, December, 1999.

[15] P. Biswas, Y. Ye. Semidefinite Programming for Ad Hoc Wireless Sensor Network Localization. *Proc. 3rd IPSN* 46–54, 2004.

[16] P. Biswas and Y. Ye. A distributed method for solving semidefinite programming arising from ad hoc wireless sensor network localization. Technical report, Dept of Management Science and Engineering, Stanford University, October 2003.

[17] S. Boyd, L. E. Ghaoui, E. Feron, and V. Balakrishnan. Linear Matrix Inequalities in System and Control Theory. SIAM, 1994.

[18] N. Bulusu, J. Heidemann, and D. Estrin. GPS-less Low-cost Outdoor Localization for Very Small Devices, IEEE Personal Communications, 7(5):28¡V34, Oct 2000.

[19] L. Doherty, L. E. Ghaoui, and S. J. Pister. Convex position estimation in wireless sensor networks. In IEEE Infocom, Vol. 3, pages 1655–1663, April 2001.

[20] T. Eren, D. K. Goldenberg, W. Whiteley, Y. R. Yang, A. S. Moore, B. D. O. Anderson, P. N. Belhumeur. Rigidity, Computation, and Randomization in Network Localization. IEEE INFOCOM, 2004.

[21] D. Goldfarb, K. Scheinberg. Interior Point Trajectories in Semidefinite Programming. *SIAM J. Opt.* 8(4):871–886, 1998.

[22] J. C. Gower. Some distance properties of latent root and vector methods in multivariate analysis. *Biometrika* 53:325–338, 1966.

[23] J. Graver, B. Servatius, H. Servatius. Combinatorial Rigidity. *AMS*, 1993.

[24] O. Güler and Y. Ye. Convergence behavior of interior point algorithms. *Math. Programming* 60:215–228, 1993.

[25] B. Hendrickson. Conditions for Unique Graph Realizations. *SIAM J. Comput.* 21(1):65–84, 1992.

[26] B. Hendrickson. The Molecule Problem: Exploiting Structure in Global Optimization. *SIAM J. Opt.* 5(4):835–857, 1995.

[27] J. Hightower and G. Borriello. Location systems for ubiquitous computing. IEEE Computer, 34(8):57–66, August 2001.

[28] A. Howard, M. Mataric, and G. Sukhatme. Relaxation on a mesh: a formalism for generalized localization. In IEEE/RSJ Int'l Conf. on Intelligent Robots and Systems, Vol. 3, pages 1055–1060, October 2001.

[29] B. Jackson, T. Jordán. Connected Rigidity Matroids and Unique Realizations of Graphs. Preprint, 2003.

[30] E. Kaplan. Understanding GPS Principles and Applications Artech House, 1996

[31] M. Laurent. Matrix completion problems. The Encyclopedia of Optimization., 3:221–229, 2001.

[32] T.C. Liang, T.C. Wang and Y. Ye. A Gradient Search Method to Round the Semidefinite Programming Relaxation Solution for Ad Hoc Wireless Sensor Network Localization. Technical report, Dept of Management Science and Engineering, Stanford University, April 2004.

[33] N. Linial, E. London, Yu. Rabinovich. The Geometry of Graphs and Some of Its Algorithmic Applications. *Combinatorica* 15(2):215–245, 1995.

[34] Z. Q. Luo, J. Sturm and S. Zhang. Superlinear convergence of a symmetric primal-dual path following algorithm for semidefinite programming. *SIAM J. Optimization* 8(1):59-81, 1998.

[35] A. Mainwaring, J. Polastre, R. Szewczyk, D. Culler and J. Anderson. Wireless Sensor Networks for Habitat Monitoring, WSNA¡¦02, September 28, 2002, Atlanta, Georgia, USA.

[36] K. V. Mardia, J. T. Kent, and J. M. Bibby. *Multivariate Analysis.* Academic Press, Orlando, 1979.

[37] J. Moré and Z. Wu. Global continuation for distance geometry problems. SIAM Journal on Optimization, 7:814–836, 1997.

[38] D. Niculescu and B. Nath. Ad hoc positioning system (APS). In IEEE GLOBECOM (1), pages 2926–2931, 2001.

[39] C. Savarese, J. Rabay, K. Langendoen. Robust Positioning Algorithms for Distributed Ad–Hoc Wireless Sensor Networks. *USENIX Annual Technical Conference*, 2002.

[40] A. Savvides, C.–C. Han, M. B. Srivastava. Dynamic Fine–Grained Localization in Ad–Hoc Networks of Sensors. *Proc. 7th MOBICOM*, 166–179, 2001.

[41] A. Savvides, H. Park, M. B. Srivastava. The Bits and Flops of the n–hop Multilateration Primitive for Node Localization Problems. *Proc. 1st WSNA*, 112–121, 2002.

[42] I. J. Schoenberg. Remarks to Maurice Fréchets article "Sur la définition axiomatique dúne classe déspaces distanciés vectoriellement applicable sur léspace de Hilbert." em Annals of Mathematics 38:724–732, 1935.

[43] Y. Shang, W. Ruml, Y. Zhang, and M. P. J. Fromherz. Localization from mere connectivity. In Proceedings of the 4th ACM international symposium on Mobile Ad Hoc Networking & Computing, pages 201–212. ACM Press, 2003.

[44] Anthony Man-Cho So and Yinyu Ye. Theory of Semidefinite Programming for Sensor Network Localization. Technical report, Dept of Management Science and Engineering, Stanford University, April 2004; to appear in SODA'05.

[45] J. F. Sturm. Let sedumi seduce you, October 2001.

[46] W. S. Torgerson. Multidimensional scaling: I. Theory and method. *Psychometrika* 17:401–419, 1952.

[47] M. W. Trosset. Distance matrix completion by numerical optimization. em Computational Optimizationand Applications 17:11–22, 2000.

[48] M. W. Trosset. Extensions of classical multidimensional scaling via variable reduction. em Computational Statistics 17(2):147–162, 2002.

[49] P. Tseng. SOCP relaxation for nonconvex optimization, ICCOPT 1, RPI, Troy, NY, August, 2004.

[50] G. Young and A. S. Householder. Discussion of a set of points in terms of their mutual distances. *Psychometrika* 3:19–22, 1938.